T0138631

Multilinear Subspace Learning

Dimensionality Reduction of Multidimensional Data

Chapman & Hall/CRC
Machine Learning & Pattern Recognition Series

SERIES EDITORS

Ralf Herbrich
Amazon Development Center
Berlin, Germany

Thore Graepel
Microsoft Research Ltd.
Cambridge, UK

AIMS AND SCOPE

This series reflects the latest advances and applications in machine learning and pattern recognition through the publication of a broad range of reference works, textbooks, and handbooks. The inclusion of concrete examples, applications, and methods is highly encouraged. The scope of the series includes, but is not limited to, titles in the areas of machine learning, pattern recognition, computational intelligence, robotics, computational/statistical learning theory, natural language processing, computer vision, game AI, game theory, neural networks, computational neuroscience, and other relevant topics, such as machine learning applied to bioinformatics or cognitive science, which might be proposed by potential contributors.

PUBLISHED TITLES

MACHINE LEARNING: An Algorithmic Perspective
Stephen Marsland

HANDBOOK OF NATURAL LANGUAGE PROCESSING,
Second Edition
Nitin Indurkhya and Fred J. Damerau

UTILITY-BASED LEARNING FROM DATA
Craig Friedman and Sven Sandow

A FIRST COURSE IN MACHINE LEARNING
Simon Rogers and Mark Girolami

COST-SENSITIVE MACHINE LEARNING
Balaji Krishnapuram, Shipeng Yu, and Bharat Rao

ENSEMBLE METHODS: FOUNDATIONS AND ALGORITHMS
Zhi-Hua Zhou

MULTI-LABEL DIMENSIONALITY REDUCTION
Liang Sun, Shuiwang Ji, and Jieping Ye

BAYESIAN PROGRAMMING
Pierre Bessière, Emmanuel Mazer, Juan-Manuel Ahuactzin, and Kamel Mekhnacha

MULTILINEAR SUBSPACE LEARNING: DIMENSIONALITY REDUCTION
OF MULTIDIMENSIONAL DATA
Haiping Lu, Konstantinos N. Plataniotis, and Anastasios N. Venetsanopoulos

Chapman & Hall/CRC
Machine Learning & Pattern Recognition Series

Multilinear Subspace Learning
Dimensionality Reduction of Multidimensional Data

Haiping Lu
Konstantinos N. Plataniotis
Anastasios N. Venetsanopoulos

CRC Press
Taylor & Francis Group
Boca Raton   London   New York

CRC Press is an imprint of the
Taylor & Francis Group, an **informa** business

A CHAPMAN & HALL BOOK

CRC Press
Taylor & Francis Group
6000 Broken Sound Parkway NW, Suite 300
Boca Raton, FL 33487-2742

© 2013 by Taylor & Francis Group, LLC
CRC Press is an imprint of Taylor & Francis Group, an Informa business

No claim to original U.S. Government works

Printed on acid-free paper
Version Date: 20150126

International Standard Book Number-13: 978-1-4398-5724-3 (Hardback)

Library of Congress Cataloging-in-Publication Data

Lu, Haiping.
   Multilinear subspace learning : dimensionality reduction of multidimensional data / Haiping Lu, K.N. Plataniotis, A.N. Venetsanopoulos.
      pages cm -- (Chapman & Hall/CRC machine learning & pattern recognition series)
   Includes bibliographical references and index.
   ISBN 978-1-4398-5724-3 (hardback)
   1. Data compression (Computer science) 2. Big data. 3. Multilinear algebra. I. Plataniotis, Konstantinos N. II. Venetsanopoulos, A. N. (Anastasios N.), 1941- III. Title.

QA76.9.D33L825 2013
005.7--dc23   2013039517

Visit the Taylor & Francis Web site at
http://www.taylorandfrancis.com

and the CRC Press Web site at
http://www.crcpress.com

To Hongxia, Dailian, and Daizhen

To Ilda

Contents

List of Figures

List of Tables

List of Algorithms

Acronyms and Symbols

Acronym	Description
AdaBoost	Adaptive boosting
ALS	Alternating least squares
APP	Alternating partial projections
BSS	Blind source separation
CANDECOMP	Canonical decomposition
CCA	Canonical correlation analysis
CRR	Correct recognition rate
DATER	Discriminant analysis with tensor representation
EMP	Elementary multilinear projection
FPT	Full projection truncation
GPCA	Generalized PCA
GTDA	General tensor discriminant analysis
HOPLS	Higher-order PLS
HOSVD	High-order SVD
IC	Independent component
ICA	Independent component analysis
LDA	Linear discriminant analysis
LSL	Linear subspace learning
MCCA	Multilinear CCA
MMICA	Multilinear modewise ICA
MPCA	Multilinear PCA
MSL	Multilinear subspace learning
NIPALS	Nonlinear iterative partial least squares
NMF	Nonnegative matrix factorization
N-PLS	N-way PLS
NTF	Nonnegative tensor factorization
PARAFAC	Parallel factors
PC	Principal component
PCA	Principle component analysis
PLS	Partial least squares
R-UMLDA	Regularized UMLDA
R-UMLDA-A	Regularized UMLDA with aggregation
SMT	Sequential mode truncation
SSS	Small sample size
SVD	Singular value decomposition
SVM	Support vector machine
TR1DA	Tensor rank-one discriminant analysis
TROD	Tensor rank-one decomposition

Acronym	Description
TTP	Tensor-to-tensor projection
TVP	Tensor-to-vector projection
UMLDA	Uncorrelated multilinear discriminant analysis
UMPCA	Uncorrelated MPCA
VVP	Vector-to-vector projection

Symbol Description

$\|\mathbf{A}\|$	Determinant of matrix $\mathbf{A}$	J_n	Mode-n dimension for the second set in CCA/PLS extensions
$\|\cdot\|_F$	Frobenius norm		
a or A	A scalar		
$\mathbf{a}$	A vector	K	Maximum number of iterations
$\mathbf{A}$	A matrix		
$\mathcal{A}$	A tensor	k	Iteration step index
$\bar{\mathbf{A}}$ or $\bar{\mathcal{A}}$	The mean of samples $\{\mathbf{A}_m\}$ or $\{\mathcal{A}_m\}$	L	Number of training samples for each class
$\mathbf{A}^T$	Transpose of matrix $\mathbf{A}$	M	Number of training samples
$\mathbf{A}^{-1}$	Inverse of matrix $\mathbf{A}$	m	Index of training sample
$\mathbf{A}(i_1, i_2)$	Entry at the i_1th row and i_2th column of $\mathbf{A}$	M_c	Number of training samples in class c
$\mathbf{A}_{(n)}$	Mode-n unfolding of tensor $\mathcal{A}$	N	Order of a tensor, number of indices/modes
$< \mathcal{A}, \mathcal{B} >$	Scalar product of $\mathcal{A}$ and $\mathcal{B}$		
$\mathcal{A} \times_n \mathbf{U}$	Mode-n product of $\mathcal{A}$ by $\mathbf{U}$	n	Mode index of a tensor
$\mathbf{a} \circ \mathbf{b}$	Outer (tensor) product of $\mathbf{a}$ and $\mathbf{b}$	P	Dimension of the output vector, also number of EMPs in a TVP, or number of latent factors in PLS
C	Number of classes		
c	Class index		
c_m	Class label for the mth training sample, the mth element of the class vector $\mathbf{c}$	P_n	Mode-n dimension in the projected (output) space of a TTP
δ_{pq}	Kronecker delta, $\delta_{pq} = 1$ iff $p = q$ and 0 otherwise	p	Index of the output vector, also index of the EMP in a TVP, or index of latent factor in PLS
$\frac{\partial f(\mathbf{x})}{\partial \mathbf{x}}$	Partial derivative of f with respect to $\mathbf{x}$		
$\mathbf{g}_p$	The pth coordinate vector	Ψ_B	Between-class scatter (measure)
g_{p_m}	$\mathbf{g}_p(m)$, the mth element of $\mathbf{g}_p$, see y_{m_p}	Ψ_T	Total scatter (measure)
$H_\mathbf{y}$	Number of selected features in MSL	Ψ_W	Within-class scatter (measure)
$\mathbf{I}$	An identity matrix	Q	Ratio of total scatter kept in each mode
I_n	Mode-n dimension or mode-n dimension for the first set in CCA/PLS extensions	$\mathbb{R}$	The set of real numbers
		$\mathbf{r}_m$	The (TVP) projection of

the first set sample $\mathbf{X}_m$ in second-order MCCA

ρ Sample Pearson correlation

$\mathbf{S}_B$ Between-class scatter matrix in LSL

$\mathbf{S}_B^{(n)}$ Mode-n between-class scatter matrix in MSL

$S_{B_{y_p}}$ Between-class scatter of pth EMP projections $\{y_{m_p}, m = 1, ..., M\}$

$\mathbf{S}_T$ Total scatter matrix in LSL

$\mathbf{S}_T^{(n)}$ Mode-n total scatter matrix in MSL

$S_{T_{y_p}}$ Total scatter of pth EMP projections $\{y_{m_p}, m = 1, ..., M\}$

$\mathbf{S}_W$ Within-class scatter matrix in LSL

$\mathbf{S}_W^{(n)}$ Mode-n within-class scatter matrix in MSL

$S_{W_{y_p}}$ Within-class scatter of the pth EMP projections $\{y_{m_p}, m = 1, ..., M\}$

$\mathbf{s}_m$ The (TVP) projection of the second set sample $\mathbf{Y}_m$ in second-order MCCA

$tr(\mathbf{A})$ The trace of matrix $\mathbf{A}$

$\mathcal{X}_m$ The mth input tensor sample

$\mathbf{x}_m$ The mth input vector sample or the mth sample in the first set in CCA/PLS

$\mathbf{U}$ Projection matrix in LSL

$\tilde{\mathbf{U}}$ or $\tilde{\mathbf{u}}$ The (sub)optimal solution of $\mathbf{U}$ or $\mathbf{u}$

$\mathbf{U}^{(n)}$ Mode-n projection matrix

$\{\mathbf{U}^{(n)}\}$ A TTP, consisting of N projection matrices

$\mathbf{u}^{(n)}$ Mode-n projection vector

$\mathbf{u}_p$ The pth projection vector in LSL, or the pth mode-2 projection vector in tri-linear PLS1 (N-PLS)

$\mathbf{u}_{x_p}$ The pth projection vector for the first set in CCA/PLS, or the pth mode-1 projection vector for the first set in second-order MCCA, or the pth latent vector in HOPLS

$\mathbf{u}_{y_p}$ The pth projection vector for the second set in CCA/PLS, or the pth mode-1 projection vector for the second set in second-order MCCA

$\{\mathbf{u}_p^{(n)}\}$ The pth EMP in a TVP, consisting of N projection vectors

$\{\mathbf{u}_p^{(n)}\}_N^P$ A TVP, consisting of P EMPs ($P \times N$ projection vectors)

$vec(\mathcal{A})$ Vectorized representation of a tensor $\mathcal{A}$

$\mathbf{v}_{x_p}$ The pth mode-2 projection vector for the first set in second-order MCCA

$\mathbf{v}_{y_p}$ The pth mode-2 projection vector for the second set in second-order MCCA

$\mathbf{w}_p$ The pth coordinate vector for the first set in CCA/PLS or second-order MCCA, or the pth latent factor in trilinear PLS1 (N-PLS)

$\mathcal{X}_m$ The mth (training/input) tensor sample

$\mathcal{Y}_m$ Projection of $\mathcal{X}_m$ on a TTP $\{\mathbf{U}^{(n)}\}$, or the mth sample in the second set in CCA/PLS extensions

$\hat{y}^{(n)}$ Mode-n partial multilinear projection of raw samples in TTP

$\hat{\mathcal{y}}^{(n)}$ Mode-n partial multilinear projection of centered (zero-mean) samples in TTP

$\mathbf{y}_m$ Vector projection of $\mathcal{X}_m$ (rearranged from TTP projection $\mathcal{Y}_m$ in TTP-based MSL or projection on a TVP in

TVP-based MSL), or the mth sample in the second set in CCA/PLS

$\acute{\mathbf{y}}_p^{(n)}$　Mode-n partial multilinear projection of raw samples in the pth EMP of a TVP

$\hat{\mathbf{y}}_p^{(n)}$　Mode-n partial multilinear projection of centered (zero-mean) samples in the pth EMP of a TVP

y_{m_p}　= $\mathbf{y}_m(p)$ = $\mathbf{g}_p(m)$, projection of $\mathcal{X}_m$ on the pth EMP $\{\mathbf{u}_p^{(n)}\}$

$\mathbf{z}_p$　The pth coordinate vector for the second set in CCA/PLS or second-order MCCA

Acknowledgments

We thank the following careful readers for pointing out typos in our book:

- Tiejun Tong, Department of Mathematics, Hong Kong Baptist University.

- Woon Cho, the University of Tennessee, Knoxville.

- Qiquan Shi, Department of Computer Science, Hong Kong Baptist University.

Preface

With the advances in sensor, storage, and networking technologies, bigger and bigger data are being generated on a daily basis in a wide range of applications, especially in emerging cloud computing, mobile Internet, and big data applications. Most real-world data, either big or small, have multidimensional representations. Two-dimensional (2D) data include gray-level images in computer vision and image processing, multichannel electroencephalography (EEG) signals in neuroscience and biomedical engineering, and gene expression data in bioinformatics. Three-dimensional (3D) data include 3D objects in generic object recognition, hyperspectral cube in remote sensing, and gray-level video sequences in activity or gesture recognition for surveillance and human–computer interaction. A functional magnetic resonance imaging (fMRI) sequence in neuroimaging is an example of four-dimensional (4D) data. Other multidimensional data appear in medical image analysis, content-based retrieval, and space-time super-resolution. In addition, many streaming data and mining data are frequently organized in multidimensional representations, such as those in social network analysis, Web data mining, sensor network analysis, and network forensics. Moreover, multiple features (e.g., different image cues) can also form higher-order tensors in feature fusion.

These multidimensional data are usually very high-dimensional, with a large amount of redundancy and occupying only a small subspace of the entire input space. Therefore, dimensionality reduction is frequently employed to map high-dimensional data to a low-dimensional space while retaining as much information as possible. Linear subspace learning (LSL) algorithms are traditional dimensionality reduction techniques that represent input data as vectors and solve for an optimal linear mapping to a lower-dimensional space. However, they often become inadequate when dealing with big multidimensional data. They result in very high-dimensional vectors, lead to the estimation of a large number of parameters, and also break the natural structure and correlation in the original data.

Due to the above challenges, especially in emerging big data applications, there has been an urgent need for more efficient dimensionality reduction schemes for big multidimensional data. Consequently, there has been a growing interest in multilinear subspace learning (MSL) that reduces the dimensionality of big data directly from their natural multidimensional representation: tensors, which refer to multidimensional arrays here. The research on MSL has progressed from heuristic exploration to systematic investigation, while recent prevalence of big data applications has increased the demand for technical developments in this emerging research field. Thus, we found that there is a strong need for a new book devoted to the fundamentals and foundations of MSL, as well as MSL algorithms and their applications.

The primary goal of this book is to give a comprehensive introduction to both theoretical and practical aspects of MSL for dimensionality reduction of multidimensional data. It expects not only to detail recent advances in MSL, but also to trace the history and explore future developments and emerging applications. In particular, the emphasis is on the fundamental concepts and system-level perspectives. This book provides a foundation upon which we can build solutions for many of today's most interesting and challenging problems in big multidimensional data processing. Specifically, it includes the following important topics in MSL: multilinear algebra fundamentals, multilinear projections, MSL framework formulation, MSL optimality criterion construction, and MSL algorithms, solutions, and applications. The MSL framework enables us to develop MSL algorithms systematically with various optimality criteria. Under this unifying MSL framework, a number of MSL algorithms are discussed and analyzed in detail. This book covers their applications in various fields, and provides their pseudocodes and implementation tips to help practitioners in further development, evaluation, and application. MATLAB$^\circledR$ source codes are made available online.

The topics covered in this book are of great relevance and importance to both theoreticians and practitioners who are interested in learning compact features from big multidimensional data in machine learning and pattern recognition. Most examples given in this book highlight our own experiences, which are directly relevant for researchers who work on applications in video surveillance, biometrics, and object recognition. This book can be a useful reference for researchers dealing with big multidimensional data in areas such as computer vision, image processing, audio and speech processing, machine learning, pattern recognition, data mining, remote sensing, neurotechnology, bioinformatics, and biomedical engineering. It can also serve as a valuable resource for advanced courses in these areas. In addition, this book can serve as a good reference for graduate students and instructors in the departments of electrical engineering, computer engineering, computer science, biomedical engineering, and bioinformatics whose orientation is in subjects where dimensionality reduction of big multidimensional data is essential.

We organize this book into two parts. The "ingredients" are in Part I while the "dishes" are in Part II. On the first page of each chapter, we include a figure serving as a "graphic abstract" for the chapter wherever possible.

In summary, this book provides a foundation for solving many dimensionality reduction problems in multidimensional data applications. It is our hope that its publication will foster more principled and successful applications of MSL in a wide range of research disciplines.

We have set up the following websites for this book:

http://www.comp.hkbu.edu.hk/~haiping/MSL.html

or
http://www.dsp.toronto.edu/~haiping/MSL.html
or
https://sites.google.com/site/tensormsl/

We will update these websites with open source software, possible corrections, and any other useful materials to distribute after publication of this book.

The authors would like to thank the Edward S. Rogers Sr. Department of Electrical and Computer Engineering, University of Toronto, for supporting this research work. H. Lu would like to thank the Institute for Infocomm Research, the Agency for Science, Technology and Research (A*STAR), in particular, How-Lung Eng, Cuntai Guan, Joo-Hwee Lim, and Yiqun Li, for hosting him for almost four years. H. Lu would also like to thank the Department of Computer Science, Hong Kong Baptist University, in particular, Pong C. Yuen, and Jiming Liu for supporting this work. We thank Dimitrios Hatzinakos, Raymond H. Kwong, and Emil M. Petriu for their help in our work on this topic. We thank Kar-Ann Toh, Constantine Kotropoulos, Andrew Teoh, and Althea Liang for reading through the draft and offering useful comments and suggestions. We would also like to thank the many anonymous reviewers of our papers who have given us tremendous help in advancing this field. This book would not have been possible without the contributions from other researchers in this field. In particular, we want to thank the following researchers whose works have been particularly inspiring and helpful to us: Lieven De Lathauwer, Tamara G. Kolda, Amnon Shashua, Jian Yang, Jieping Ye, Xiaofei He, Deng Cai, Dacheng Tao, Shuicheng Yan, Dong Xu, and Xuelong Li. We also thank editor Randi Cohen and the staff at CRC Press, Taylor & Francis Group, for their support during the writing of this book.

Haiping Lu
Hong Kong

Konstantinos N. Plataniotis
Anastasios N. Venetsanopoulos
Toronto

Chapter 1

Introduction

With the advances in sensor, storage, and networking technologies, bigger and bigger data are being generated daily in a wide range of applications. Figures 1.1 through 1.4 show some examples in computer vision, audio processing, neuroscience, remote sensing, and data mining. To succeed in this era of *big data* [Howe et al., 2008], it becomes more and more important to learn *compact features* for efficient processing. Most big data are *multidimensional* and they can often be represented as *multidimensional arrays*, which are referred to as *tensors* in mathematics [Kolda and Bader, 2009]. Thus, *tensor-based computation* is emerging, especially with the growth of mobile Internet [Lenhart et al., 2010], cloud computing [Armbrust et al., 2010], and big data such as the MapReduce model [Dean and Ghemawat, 2008; Kang et al., 2012].

This book deals with tensor-based learning of compact features from multidimensional data. In particular, we focus on *multilinear subspace learning* (MSL) [Lu et al., 2011], a dimensionality reduction [Burges, 2010] method developed for tensor data. The objective of MSL is to learn a *direct mapping* from high-dimensional tensor representations to low-dimensional vector/tensor representations.

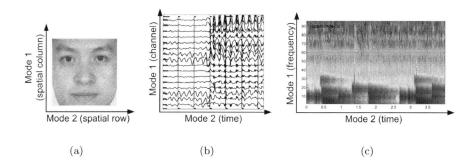

(a)   (b)   (c)

FIGURE 1.1: Examples of second-order tensor (matrix) data: (a) a gray-level image, (b) multichannel electroencephalography (EEG) signals ("Electroencephalography," Wikipedia, the free encyclopedia, http://en.wikipedia.org/wiki/Electroencephalography), (c) an auditory spectrogram.

1.1 Tensor Representation of Multidimensional Data

Multidimensional data can be naturally represented as multidimensional (multiway) arrays, which are referred to as *tensors* in mathematics [Lang, 1984; Kolda and Bader, 2009]. The number of dimensions (ways) N defines the *order* of a tensor, and the elements (entries) of a tensor are addressed by N indices. Each index defines one *mode*. Tensor is a generalization of vector and matrix. Scalars are zero-order tensors, vectors are first-order tensors, matrices are second-order tensors, and tensors of order three or higher ($N \geq 3$) are called *higher-order tensors* [De Lathauwer et al., 2000a; Kolda and Bader, 2009].

> **Tensor terminology:** The term *tensor* has different meanings in mathematics and physics. The usage in this book refers to its meaning in mathematics, in particular multilinear algebra [De Lathauwer et al., 2000b,a; Greub, 1967; Lang, 1984]. In physics, the same term generally refers to a *tensor field* [Lebedev and Cloud, 2003], a generalization of a vector field. It is an association of a tensor with each point of a geometric space and it varies continuously with position.

Second-order tensor (matrix) data are two-dimensional (2D), with some examples shown in Figure 1.1. Figure 1.1(a) shows a gray-level face image in computer vision applications, with spatial column and row modes. Figure 1.1(b) depicts multichannel electroencephalography (EEG) signals in neuroscience, where the two modes consist of channel and time. Figure 1.1(c) shows an audio spectrogram in audio and speech processing with frequency and time modes.

Third-order tensor data are three-dimensional (3D), with some examples shown in Figure 1.2. Figure 1.2(a) is a 3D face object in computer vision or computer graphics [Sahambi and Khorasani, 2003], with three modes of (spatial) column, (spatial) row, and depth. Figure 1.2(b) shows a hyperspectral cube in remote sensing [Renard and Bourennane, 2009], with three modes of column, row, and spectral wavelength. Figure 1.2(c) depicts a binary gait video sequence for activity or gesture recognition in computer vision or human-computer interaction (HCI) [Chellappa et al., 2005; Green and Guan, 2004], with the column, row, and time modes. Figure 1.2(d) illustrates social network analysis data organized in three modes of conference, author, and keyword [Sun et al., 2006]. Figures 1.2(e) and 1.2(f) demonstrate web graph mining data organized in three modes of source, destination, and text, and environmental sensor monitoring data organized in three modes of type, location, and time [Faloutsos et al., 2007].

Similarly, fourth-order tensor data are four-dimensional (4D). Figure 1.3(a) depicts a functional magnetic resonance imaging (fMRI) scan sequence in brain mapping research [van de Ven et al., 2004]. It is a 4D object with four modes: three spatial modes (column, row, and depth) and one temporal mode. Another fourth-order tensor example is network traffic data with four modes: source IP, destination IP, port number, and time [Kolda and Sun, 2008], as illustrated in Figure 1.3(b).

Our tour through tensor data examples is not meant to be exhaustive. Many other interesting tensor data have appeared and are emerging in a broad spectrum of application domains including computational biology, chemistry, physics, quantum computing, climate modeling, and control engineering [NSF, 2009].

Tensor for feature fusion: Moreover, multiple features of an image (and

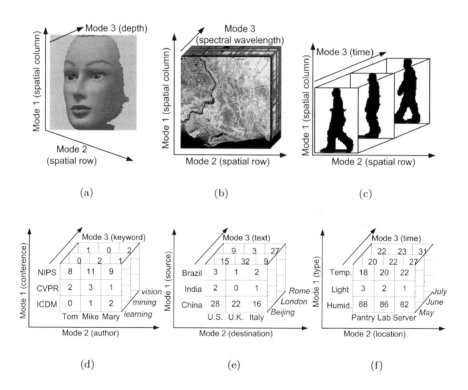

FIGURE 1.2: Examples of third-order tensor data: (a) a 3D face image (Source: www.dirk.colbry.com by Dr. Dirk Colbry), (b) a hyperspectral cube ("Hyperspectral imaging," Wikipedia, the free encyclopedia, http://en.wikipedia.org/wiki/Hyperspectral_imaging), (c) a video sequence, (d) social networks organized in conference×author×keyword, (e) web graphs organized in source×destination×text, (f) environmental sensor monitoring data organized in type×location×time.

other data as well) can be represented as a third-order tensor where the first two modes are column and row, and the third mode indexes different features such that tensor is used as a feature combination/fusion scheme. For example, local descriptors such as the Scale-Invariant Feature Transform (SIFT) [Lowe, 2004] and Histogram of Oriented Gradients (HOG) [Dalal and Triggs, 2005] form a local descriptor tensor in [Han et al., 2012], which is shown to be more efficient than the bag-of-feature (BOF) model [Sivic and Zisserman, 2003]. Local binary patterns [Ojala et al., 2002] on a Gaussian pyramid [Lindeberg, 1994] are employed to form feature tensors in [Ruiz-Hernandez et al., 2010a,b]. Gradient-based appearance cues are combined in a tensor form in [Wang et al., 2011a], and wavelet transform [Antonini et al., 1992] and Gabor filters [Jain and Farrokhnia, 1991] are used to generate higher-order tensors in [Li et al., 2009a; Barnathan et al., 2010], and [Tao et al., 2007b], respectively. Figure 1.4

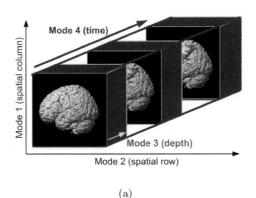

(a)

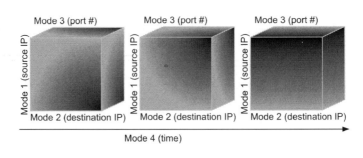

(b)

FIGURE 1.3: Examples of fourth-order tensor data: (a) a functional magnetic resonance imaging (fMRI) scan sequence with three spatial modes and one temporal mode [Pantano et al., 2005], (b) network traffic data organized in source IP×destination IP×port number×time.

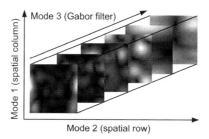

FIGURE 1.4: A third-order tensor formed by the Gabor filter outputs of a gray-level face image. Here, tensor is used as a feature fusion scheme.

shows an example of a third-order tensor formed by the Gabor filter outputs of a gray-level face image.

1.2   Dimensionality Reduction via Subspace Learning

Real-world tensor data are commonly specified in a high-dimensional space. Direct operation on this space suffers from the so-called *curse of dimensionality*:

- Handling high-dimensional data puts a high demand on processing power and resources so it is computationally expensive [Shakhnarovich and Moghaddam, 2004].

- When the number of data samples available is small compared to their high dimensionality, that is, in the *small sample size* (SSS) scenario, conventional tools become inadequate and many problems become ill-posed or poorly conditioned [Ma et al., 2011].

Fortunately, these tensor data do not lie randomly in the high-dimensional space; rather, they are highly constrained and confined to a *subspace* [Shakhnarovich and Moghaddam, 2004; Zhang et al., 2004]. For example, as shown in Figure 1.5, a 256-level facial image of 100×100 (on the left) is only one of the $256^{10,000}$ points in the corresponding image space. As faces are constrained with certain specific characteristics, all 256-level facial images of 100×100 will occupy only a very small portion, that is, a subspace, of the corresponding image space. Thus, they are intrinsically low-dimensional and there is lots of redundancy.

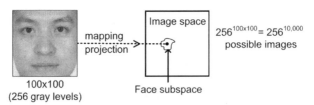

FIGURE 1.5: A face can be represented as one point in an image space of the same size. All 8-bit 100×100 faces occupy only a small portion, that is, a subspace, of the whole image space of size 100×100 with 8 bits per pixel.

Dimensionality reduction[1] is an attempt to transform a high-dimensional dataset into a low-dimensional representation while retaining most of the information regarding the underlying structure or the actual physical phenomenon [Law and Jain, 2006]. In other words, in dimensionality reduction, we are learning a *mapping* from high-dimensional input space to low-dimensional output space that is a subspace of the input space, that is, we are doing *subspace learning*. We can view the low-dimensional representation as *latent variables* to estimate. Also, we can view this as a *feature extraction* process and the low-dimensional representation as the features learned. These features can then be used to perform various tasks, for example, they can be fed into a classifier to identify its class label.

"In an information-rich world, the wealth of information means a dearth of something else: a scarcity of whatever it is that information consumes. What information consumes is rather obvious: it consumes the attention of its recipients. Hence a wealth of information creates a poverty of attention and a need to allocate that attention efficiently among the overabundance of information sources that might consume it."

Herbert Simon (1916–2001)
Economist, Turing Award Winner, and Nobel Laureate

Traditional subspace learning algorithms are linear ones operating on vectors, that is, first-order tensors, including principal component analysis (PCA) [Jolliffe, 2002], independent component analysis (ICA) [Hyvärinen et al.,

---

[1]Dimensionality reduction is also known as *dimension reduction* or *dimensional reduction* [Burges, 2010]. Here, we adopt the name most commonly known in the machine learning and pattern recognition literature.

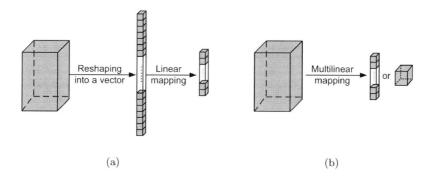

(a)   (b)

FIGURE 1.6: Linear versus multilinear mapping: (a) a tensor needs to be reshaped before a linear mapping to a low-dimensional vector, (b) direct multilinear mapping from a tensor to a vector or another tensor of lower dimension.

FIGURE 1.7: Reshaping (vectorization) of a 32×32 face image to a 1024×1 vector breaks the natural structure and correlation in the original face image. The 2D and 1D representations are shown on the same scale here.

2001], linear discriminant analysis (LDA) [Duda et al., 2001], canonical correlation analysis (CCA) [Hotelling, 1936], and partial least squares (PLS) analysis [Wold et al., 2001]. To apply these *linear subspace learning* (LSL) methods on tensor data of order higher than one, such as images and videos, we have to *reshape* (*vectorize*) tensors into vectors first, that is, to convert N-dimensional arrays ($N > 1$) to one-dimensional arrays, as depicted in Figure 1.6(a). Thus, LSL only partly alleviates the curse of dimensionality while such *reshaping*, (i.e., *vectorization*), has two fundamental limitations:

- Vectorization breaks the natural structure and correlation in the original data, reduces redundancies and/or higher order dependencies present in the original dataset, and loses potentially more compact or useful representations that can be obtained in the original tensor forms. For example, Figure 1.7 shows a 2D face image of size 32×32 and its corresponding reshaped vector with size 1024×1 on the same scale. From the 2D image (matrix) representation above, we can tell it is a face. However, from the 1D vector representation below, we cannot tell what it is.

- For higher-order tensor data of large size such as video sequences, the reshaped vectors are very high dimensional. Analysis of these vectors

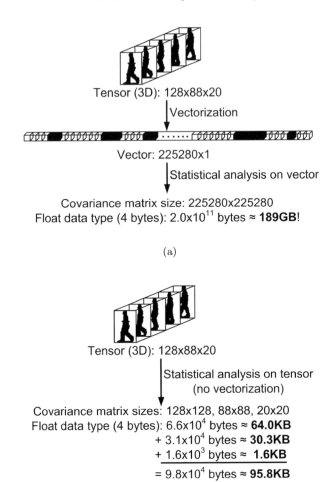

(a)

(b)

FIGURE 1.8: Vector-based versus tensor-based analysis of a 3D object: (a) reshaping (vectorization) of a $128 \times 88 \times 20$ gait silhouette sequence to a $225,280 \times 1$ vector will lead to a covariance (or scatter) matrix of size $225,280 \times 225,280$, which is about 189GB using floating-point data type; (b) tensor-based processing of the same gait sequence will lead to three covariance matrices of size 128×128, 88×88, and 20×20. The total size of these three covariance matrices will be about 95.8KB using floating-point data type.

often leads to high (sometimes impractical) computational and memory demand, and results in the SSS difficulties due to a large number of parameters to be estimated. For example, Figure 1.8(a) shows a 3D gait silhouette sequence of size $128 \times 88 \times 20$ on the top. Its vectorized version has a size of $225,280 \times 1$. In statistical analysis, we often need to calculate the covariance (or scatter) matrix, which will have a size of $225,280 \times 225,280$ for the vectorized gait sequence. If we use a floating-point data type for the covariance matrix, it will need about 2.0×10^{11} bytes of memory, which is about **189GB**. Although there are tricks to avoid such a big covariance matrix when it is of low-rank [Turk and Pentland, 1991], these tricks may have limited applicability or usefulness as more and more data are available in an era of big data.

Therefore, there has been a surging interest in more effective and efficient dimensionality reduction schemes for tensor data. Methods working directly on tensor representations have emerged as a promising approach. When we use tensor-based analysis, tensors are processed directly without vectorization. Figure 1.8(b) shows tensor-based analysis of the same 3D object in Figure 1.8(a), leading to three covariance matrices of size 128×128, 88×88, and 20×20. The total size of these three covariance matrices will be only about 95.8KB, which is several orders of magnitude smaller than the size of the covariance matrix in vector-based processing ($95.8\text{KB}/189\text{GB} \approx 4.8 \times 10^{-7}$). In other words, vector-based processing will need about **2 millions times more** memory for the covariance matrix than that of tensor-based processing in this case.

1.3 Multilinear Mapping for Subspace Learning

Driven by a proliferation of data-intensive applications, researchers have been working on broadening and generalizing successful statistical and computational tools based on linear algebra. Recently, the National Science Foundation (NSF) is calling for advancement in "computational thinking." In the report of the 2009 NSF Workshop on "Future Directions in Tensor-Based Computation and Modeling," it was pointed out that in matrix computations, the level of thinking seems to "kick up" about every twenty years as shown in Figure 1.9 [NSF, 2009]. We have progressed from scalar-level thinking to matrix level thinking, and then to *block matrix*-level thinking. Now we are in a transition from matrix-based to tensor-based computational thinking.

From a machine learning perspective, the goal of this book is to *advance tensor-level computational thinking in dimensionality reduction via subspace learning*. We investigate *multilinear subspace learning (MSL) of compact representations for multidimensional data*.

MSL solves for a *multilinear mapping*, which transforms input tensor data directly to low-dimensional tensors of the same or lower order (e.g., vectors), as

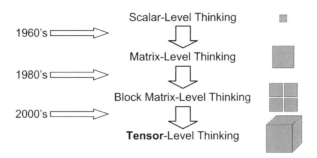

FIGURE 1.9: The field of matrix computations seems to "kick up" its level of thinking about every 20 years. (Adapted from the report of the NSF Workshop on "Future Directions in Tensor-Based Computation and Modeling," 2009 [NSF, 2009].)

shown in Figure 1.6(b). In contrast with linear mapping, multilinear mapping does not reshape tensors into vectors. Each *basis* in a linear mapping of tensor data is often specified by a large number of parameters, which equals to the input dimensionality, that is, *product* of dimensions in each mode. In contrast, each basis in a multilinear mapping of the same tensor data is often specified by a much smaller number of parameters, which equals to the *sum* of dimensions in each mode. Thus, bases in linear mapping have large degrees of freedom so they are able to capture richer representations, while bases in multilinear mapping are more constrained so the learned representations are sparser with more "grid-like" structures or regularities. This brings MSL three key benefits [Lu et al., 2011]:

- It preserves data structure *before mapping* by taking tensors directly as input.

- It can learn more compact and potentially more useful representations than linear subspace learning. With the same amount of data, MSL has no or a much less severe small sample size problem than LSL.

- It can handle big tensor data more efficiently with computations in much lower dimensions than linear methods.

Next, we demonstrate the different characteristics of features learned through linear and multilinear subspace learning with two examples. Figure 1.10 shows a second-order example of subspace learning on 1,360 face images. The input face images are of size $80 \times 80 = 6,400$, with a sample shown in Figure 1.10(a). Figures 1.10(b) and 1.10(c) depict three most discriminative face bases learned through linear and multilinear subspace learning using the same principle, respectively. A linear basis requires 6,400 parameters to specify while a multilinear basis needs only 160 parameters to specify, which is more compact ($\frac{160}{6400} = \frac{1}{40}$). Consequently, the linear bases look like "ghost-faces" while the multilinear bases have simpler "grid-like" structures. Despite being simpler, the multilinear bases have been shown to have higher face recognition accuracy in [Ye, 2005a].

Figure 1.11 is a third-order example on 731 gait silhouette sequences of size $64 \times 44 \times 20 = 56,320$, with a sample shown in Figure 1.11(a). Figures 1.11(b) and 1.11(c) show three most discriminative gait bases learned through linear and multilinear subspace learning using the same principle, respectively. In this case, each linear basis requires 56,320 parameters to specify, while a multilinear basis needs only 128 parameters to specify, which is more compact ($\frac{128}{56320} = \frac{1}{440}$). Similar to the second-order case in Figure 1.10, the linear bases closely resemble a real gait sequence in Figure 1.11(a) while the multilinear bases look more like filter banks with simpler "grid-like" structures. These multilinear bases have been shown to give higher gait recognition accuracy in [Lu, 2008; Lu et al., 2008b].

1.4 Roadmap

This book aims to provide a systematic and unified treatment of multilinear subspace learning. As Hamming [1986] suggested in his inspiring talk "You and Your Research," we want to provide the essence of this specific field. This first chapter has provided the motivation for MSL and a brief introduction to it. We started with the tensor representation of multidimensional data

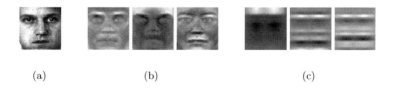

(a)   (b)   (c)

FIGURE 1.10: Illustration of second-order feature characteristics: (a) a sample face image of size 80×80, (b) three most discriminative face bases learned by linear subspace learning where each basis is specified by 6,400 parameters, and (c) three most discriminative face bases learned by multilinear subspace learning using a similar optimality criterion, where each basis is specified by 160 parameters. These simpler representations are shown to have higher face recognition accuracy in [Ye, 2005a].

and the need for dimensionality reduction to deal with the curse of dimensionality. Then, we discussed how conventional linear subspace learning for dimensionality reduction becomes inadequate for big tensor data as it needs to reshape tensors into high-dimensional vectors. From tensor-level computational thinking, MSL was next introduced to learn compact representations through direction multilinear mapping of tensors to alleviate those difficulties encountered by their linear counterparts.

The rest of this book consists of two parts:

Part I covers the **fundamentals** and **foundations** of MSL. **Chapter 2** reviews five basic LSL algorithms: PCA, ICA, LDA, CCA, and PLS. It

(a)

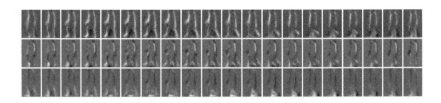

(b)

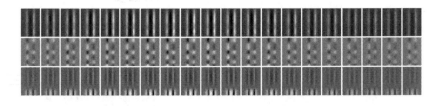

(c)

FIGURE 1.11: Illustration of third-order feature characteristics: (a) a sample gait silhouette sequence of size $64 \times 44 \times 20$, (b) three most discriminative gait bases (one in each row) learned by linear subspace learning where each basis is specified by 56,320 parameters, (c) three most discriminative gait bases (one in each row) learned by multilinear subspace learning using a similar optimality criterion, where each basis is specified by 128 parameters. These more compact representations are shown to give higher gait recognition accuracy in [Lu et al., 2008b].

also discusses closely-related topics including regularization, model selection, and ensemble-based learning. **Chapter 3** presents multilinear algebra preliminaries and two popular tensor decompositions in the beginning. Then, it introduces multilinear projections and scatter measures commonly used for constructing optimization criteria. These two chapters lay the groundwork necessary for the MSL algorithms in Part II. **Chapter 4** develops the MSL framework and gives an overview of PCA-based and LDA-based MSL algorithms. It also traces the history of MSL and related works, and points out future research directions. **Chapter 5** considers the algorithmic and computational aspects including typical solutions for MSL and related issues such as initialization and termination, synthetic data generation, feature selection strategies, resource demand, computational complexity, and some implementation tips for large datasets. It is an important chapter for practitioners who wants to employ or further develop MSL algorithms in their applications of interest.

"In this day of practically infinite knowledge, we need orientation to find our way. ... And we cope with that, essentially, by specialization. ... The present growth of knowledge will choke itself off until we get different tools. I believe that books which try to digest, coordinate, get rid of the duplication, get rid of the less fruitful methods and present the underlying ideas clearly of what we know now, will be the things the future generations will value. ... But I am inclined to believe that, in the long-haul, books which leave out what's not essential are more important than books which tell you everything because you don't want to know everything. ... You just want to know the essence."

Richard Hamming (1915–1998)
Mathematician, Turing Award Winner

Part II presents specific MSL **algorithms** and **applications**. **Chapters 6** and **7** are devoted to multilinear extensions of PCA and LDA, respectively. **Chapter 8** covers other MSL algorithms including multilinear extensions of ICA, CCA, and PLS. Within each chapter, five selected algorithms will be treated with more details for a deeper understanding of the subject. Lastly, **Chapter 9** describes a typical pattern recognition system and examines various applications of MSL. This chapter includes some experimental performance comparisons of MSL algorithms on popular face and gait recognition problems.

Appendix A gives some mathematical background for reference. **Appendix B** provides details on popular face and gait databases for MSL and the necessary preprocessing steps involved. Finally, **Appendix C** points the readers to related *open source software* and discusses the motivations, benefits, and common concerns of open source, as well as some software development tips.

On the first page of each chapter, we try to provide a figure that can (hopefully) act as a graphical abstract[2] to illustrate the most important concept in the chapter. At the end of each chapter in Part I, we summarize the key points and provide further readings. As Part II deals with specific algorithms and applications, further readings will be the respective references for chapters in this part. In addition, the contents of Chapter 2 are covered in many existing books on machine learning with varying levels of detail so knowledgeable readers may skip this chapter on first read.

This book focuses on machine learning. However, machine learning algorithms alone are often not enough to make a real impact in real-world applications, as pointed out in [Wagstaff, 2012]. To build a working system based on machine learning, it is important to formulate the real-world problems into machine learning tasks, collect data to learn from, and preprocess raw data into a form suitable for learning. Furthermore, it is also important to evaluate the learning system properly, give insightful interpretations, communicate the results to relevant scientific community, and persuade potential users to adopt the methods or systems.

1.5   Summary

- Most big data are multidimensional.

- Multidimensional data are called *tensors*, and each dimension is called a *mode*.

- Dimensionality reduction is often needed to transform tensors into lower dimension by learning a mapping to a *subspace*.

- Linear subspace learning needs to reshape input tensors into vectors, requiring lots of parameters to be estimated. This leads to difficulties when dealing with big tensor data.

---

[2]This is inspired by the graphical abstracts introduced by Elsevier: http://www.elsevier.com/authors/graphical-abstract

- Multilinear subspace learning directly maps tensor data to lower dimension. It can preserve data structure, result in more compact representations, and lead to lower computational demand than its linear counterpart.

Part I

Fundamentals and Foundations

Chapter 2

Linear Subspace Learning for Dimensionality Reduction

In Chapter 1, we pointed out why *dimensionality reduction* is commonly needed in practice. On one hand, real-world data are often specified in a high-dimensional space. Direct processing of high-dimensional data is computationally expensive and the number of samples available is often small compared to their high dimensionality. On the other hand, real-world data are often highly constrained to a low-dimensional *subspace*. Thus, dimensionality reduction attempts to learn a mapping of high-dimensional data to a low-dimensional space, that is, a subspace.

This chapter introduces the basics of linear subspace learning (LSL) for dimensionality reduction and serves as the foundation for Chapters 6, 7, and 8. LSL solves for a linear mapping to a subspace by optimizing a criterion [Duda et al., 2001; Shakhnarovich and Moghaddam, 2004]. It takes a high-dimensional random vector $\mathbf{x}$ as input and transforms it to a low-dimensional vector $\mathbf{y}$ through a *projection matrix* $\mathbf{U}$, as shown in Figure 2.1. When the input data are multidimensional, they need to be reshaped into vectors first.

LSL captures some *population* properties of the random vector $\mathbf{x}$ [Jolliffe, 2002]. In practice, the mapping or projection is learned from only a *sample* (a selected/collected subset) of the (whole) population [Anderson, 2003], e.g., a dataset $\{\mathbf{x}_1, \mathbf{x}_2, \mathbf{x}_3, ...\}$. Throughout this book, learning is with respect to the **sample** rather than the **population**.

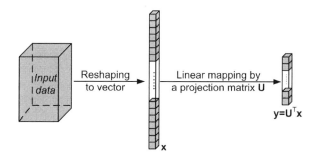

FIGURE 2.1: Linear subspace learning solves for a projection matrix $\mathbf{U}$, which maps a high-dimensional vector $\mathbf{x}$ to a low-dimensional vector $\mathbf{y}$.

We first consider five basic LSL algorithms, namely principal component analysis (PCA), independent component analysis (ICA), linear discriminant analysis (LDA), canonical correlation analysis (CCA), and partial least squares (PLS) analysis, in Sections. 2.1 to 2.5. PCA and ICA are *unsupervised learning* algorithms, which take unlabeled data as the input. LDA is a *supervised learning* algorithm that requires labeled data as input. While PCA, ICA, and LDA learn a subspace for one dataset, CCA and PLS learn two subspaces for two (paired) datasets. CCA and PLS are *unsupervised learning* algorithms by design. However, they are often used as *supervised learning* algorithms where one set of data is simply the labels.

After presenting these five algorithms, we give a unified view of PCA, LDA, CCA, and PLS as generalized eigenvalue problems. Next, we discuss several closely related topics, including regularization for reducing overfitting and improving generalization, cross-validation for model selection, and ensemble methods, where multiple models or learners are combined to obtain better performance.

2.1 Principal Component Analysis

PCA reduces the dimensionality of an input dataset consisting of correlated variables into an output set of linearly *uncorrelated* variables, the principal components (PCs). The output set, with reduced dimensionality, retains as much as possible the variation (or energy) present in the original input set[1]. The PCs are usually ordered by the variation captured in descending order and the tail values can be truncated for a desirable amount of dimensionality reduction. PCs are *independent* only if the dataset is jointly normally distributed [Jolliffe, 2002]. PCA is an unsupervised learning method. In the following, we show a standard derivation of PCA.

Suppose that the input dataset has M samples $\{\mathbf{x}_1, \mathbf{x}_2, ..., \mathbf{x}_M\}$, where each sample $\mathbf{x}_m \in \mathbb{R}^I$ is a vector, $m = 1, ..., M$. We measure the variation through the *total scatter matrix* $\mathbf{S}_T$ simply as $(m - 1)$ times the sample covariance matrix [Duda et al., 2001],

$$\mathbf{S}_T = \sum_{m=1}^{M} (\mathbf{x}_m - \bar{\mathbf{x}})(\mathbf{x}_m - \bar{\mathbf{x}})^T, \qquad (2.1)$$

---

[1] One of the rationales behind capturing the maximal variation is that noise can be assumed to be uniformly spread so directions of high variation tend to have a higher signal-to-noise ratio [Bie et al., 2005].

where $\bar{\mathbf{x}}$ is the *sample mean* defined as

$$\bar{\mathbf{x}} = \frac{1}{M} \sum_{m=1}^{M} \mathbf{x}_m. \tag{2.2}$$

In PCA, we want to find a linear mapping of the centered samples $\{\mathbf{x}_m - \bar{\mathbf{x}}\}$ to a low-dimensional representation $\{\mathbf{y}_m \in \mathbb{R}^P\}$, $P < I$, through a projection matrix $\mathbf{U} \in \mathbb{R}^{I \times P}$ as

$$\mathbf{y}_m = \mathbf{U}^T (\mathbf{x}_m - \bar{\mathbf{x}}). \tag{2.3}$$

The objective of PCA is to maximize the variation captured by the projected samples (or extracted features) $\{\mathbf{y}_m\}$. The projection matrix $\mathbf{U}$ can be considered to consist of P projection directions $\{\mathbf{u}_1, \mathbf{u}_2, ..., \mathbf{u}_P\}$. Then the projection in each direction is

$$y_{m_p} = \mathbf{u}_p^T (\mathbf{x}_m - \bar{\mathbf{x}}), \tag{2.4}$$

where $p = 1, ..., P$. Let us define a *coordinate vector* $\mathbf{g}_p \in \mathbb{R}^M$ for each p, where the mth element of $\mathbf{g}_p$ is defined as

$$g_{p_m} = \mathbf{u}_p^T (\mathbf{x}_m - \bar{\mathbf{x}}) = y_{m_p}. \tag{2.5}$$

PCA requires that each direction maximizes $\mathbf{u}_p^T \mathbf{S}_T \mathbf{u}_p$, and $\mathbf{g}_p$ and $\mathbf{g}_q$ are uncorrelated for $q \neq p$, $p, q = 1, ..., P$.

Centering: The subtraction of the mean in Equations (2.1) and (2.3) is called *centering*. It is often carried out as a preprocessing procedure. When the mean is zero ($\bar{\mathbf{x}} = \mathbf{0}$), the total scatter matrix becomes

$$\mathbf{S}_T = \sum_{m=1}^{M} \mathbf{x}_m \mathbf{x}_m^T = \mathbf{X}\mathbf{X}^T, \tag{2.6}$$

where $\mathbf{X} \in \mathbb{R}^{I \times M}$ is a data matrix formed with the M input samples $\{\mathbf{x}_m\}$ as its columns. Sometimes, Equation (2.6) is used directly without centering even when the mean is nonzero ($\bar{\mathbf{x}} \neq \mathbf{0}$). It should be noted that centering is preferred in PCA to ensure that the first PC describes the direction of maximum variation. If centering is not performed, the first PC often represents the mean of the data and the variation captured by the first PC often accounts for over 99% of the total variation, as pointed out in Section 3.4.3 of [Jackson, 1991]. Moreover, the further from the origin is the mean, the larger will be the variation captured by the first PC relative to that captured by the other PCs.

The first projection: To derive PCA, consider $p = 1$ to solve for $\mathbf{u}_1$ by maximizing $\mathbf{u}_1^T \mathbf{S}_T \mathbf{u}_1$ first. We need to impose a normalization constraint

$\mathbf{u}_1^T\mathbf{u}_1 = 1$ since the maximum will not be achieved for finite $\mathbf{u}_1$. The objective then becomes

$$\tilde{\mathbf{u}}_1 = \arg\max_{\mathbf{u}_1} \mathbf{u}_1^T\mathbf{S}_T\mathbf{u}_1 \quad \text{subject to} \quad \mathbf{u}_1^T\mathbf{u}_1 = 1. \tag{2.7}$$

Using the technique of *Lagrange multipliers* (Section A.3), we need to maximize

$$\psi_1 = \mathbf{u}_1^T\mathbf{S}_T\mathbf{u}_1 - \lambda(\mathbf{u}_1^T\mathbf{u}_1 - 1), \tag{2.8}$$

where λ is a Lagrange multiplier. We differentiate Equation (2.8) with respect to $\mathbf{u}_1$ and set the result to zero as

$$\frac{\partial\psi_1}{\partial\mathbf{u}_1} = \mathbf{S}_T\mathbf{u}_1 - \lambda\mathbf{u}_1 = (\mathbf{S}_T - \lambda\mathbf{I})\mathbf{u}_1 = \mathbf{0}, \tag{2.9}$$

where $\mathbf{I}$ is an identity matrix[2]. Therefore, λ and $\mathbf{u}_1$ are an eigenvalue and its corresponding eigenvector of $\mathbf{S}_T$, respectively (Section A.1.8). In addition, the quantity to be maximized is

$$\mathbf{u}_1^T\mathbf{S}_T\mathbf{u}_1 = \mathbf{u}_1^T\lambda\mathbf{u}_1 = \lambda\mathbf{u}_1^T\mathbf{u}_1 = \lambda. \tag{2.10}$$

Thus, we obtain $\tilde{\mathbf{u}}_1$ as the eigenvector associated with the largest eigenvalue λ_1 of $\mathbf{S}_T$.

The second projection: Next, we solve for the second projection vector $\mathbf{u}_2$ that maximizes $\mathbf{u}_2^T\mathbf{S}_T\mathbf{u}_2$ subject to the constraint that the projections by $\mathbf{u}_2$ (i.e., $\mathbf{g}_2$) are uncorrelated with those by $\mathbf{u}_1$ (i.e., $\mathbf{g}_1$). This *zero-correlation* constraint can be written as

$$\mathbf{u}_1^T\mathbf{S}_T\mathbf{u}_2 = \mathbf{u}_2^T\mathbf{S}_T\mathbf{u}_1 = \mathbf{u}_2^T\lambda_1\mathbf{u}_1 = \lambda_1\mathbf{u}_2^T\mathbf{u}_1 = 0 \Rightarrow \mathbf{u}_2^T\mathbf{u}_1 = 0. \tag{2.11}$$

Similar to the derivation of $\mathbf{u}_1$, we maximize the following quantity by incorporating the normalization and zero-correlation constraints:

$$\psi_2 = \mathbf{u}_2^T\mathbf{S}_T\mathbf{u}_2 - \lambda(\mathbf{u}_2^T\mathbf{u}_2 - 1) - \mu\mathbf{u}_2^T\mathbf{u}_1, \tag{2.12}$$

where λ and μ are Lagrange multipliers. We differentiate Equation (2.12) with respect to $\mathbf{u}_2$ and set the result to zero as

$$\frac{\partial\psi_2}{\partial\mathbf{u}_2} = 2\mathbf{S}_T\mathbf{u}_2 - 2\lambda\mathbf{u}_2 - \mu\mathbf{u}_1 = \mathbf{0}. \tag{2.13}$$

Multiplication of Equation (2.13) on the left by $\mathbf{u}_1^T$ gives

$$2\mathbf{u}_1^T\mathbf{S}_T\mathbf{u}_2 - 2\lambda\mathbf{u}_1^T\mathbf{u}_2 - \mu\mathbf{u}_1^T\mathbf{u}_1 = 0 \Rightarrow \mu = 0, \tag{2.14}$$

where the first two terms are zero by Equation (2.11) and $\mathbf{u}_1^T\mathbf{u}_1 = 1$. Again, from Equations (2.13) and (2.14) we have

$$\mathbf{S}_T\mathbf{u}_2 - \lambda\mathbf{u}_2 = (\mathbf{S}_T - \lambda\mathbf{I})\mathbf{u}_2 = \mathbf{0}, \tag{2.15}$$

---

[2]As the dimension of the identity matrix should be clear from the context, we do not indicate its dimension using subscript for simpler notation.

Algorithm 2.1 Principal component analysis (PCA)

Input: A set of M training samples $\{\mathbf{x}_1, \mathbf{x}_2, ..., \mathbf{x}_M\}$, and the number of PCs to estimate P or percentage of captured variation η to determine P by Equation (2.17).

Process:
 1: Calculate the mean $\bar{\mathbf{x}}$ from Equation (2.2).
 2: Calculate the total scatter matrix $\mathbf{S}_T$ from Equation (2.1).
 3: Get the P eigenvectors associated with the P largest eigenvalues of $\mathbf{S}_T$ to form $\tilde{\mathbf{U}}$.

Output: The projection matrix $\tilde{\mathbf{U}}$.

---

and λ and $\mathbf{u}_2$ are an eigenvalue and its corresponding eigenvector of $\mathbf{S}_T$, respectively. Following Equation (2.10), we are maximizing $\mathbf{u}_2^T \mathbf{S}_T \mathbf{u}_2 = \lambda$. Assuming that $\mathbf{S}_T$ does not have repeated eigenvalues[3], λ should not equal to λ_1 due to the constraint $\mathbf{u}_2^T \mathbf{u}_1 = 0$. Therefore, we obtain $\tilde{\mathbf{u}}_2$ as the eigenvector associated with the second largest eigenvalue λ_2 of $\mathbf{S}_T$.

Following a similar derivation, we can get the other projection vectors $\tilde{\mathbf{u}}_p$ $(p = 3, ..., P)$ as the pth eigenvector associated with the pth largest eigenvalue λ_p of $\mathbf{S}_T$. Thus, the projection matrix $\tilde{\mathbf{U}}$ is composed of the eigenvectors corresponding to the P $(P < I)$ largest eigenvalues of $\mathbf{S}_T$ and the variation captured by the pth PC is the pth eigenvalue λ_p. The projected PCA features have the following properties [Fukunaga, 1990]:

1. The effectiveness of each feature, in terms of representing the input, is determined by its corresponding eigenvalue.

2. The projected features are uncorrelated. If the input data is Gaussian (i.e., normally distributed), the projected features are mutually *independent* (i.e., their *mutual information* is maximized).

3. The P eigenvectors minimize the mean squared error in reconstruction over all choices of P *orthonormal* basis vectors.

The projection of a test sample $\mathbf{x}$ in the PCA subspace is obtained as

$$\mathbf{y} = \tilde{\mathbf{U}}^T(\mathbf{x} - \bar{\mathbf{x}}). \tag{2.16}$$

The number of PCs, P, is commonly determined by specifying the total variation (energy) to be captured. For example, suppose $\mathbf{S}_T$ has β (non-zero) eigenvalues $\{\lambda_1, ..., \lambda_\beta\}$ in total, the number of PCs to be kept to capture at least $\eta\%$ of the total variation (energy) is

$$P^* = \arg\min_P \frac{\sum_{p=1}^{P} \lambda_p}{\sum_{p=1}^{\beta} \lambda_p} > \frac{\eta}{100}. \tag{2.17}$$

---

[3] Please refer to Section 2.4 of [Jolliffe, 2002] for the case of repeated eigenvalues.

Algorithm 2.1 provides the pseudocode for PCA. It should be noted that when $M < I$, especially when $M \ll I$, the computation of eigenvectors for an $I \times I$ matrix could be computationally expensive (sometimes infeasible). However, there is an efficient way to compute them through the transposed scatter matrix as described in [Turk and Pentland, 1991].

2.2 Independent Component Analysis

ICA is an important unsupervised learning method for finding representational components of data with maximum statistical independence [Hyvärinen et al., 2001]. Statistically independent components mean that the value of any one of the components gives no information on the values of the other components [Hyvärinen et al., 2001]. PCA considers only the *first-order* and *second-order* statistics and it gives *independent components* (ICs)[4] only for *Gaussian* data. In constrast, ICA finds ICs for the general case of *non-Gaussian* data. Thus, many ICA estimation methods aim to maximize the *non-Gaussianity* of data [Hyvärinen and Oja, 2000] and make use of *higher-order* statistics.

> **The cocktail-party problem:** Imagine we have two microphones at two different locations in a room where two people are speaking simultaneously. The two microphones record two time signals, $x_1(t)$ and $x_2(t)$, where t is the time index. Each recorded signal is then a weighted sum of the speech signals from the two speakers, denoted as $s_1(t)$ and $s_2(t)$. Thus, we could write $x_1(t)$ and $x_2(t)$ as the following linear equations:
>
> $$x_1(t) = a_{11}s_1(t) + a_{12}s_2(t), \tag{2.18}$$
> $$x_2(t) = a_{21}s_1(t) + a_{22}s_2(t), \tag{2.19}$$
>
> where a_{11}, a_{12}, a_{21}, and a_{22} are the *weight parameters* dependent on the distances between the microphones and the speakers. We would like to estimate the two original speech signals $s_1(t)$ and $s_2(t)$ from only the recorded signals $x_1(t)$ and $x_2(t)$. This is the well-known *cocktail-party problem*, a typical example of the *blind source separation* problem where ICA can be very useful [Hyvärinen and Oja, 2000].

The ICA model: ICA is an important tool for blind source separation (BSS). A typical BSS problem is the "cocktail party problem," where the underlying speech signals are separated from recorded signals of people talking simultaneously in a room. In this case, the recorded signals are the *mixtures*

---

[4]ICs are also known as factors, latent variables, or sources.

and the speech signals are the *sources* or *latent variables*. The problem is usually simplified by assuming no time delays or echoes. The (simplified, noise-free) *ICA model* assumes that we observe P linear mixtures $x_1, ..., x_P$ of P ICs[5] with weight parameters $\{a_{qp}\}$ [Hyvärinen and Oja, 2000].

$$x_q = a_{q_1}s_1 + a_{q_2}s_2 + ... + a_{q_P}s_P, \quad \text{for } q = 1, ..., P. \tag{2.20}$$

Let $\mathbf{x}$ denote the vector with elements $x_1, x_2, ..., x_P$. Let $\mathbf{s}$ denote the vector with elements $s_1, s_2, ..., s_P$. Collecting the coefficients a_{q_p} in a matrix $\mathbf{A}$, the mixing model in Equation (2.20) then becomes

$$\mathbf{x} = \mathbf{As}. \tag{2.21}$$

The ICA model is a *generative model* describing how the observed data are generated by mixing the components $\{s_p\}$. Matrix $\mathbf{A}$ is called the *mixing matrix*. In the ICA problem, all we observe are the mixtures $\mathbf{x}$ and we need to estimate $\mathbf{A}$ and $\mathbf{s}$ using only $\mathbf{x}$.

To solve the ICA estimation problem, we assume that the components $\{s_p\}$ are statistically *independent*. Another commonly made assumption is that the unknown mixing matrix $\mathbf{A}$ is square ($\mathbf{A} \in \mathbb{R}^{P \times P}$), as in Equation (2.20). Thus, after the estimation of $\mathbf{A}$, its inverse $\mathbf{W}$ can be computed to obtain the ICs as

$$\mathbf{s} = \mathbf{Wx}. \tag{2.22}$$

The matrix $\mathbf{W}$ is then called the *unmixing matrix* or *separating matrix*. There are a few points to note regarding ICA [Hyvärinen and Oja, 2000]:

1. Similar to PCA, ICA can only identify ICs up to scaling (including sign).

2. Unlike PCA, the ICs estimated by ICA are not ordered.

3. ICs are identifiable only if at most one of the sources is Gaussian and the number of observed mixtures is not smaller than the number of estimated components.

4. ICA cannot identify the actual number of source signals.

Estimation principles for ICA: The ICA estimation problem is solved by maximizing the statistical independence of the estimated ICs and it is much harder than the estimation problem in PCA. There are two popular ways to define an independence measure to be optimized. One approach estimates ICs through minimization of *mutual information* [Bell and Sejnowski, 2000], based on information theory. Mutual information is a natural measure of the dependence between random variables and it is equivalent to the *Kullback–Leibler*

---

[5]In the ICA model, it is often assumed that the number of mixtures is equal to the number of ICs for convenience. In practice, ICA can be performed as long as the number of mixtures is not less than the number of ICs.

divergence. The other approach solves the ICA problem by maximization of *non-Gaussianity* [Hyvärinen and Oja, 2000], motivated by the central limit theorem. Non-Gaussianity can be measured by either *kurtosis*, that is, the fourth-order *cumulant*[6], or *negentropy*, which is based on the information-theoretic quantity of (differential) entropy. Unlike PCA, closed-form solutions do not exist in general for ICA. Thus, ICA estimation algorithms are iterative and computationally more expensive.

Entropy is a fundamental concept in information theory. It measures the uncertainty associated with a random variable, or the amount of information contained in the observed variable. The more random the variable is, the larger its entropy. Entropy is usually quantified as the expected value (e.g., coding length in bits) of the information contained in a message. A single toss of a fair coin has an entropy of one bit.

Preprocessing and dimensionality reduction: Before an ICA algorithm is applied to data, several preprocessing steps are usually very helpful to make the estimation problem simpler and better conditioned. Typical preprocessing includes centering, *whitening*, and dimensionality reduction. Centering makes $\mathbf{x}$ a zero-mean variable, and $\mathbf{s}$ becomes zero-mean too. This reduces the complexity of the ICA estimation problem. After centering, whitening transforms the observed vector $\mathbf{x}$ linearly to obtain a new vector with uncorrelated components and unit variances. Furthermore, the number of ICs found by ICA algorithms typically corresponds to the dimension of the input so dimensionality reduction is often needed before applying ICA. Therefore, PCA, which does centering, whitening, and dimensionality reduction, is a frequently used preprocessing step before ICA [Bartlett et al., 2002].

Well-known ICA algorithms include the infomax algorithm [Bell and Sejnowski, 2000], the FastICA algorithm [Hyvärinen, 1999], and the joint approximate diagonalization of eigenmatrices (JADE) [Cardoso and Souloumiac, 1993; Cardoso, 1998]. They are more complex compared to the PCA algorithm so they are not described and their pseudocodes are not provided here. Interested readers can refer to the references provided.

Two ICA architectures for image data: For image representation and recognition, two ICA architectures can be employed [Bartlett et al., 2002]. *Architecture I* treats images as random variables and pixels as random trials to find spatially local basis images that are statistically independent. *Architecture II* treats pixels as random variables and images as random trials to find factorial code that reflects global properties.

We first extend the ICA model for scalar variables in Equation (2.20) to vector-valued variables as

$$\mathbf{x}_q = a_{q_1}\mathbf{s}_1 + a_{q_2}\mathbf{s}_2 + ... + a_{q_P}\mathbf{s}_P, \quad \text{for } q = 1, ..., P, \qquad (2.23)$$

---

[6]Therefore, ICA makes use of higher-order statistics.

that is, we observe P linear mixtures $\mathbf{x}_1, ..., \mathbf{x}_P$ of P independent vector-valued sources $\mathbf{s}_1, ..., \mathbf{s}_P$ (the ICs), where $\mathbf{x}_p \in \mathbb{R}^I$ and $\mathbf{s}_p \in \mathbb{R}^I$. Let $\mathbf{X}$ denote the mixture matrix with its columns as $\mathbf{x}_1, \mathbf{x}_2, ..., \mathbf{x}_P$. Let $\mathbf{S}$ denote the source matrix with its columns as $\mathbf{s}_1, \mathbf{s}_2, ..., \mathbf{s}_P$. Form the mixing matrix $\mathbf{A}$ with the coefficents a_{q_p}, the mixing model in Equation (2.23) then becomes

$$\mathbf{X} = \mathbf{AS}. \tag{2.24}$$

To find a good set of basis images to represent a set of M images through ICA, each image is reshaped into a vector of $I \times 1$, and the image set can be organized into a matrix $\mathbf{X}$ in two ways to perform ICA [Bartlett et al., 2002].

1. In Architecture I, each row vector of $\mathbf{X}$ is a different image. Each mixture is an M-dimensional vector consisting of pixel values of the same pixel location from all M images. There are I such mixtures. Thus, in this architecture, ICA estimates the independence of images or functions of images. Two images are independent if when moving across pixels, it is not possible to predict the value taken by the pixel on one image based on the value taken by the same pixel on the other image. This is the typical usage of ICA for blind source separation.

2. In Architecture II, each column vector of $\mathbf{X}$ is a different image. Each mixture is an I-dimensional vector corresponding to an image. There are M such mixtures. Thus, in this architecture, ICA estimates the independence of pixels or functions of pixels. Two pixels would be independent if when moving across the entire set of images it is not possible to predict the value taken by one pixel based on the corresponding value taken by the other pixel on the same image.

2.3   Linear Discriminant Analysis

LDA is a classical approach to dimensionality reduction for labeled data, that is, it is a classical *supervised* dimensionality reduction algorithm. Frequently, LDA for binary class labels is called *Fisher discriminant analysis* (FDA), and LDA for multiple classes is called *multiple discriminant analysis* (MDA). LDA may be viewed as a natural extension of PCA to the case of labeled data. PCA finds components that are useful for representing data, while LDA seeks components that are useful for discriminating data. In other words, PCA is designed for *representation* while LDA is designed for *discrimination*, although they both look for linear combinations of input variables that best explain the data.

In supervised learning, the M input samples $\{\mathbf{x}_m \in \mathbb{R}^I\}$ are provided with class labels. Let us define C as the number of classes, and $c = 1, ..., C$ as the class index.

In LDA[7], a class separability criterion is formulated through the definition of two scatter matrices. The *within-class scatter* matrix captures the scatter of samples around their respective class means as [Duda et al., 2001]

$$\mathbf{S}_W = \sum_{m=1}^{M} (\mathbf{x}_m - \bar{\mathbf{x}}_{c_m})(\mathbf{x}_m - \bar{\mathbf{x}}_{c_m})^T, \qquad (2.25)$$

where c_m is the class label c for the mth training sample $\mathbf{x}_m$, and $\bar{\mathbf{x}}_c$ is the mean of the M_c training samples in class c:

$$\bar{\mathbf{x}}_c = \frac{1}{M_c} \sum_{m,c_m=c} \mathbf{x}_m. \qquad (2.26)$$

On the other hand, the *between-class scatter* matrix captures the scatter of class means around the overall data mean as [Duda et al., 2001]

$$\mathbf{S}_B = \sum_{c=1}^{C} M_c (\bar{\mathbf{x}}_c - \bar{\mathbf{x}})(\bar{\mathbf{x}}_c - \bar{\mathbf{x}})^T. \qquad (2.27)$$

The *total scatter* matrix defined in Equation (2.1) equals the sum of the within-class and between-class scatter matrices:

$$\mathbf{S}_T = \mathbf{S}_W + \mathbf{S}_B. \qquad (2.28)$$

While the scatter matrices above are defined in the input space, in LDA, we are interested in the scatter in the output (feature) space defined through a projection matrix $\mathbf{U}$. The within-class scatter matrix $\mathbf{S}_{WY}$ and the between-class scatter matrix $\mathbf{S}_{BY}$ in the output space are related to those in the input space by

$$\mathbf{S}_{WY} = \mathbf{U}^T \mathbf{S}_W \mathbf{U}, \;\; \mathbf{S}_{BY} = \mathbf{U}^T \mathbf{S}_B \mathbf{U}. \qquad (2.29)$$

Similarly, the total scatter matrix $\mathbf{S}_{TY}$ in the output space is related to $\mathbf{S}_T$ as

$$\mathbf{S}_{TY} = \mathbf{U}^T \mathbf{S}_T \mathbf{U}. \qquad (2.30)$$

To find the most discriminative features, LDA seeks a projection $\mathbf{U}$ that maximizes the between-class scatter $\mathbf{S}_{BY}$ while minimizing the within-class scatter $\mathbf{S}_{WY}$ in the output space [Belhumeur et al., 1997]. Thus, the LDA criterion function can be defined for nonsingular $\mathbf{S}_W$ as

$$\tilde{\mathbf{U}} = \arg\max_{\mathbf{U}} \mathrm{tr}(\mathbf{S}_{WY}^{-1}\mathbf{S}_{BY}) = \arg\max_{\mathbf{U}} \mathrm{tr}\left((\mathbf{U}^T\mathbf{S}_W\mathbf{U})^{-1}\mathbf{U}^T\mathbf{S}_B\mathbf{U}\right), \quad (2.31)$$

where $\mathrm{tr}(\cdot)$ is the trace of a matrix, which is the sum of the eigenvalues measuring the variations in principal directions of the matrix [Duda et al., 2001].

---

[7]The mapping in LDA is to a space of dimension $C-1$. Thus, it is tacitly assumed that the input dimensionality I is no smaller than C: $I \geq C$ [Duda et al., 2001].

This criterion corresponds to the J_1 criterion in Section 10.2 of [Fukunaga, 1990]. It should be noted that besides $\{\mathbf{S}_{WY}, \mathbf{S}_{BY}\}$ used in Equation (2.31), $\{\mathbf{S}_{WY}, \mathbf{S}_{TY}\}$ or $\{\mathbf{S}_{TY}, \mathbf{S}_{BY}\}$ can also be used, with the same set of features produced [Fukunaga, 1990].

Another commonly seen LDA criterion is the ratio of the scatter matrix determinants:

$$\tilde{\mathbf{U}} = \arg\max_{\mathbf{U}} \frac{|\mathbf{S}_{BY}|}{|\mathbf{S}_{WY}|} = \arg\max_{\mathbf{U}} \frac{|\mathbf{U}^T \mathbf{S}_B \mathbf{U}|}{|\mathbf{U}^T \mathbf{S}_W \mathbf{U}|}, \qquad (2.32)$$

where $|\cdot|$ is the determinant of a matrix, which is the product of the eigenvalues. This criterion corresponds to the J_2 criterion in Section 10.2 of [Fukunaga, 1990]. It should be noted, however, that this criterion may not be properly defined in general because the rank of $\mathbf{S}_B$ (and $\mathbf{S}_{BY}$ in turn) is no greater than $C - 1$ from Equation (2.27), $|\mathbf{S}_{BY}| = 0$ for $I \geq C$ [Fukunaga, 1990]. For LDA criterion defined through the determinant ratio, $\frac{|\mathbf{S}_{TY}|}{|\mathbf{S}_{WY}|}$ should be used, which produces the same features as Equation (2.31) nonetheless [Fukunaga, 1990].

LDA for two classes: When there are only two classes, that is, $C = 2$, (or when we are interested in only the most discriminative projection for $C > 2$), only one projection is to be solved and Equation (2.32) becomes

$$\tilde{\mathbf{u}} = \arg\max_{\mathbf{u}} \frac{\mathbf{u}^T \mathbf{S}_B \mathbf{u}}{\mathbf{u}^T \mathbf{S}_W \mathbf{u}}. \qquad (2.33)$$

We can solve for $\mathbf{u}$ above through the constrained optimization problem below:

$$\tilde{\mathbf{u}} = \arg\max_{\mathbf{u}} \mathbf{u}^T \mathbf{S}_B \mathbf{u} \text{ subject to } \mathbf{u}^T \mathbf{S}_W \mathbf{u} = 1. \qquad (2.34)$$

We form the Lagrangian as

$$\psi = \mathbf{u}^T \mathbf{S}_B \mathbf{u} - \lambda(\mathbf{u}^T \mathbf{S}_W \mathbf{u} - 1). \qquad (2.35)$$

Differentiate ψ with respect to $\mathbf{u}$ and set to zero gives

$$\frac{\partial \psi}{\partial \mathbf{u}} = \mathbf{S}_B \mathbf{u} - \lambda \mathbf{S}_W \mathbf{u} = \mathbf{0} \implies \mathbf{S}_B \mathbf{u} = \lambda \mathbf{S}_W \mathbf{u}, \qquad (2.36)$$

which is a generalized eigenvalue problem (Section A.1.9). The quantity to be maximized is

$$\mathbf{u}^T \mathbf{S}_B \mathbf{u} = \lambda \mathbf{u}^T \mathbf{S}_W \mathbf{u} = \lambda. \qquad (2.37)$$

Thus, we obtain the solution $\tilde{\mathbf{u}}$ as the generalized eigenvector corresponding to the largest generalized eigenvalue λ of the generalized eigenvalue problem $\mathbf{S}_B \mathbf{u} = \lambda \mathbf{S}_W \mathbf{u}$.

For general $C > 2$, the maximization of Equation (2.31) leads to the following generalized eigenvalue problem:

$$\mathbf{S}_B \mathbf{u}_p = \lambda_p \mathbf{S}_W \mathbf{u}_p, \qquad (2.38)$$

Algorithm 2.2 Linear discriminant analysis (LDA)

Input: A set of M training samples $\{\mathbf{x}_1, \mathbf{x}_2, ..., \mathbf{x}_M\}$, with class labels $\mathbf{c} \in \mathbb{R}^M$.

Process:

1: Calculate the within-class scatter matrix $\mathbf{S}_W$ from Equation (2.25).
2: Calculate the between-class scatter matrix $\mathbf{S}_B$ from Equation (2.27).
3: Get the first $C - 1$ generalized eigenvectors of the generalized eigenvalue problem in Equation (2.38) to form $\tilde{\mathbf{U}}$.

Output: The projection matrix $\tilde{\mathbf{U}}$.

---

where $\mathbf{u}_p$ is the pth column of $\mathbf{U} \in \mathbb{R}^{I \times P}$. The LDA projection matrix $\tilde{\mathbf{U}}$ is comprised of the generalized eigenvectors corresponding to the largest P generalized eigenvalues of Equation (2.38). As the rank of $\mathbf{S}_B$ is no greater than $C - 1$, LDA can extract at most $C - 1$ features, that is, $P \leq C - 1$. When $\mathbf{S}_W$ is nonsingular, $\tilde{\mathbf{U}}$ can be obtained as the eigenvectors corresponding to the largest P eigenvalues of $\mathbf{S}_W^{-1}\mathbf{S}_B$. There are two important observations regarding LDA projection matrix and the projected features:

1. The projection matrix $\tilde{\mathbf{U}}$ is not unique, as the criterion to be maximized (trace/determinant ratio) is invariant under any nonsingular linear transformations [Fukunaga, 1990].

2. For nonsingular $\mathbf{S}_W$, LDA is equivalent to one of its variant named as uncorrelated LDA (ULDA) [Jin et al., 2001a], which produces uncorrelated features. This is proved in [Jin et al., 2001b; Ye et al., 2006]. Thus, the projected features by LDA are uncorrelated.

Similar to PCA, the projection of a test sample $\mathbf{x}$ in the LDA space is obtained as

$$\mathbf{y} = \tilde{\mathbf{U}}^T \mathbf{x}. \tag{2.39}$$

Algorithm 2.2 provides the pseudocode for LDA. When applied to real data, LDA often has problems due to an insufficient number of training samples, that is, the small sample size (SSS) problem, which results in singular $\mathbf{S}_W$. PCA is routinely employed before LDA to reduce the dimensionality before LDA to avoid this difficulty, leading to the PCA+LDA approach, for example, the *Fisherface* method for face recognition [Belhumeur et al., 1997]. Another popular approach is to regularize the scatter matrices [Friedman, 1989], which will be briefly discussed in Section 2.7.

2.4   Canonical Correlation Analysis

The three LSL algorithms discussed so far, PCA, ICA, and LDA, all consider the linear mapping of one set of input data, either labeled or unlabeled. In contrast, CCA considers the mappings of two sets of *paired data* [Hotelling, 1936] and aims to find correlations between them through linear projections. It is designed to capture the correlations between two sets of vector-valued variables that are assumed to be different representations of the same set of objects [Anderson, 2003]. CCA has various applications such as information retrieval [Hardoon et al., 2004], multi-label learning [Sun et al., 2008c; Rai and Daumé III, 2009], and multi-view learning [Chaudhuri et al., 2009; Dhillon et al., 2011].

Consider two paired datasets $\{\mathbf{x}_m \in \mathbb{R}^I\}$, $\{\mathbf{y}_m \in \mathbb{R}^J\}$, $m = 1, ..., M$, $I \neq J$ in general. It is assumed that $\{\mathbf{x}_m\}$ and $\{\mathbf{y}_m\}$ are different views of the same objects. The pth pair of projection $(\mathbf{u}_{x_p}, \mathbf{u}_{y_p})$, $\mathbf{u}_{x_p} \in \mathbb{R}^I$, $\mathbf{u}_{y_p} \in \mathbb{R}^J$, projects the two datasets $\{\mathbf{x}_m\}$ and $\{\mathbf{y}_m\}$ to $\{r_{m_p}\}$ and $\{s_{m_p}\}$ as

$$r_{m_p} = \mathbf{u}_{x_p}^T \mathbf{x}_m, \quad s_{m_p} = \mathbf{u}_{y_p}^T \mathbf{y}_m. \tag{2.40}$$

The pth coordinates form the coordinate vectors $\mathbf{w}_p$ and $\mathbf{z}_p$, where

$$w_{p_m} = r_{m_p}, \quad z_{p_m} = s_{m_p}, \tag{2.41}$$

as defined in Equation (2.5). CCA seeks paired projections $\{\mathbf{u}_{x_p}, \mathbf{u}_{y_p}\}$ such that $\mathbf{w}_p$ is maximally correlated with $\mathbf{z}_p$. In addition, for $p \neq q$, $\mathbf{w}_p$ and $\mathbf{w}_q$, $\mathbf{z}_p$ and $\mathbf{z}_q$, and $\mathbf{w}_p$ and $\mathbf{z}_q$, respectively, are all uncorrelated. $\{\mathbf{w}_p, \mathbf{z}_p\}$ are the pth pair of *canonical variates*, and their correlation is the pth *canonical correlation* [Anderson, 2003].

Let us define the *sample covariance matrices* and the *sample cross-covariance matrices* first as [Duda et al., 2001; Anderson, 2003]

$$\boldsymbol{\Sigma}_{\mathbf{xx}} = \frac{1}{M-1} \sum_{m=1}^{M} (\mathbf{x}_m - \bar{\mathbf{x}})(\mathbf{x}_m - \bar{\mathbf{x}})^T, \tag{2.42}$$

$$\boldsymbol{\Sigma}_{\mathbf{yy}} = \frac{1}{M-1} \sum_{m=1}^{M} (\mathbf{y}_m - \bar{\mathbf{y}})(\mathbf{y}_m - \bar{\mathbf{y}})^T, \tag{2.43}$$

$$\boldsymbol{\Sigma}_{\mathbf{xy}} = \frac{1}{M-1} \sum_{m=1}^{M} (\mathbf{x}_m - \bar{\mathbf{x}})(\mathbf{y}_m - \bar{\mathbf{y}})^T, \tag{2.44}$$

$$\boldsymbol{\Sigma}_{\mathbf{yx}} = \frac{1}{M-1} \sum_{m=1}^{M} (\mathbf{y}_m - \bar{\mathbf{y}})(\mathbf{x}_m - \bar{\mathbf{x}})^T, \tag{2.45}$$

where $\bar{\mathbf{x}}$ and $\bar{\mathbf{y}}$ are the sample means as defined in Equation (2.2) and note that $\boldsymbol{\Sigma}_{\mathbf{yx}} = \boldsymbol{\Sigma}_{\mathbf{xy}}^T$. The correlation ρ_p between $\mathbf{w}_p$ and $\mathbf{z}_p$ is then given by

$$\rho_p = \frac{\mathbf{u}_{x_p}^T \boldsymbol{\Sigma}_{\mathbf{xy}} \mathbf{u}_{y_p}}{\sqrt{(\mathbf{u}_{x_p}^T \boldsymbol{\Sigma}_{\mathbf{xx}} \mathbf{u}_{x_p})(\mathbf{u}_{y_p}^T \boldsymbol{\Sigma}_{\mathbf{yy}} \mathbf{u}_{y_p})}}. \tag{2.46}$$

To solve the problem, we follow the derivation in [Anderson, 2003], which is a similar procedure as that in the derivation of PCA in Section 2.1. We find the first projection pair giving the maximum correlation first, and then solve for other projection pairs with zero-correlation constraints.

The first projection pair: We first find a pair of projection vectors $\{\mathbf{u}_{x_1}, \mathbf{u}_{y_1}\}$ such that ρ_1 is maximized. Because rescaling of these projection vectors does not change the correlation ρ_1, we can maximize $\mathbf{u}_{x_1}^T \boldsymbol{\Sigma}_{\mathbf{xy}} \mathbf{u}_{y_1}$ subject to

$$\mathbf{u}_{x_1}^T \boldsymbol{\Sigma}_{\mathbf{xx}} \mathbf{u}_{x_1} = 1, \quad \mathbf{u}_{y_1}^T \boldsymbol{\Sigma}_{\mathbf{yy}} \mathbf{u}_{y_1} = 1. \tag{2.47}$$

Thus, we write the objective function as

$$\begin{aligned}
(\tilde{\mathbf{u}}_{x_1}, \tilde{\mathbf{u}}_{y_1}) &= \arg \max_{(\mathbf{u}_{x_1}, \mathbf{u}_{y_1})} \mathbf{u}_{x_1}^T \boldsymbol{\Sigma}_{\mathbf{xy}} \mathbf{u}_{y_1}, \\
\text{subject to} \quad &\mathbf{u}_{x_1}^T \boldsymbol{\Sigma}_{\mathbf{xx}} \mathbf{u}_{x_1} = \mathbf{u}_{y_1}^T \boldsymbol{\Sigma}_{\mathbf{yy}} \mathbf{u}_{y_1} = 1.
\end{aligned} \tag{2.48}$$

Now we write the problem above in Lagrangian form as

$$\psi_1 = \mathbf{u}_{x_1}^T \boldsymbol{\Sigma}_{\mathbf{xy}} \mathbf{u}_{y_1} - \frac{1}{2}\lambda(\mathbf{u}_{x_1}^T \boldsymbol{\Sigma}_{\mathbf{xx}} \mathbf{u}_{x_1} - 1) - \frac{1}{2}\mu(\mathbf{u}_{y_1}^T \boldsymbol{\Sigma}_{\mathbf{yy}} \mathbf{u}_{y_1} - 1), \tag{2.49}$$

where λ and μ are Lagrange multipliers. We then take the derivatives with respect to $\mathbf{u}_{x_1}$ and $\mathbf{u}_{y_1}$ and set to zero:

$$\frac{\partial \psi_1}{\partial \mathbf{u}_{x_1}} = \boldsymbol{\Sigma}_{\mathbf{xy}} \mathbf{u}_{y_1} - \lambda \boldsymbol{\Sigma}_{\mathbf{xx}} \mathbf{u}_{x_1} = \mathbf{0}, \tag{2.50}$$

$$\frac{\partial \psi_1}{\partial \mathbf{u}_{y_1}} = \boldsymbol{\Sigma}_{\mathbf{yx}} \mathbf{u}_{x_1} - \mu \boldsymbol{\Sigma}_{\mathbf{yy}} \mathbf{u}_{y_1} = \mathbf{0}. \tag{2.51}$$

Now multiply Equation (2.50) on the left by $\mathbf{u}_{x_1}^T$ and Equation (2.51) on the left by $\mathbf{u}_{y_1}^T$, we have

$$\mathbf{u}_{x_1}^T \boldsymbol{\Sigma}_{\mathbf{xy}} \mathbf{u}_{y_1} - \lambda \mathbf{u}_{x_1}^T \boldsymbol{\Sigma}_{\mathbf{xx}} \mathbf{u}_{x_1} = 0, \tag{2.52}$$

$$\mathbf{u}_{y_1}^T \boldsymbol{\Sigma}_{\mathbf{yx}} \mathbf{u}_{x_1} - \mu \mathbf{u}_{y_1}^T \boldsymbol{\Sigma}_{\mathbf{yy}} \mathbf{u}_{y_1} = 0. \tag{2.53}$$

Because we have $\mathbf{u}_{x_1}^T \boldsymbol{\Sigma}_{\mathbf{xx}} \mathbf{u}_{x_1} = 1$ and $\mathbf{u}_{y_1}^T \boldsymbol{\Sigma}_{\mathbf{yy}} \mathbf{u}_{y_1} = 1$, Equations (2.52) and

(2.53) indicate that $\lambda = \mu = \mathbf{u}_{x_1}^T \Sigma_{\mathbf{xy}} \mathbf{u}_{y_1}$ is the criterion to be maximized. From Equation (2.52), assuming $\Sigma_{\mathbf{xx}}$ is nonsingular, we have

$$\mathbf{u}_{x_1} = \frac{\Sigma_{\mathbf{xx}}^{-1} \Sigma_{\mathbf{xy}}}{\lambda} \mathbf{u}_{y_1}. \tag{2.54}$$

Substituting Equation (2.54) into Equation (2.51) and noting $\lambda = \mu$, we have

$$\Sigma_{\mathbf{yx}} \Sigma_{\mathbf{xx}}^{-1} \Sigma_{\mathbf{xy}} \mathbf{u}_{y_1} = \lambda^2 \Sigma_{\mathbf{yy}} \mathbf{u}_{y_1}, \tag{2.55}$$

which is a generalized eigenvalue problem and $\tilde{\mathbf{u}}_{y_1}$ corresponds to the generalized eigenvector associated with the largest generalized eigenvalue. Then, $\tilde{\mathbf{u}}_{x_1}$ can be obtained from Equation (2.54) or in an analogous way from

$$\Sigma_{\mathbf{xy}} \Sigma_{\mathbf{yy}}^{-1} \Sigma_{\mathbf{yx}} \mathbf{u}_{x_1} = \lambda^2 \Sigma_{\mathbf{xx}} \mathbf{u}_{x_1}, \tag{2.56}$$

assuming that $\Sigma_{\mathbf{yy}}$ is nonsingular.

For nonsingular $\Sigma_{\mathbf{xx}}$ and $\Sigma_{\mathbf{yy}}$, we can also write Equations (2.55) and (2.56) as

$$\Sigma_{\mathbf{yy}}^{-1} \Sigma_{\mathbf{yx}} \Sigma_{\mathbf{xx}}^{-1} \Sigma_{\mathbf{xy}} \mathbf{u}_{y_1} = \lambda^2 \mathbf{u}_{y_1}, \tag{2.57}$$

$$\Sigma_{\mathbf{xx}}^{-1} \Sigma_{\mathbf{xy}} \Sigma_{\mathbf{yy}}^{-1} \Sigma_{\mathbf{yx}} \mathbf{u}_{x_1} = \lambda^2 \mathbf{u}_{x_1}. \tag{2.58}$$

The subsequent pairs: Next, we find a second pair of projection vectors $\{\mathbf{u}_{x_2}, \mathbf{u}_{y_2}\}$ such that ρ_2 is maximized while $\mathbf{w}_2$ and $\mathbf{z}_2$ are uncorrelated with $\mathbf{w}_1$ and $\mathbf{z}_1$. Furthermore, at the pth step, we find the pth pair of projection vectors $\{\mathbf{u}_{x_p}, \mathbf{u}_{y_p}\}$ such that ρ_p is maximized while $\mathbf{w}_p$ and $\mathbf{z}_p$ are uncorrelated with $\mathbf{w}_q$ and $\mathbf{z}_q$ for $q = 1, ..., p - 1$, that is,

$$\mathbf{u}_{x_p}^T \Sigma_{\mathbf{xx}} \mathbf{u}_{x_q} = \mathbf{u}_{y_p}^T \Sigma_{\mathbf{yy}} \mathbf{u}_{y_q} = \mathbf{u}_{x_p}^T \Sigma_{\mathbf{xy}} \mathbf{u}_{y_q} = \mathbf{u}_{y_p}^T \Sigma_{\mathbf{yx}} \mathbf{u}_{x_q} = 0. \tag{2.59}$$

From Equation (2.50), we have

$$\Sigma_{\mathbf{xy}} \mathbf{u}_{y_1} = \lambda \Sigma_{\mathbf{xx}} \mathbf{u}_{x_1}. \tag{2.60}$$

We assume that this is true for $q = 1, ..., p - 1$:

$$\Sigma_{\mathbf{xy}} \mathbf{u}_{y_q} = \lambda_q \Sigma_{\mathbf{xx}} \mathbf{u}_{x_q}. \tag{2.61}$$

Thus, we have

$$\mathbf{u}_{x_p}^T \Sigma_{\mathbf{xx}} \mathbf{u}_{x_q} = 0 \implies \mathbf{u}_{x_p}^T \Sigma_{\mathbf{xy}} \mathbf{u}_{y_q} = \lambda_q \mathbf{u}_{x_p}^T \Sigma_{\mathbf{xx}} \mathbf{u}_{x_q} = 0. \tag{2.62}$$

Similarly, from Equation (2.51), we have

$$\Sigma_{\mathbf{yx}} \mathbf{u}_{x_1} = \lambda \Sigma_{\mathbf{yy}} \mathbf{u}_{y_1}, \tag{2.63}$$

where $\mu = \lambda$ is used. We assume that this is true for $q = 1, ..., p - 1$:

$$\Sigma_{\mathbf{yx}} \mathbf{u}_{x_q} = \lambda_q \Sigma_{\mathbf{yy}} \mathbf{u}_{y_q}. \tag{2.64}$$

Thus,

$$\mathbf{u}_{y_p}^T \boldsymbol{\Sigma}_{\mathbf{yy}} \mathbf{u}_{y_q} = 0 \implies \mathbf{u}_{y_p}^T \boldsymbol{\Sigma}_{\mathbf{yx}} \mathbf{u}_{x_q} = \lambda_q \mathbf{u}_{y_p}^T \boldsymbol{\Sigma}_{\mathbf{yy}} \mathbf{u}_{y_1} = 0. \tag{2.65}$$

Therefore, we need to impose only two constraints (instead of four constraints in Equation (2.59)) to have uncorrelated features:

$$\mathbf{u}_{x_p}^T \boldsymbol{\Sigma}_{\mathbf{xx}} \mathbf{u}_{x_q} = \mathbf{u}_{y_p}^T \boldsymbol{\Sigma}_{\mathbf{yy}} \mathbf{u}_{y_q} = 0. \tag{2.66}$$

As for the first pair, we form the Lagrangian as

$$\psi_p = \mathbf{u}_{x_p}^T \boldsymbol{\Sigma}_{\mathbf{xy}} \mathbf{u}_{y_p} - \frac{1}{2} \lambda_p (\mathbf{u}_{x_p}^T \boldsymbol{\Sigma}_{\mathbf{xx}} \mathbf{u}_{x_p} - 1) - \frac{1}{2} \mu_p (\mathbf{u}_{y_p}^T \boldsymbol{\Sigma}_{\mathbf{yy}} \mathbf{u}_{y_p} - 1)$$
$$+ \sum_{q=1}^{p-1} \theta_q \mathbf{u}_{x_p}^T \boldsymbol{\Sigma}_{\mathbf{xx}} \mathbf{u}_{x_q} + \sum_{q=1}^{p-1} \omega_q \mathbf{u}_{y_p}^T \boldsymbol{\Sigma}_{\mathbf{yy}} \mathbf{u}_{y_q}, \tag{2.67}$$

where λ_p, μ_p, $\{\theta_q\}$ and $\{\omega_q\}$ are Lagrange multipliers. We then take the derivatives with respect to $\mathbf{u}_{x_p}$ and $\mathbf{u}_{y_p}$ and set to zero:

$$\frac{\partial \psi_p}{\partial \mathbf{u}_{x_p}} = \boldsymbol{\Sigma}_{\mathbf{xy}} \mathbf{u}_{y_p} - \lambda_p \boldsymbol{\Sigma}_{\mathbf{xx}} \mathbf{u}_{x_p} + \sum_{q=1}^{p-1} \theta_q \boldsymbol{\Sigma}_{\mathbf{xx}} \mathbf{u}_{x_q} = \mathbf{0}, \tag{2.68}$$

$$\frac{\partial \psi_p}{\partial \mathbf{u}_{y_p}} = \boldsymbol{\Sigma}_{\mathbf{yx}} \mathbf{u}_{x_p} - \mu_p \boldsymbol{\Sigma}_{\mathbf{yy}} \mathbf{u}_{y_p} + \sum_{q=1}^{p-1} \omega_q \boldsymbol{\Sigma}_{\mathbf{yy}} \mathbf{u}_{y_q} = \mathbf{0}. \tag{2.69}$$

Now multiply Equation (2.68) on the left by $\mathbf{u}_{x_\gamma}^T$ and Equation (2.69) on the left by $\mathbf{u}_{y_\gamma}^T$, $\gamma = 1, ..., p-1$, and note that $\mathbf{u}_{x_\gamma}^T \boldsymbol{\Sigma}_{\mathbf{xy}} \mathbf{u}_{y_p} = \mathbf{u}_{x_\gamma}^T \boldsymbol{\Sigma}_{\mathbf{xx}} \mathbf{u}_{x_p} = \mathbf{u}_{y_\gamma}^T \boldsymbol{\Sigma}_{\mathbf{yx}} \mathbf{u}_{x_p} = \mathbf{u}_{y_\gamma}^T \boldsymbol{\Sigma}_{\mathbf{yy}} \mathbf{u}_{y_p} = 0$, we have

$$\sum_{q=1}^{p-1} \theta_q \mathbf{u}_{x_\gamma}^T \boldsymbol{\Sigma}_{\mathbf{xx}} \mathbf{u}_{x_q} = \theta_\gamma \mathbf{u}_{x_\gamma}^T \boldsymbol{\Sigma}_{\mathbf{xx}} \mathbf{u}_{x_\gamma} = \theta_\gamma = 0, \tag{2.70}$$

$$\sum_{q=1}^{p-1} \omega_q \mathbf{u}_{y_\gamma}^T \boldsymbol{\Sigma}_{\mathbf{yy}} \mathbf{u}_{y_q} = \omega_\gamma \mathbf{u}_{y_\gamma}^T \boldsymbol{\Sigma}_{\mathbf{yy}} \mathbf{u}_{y_\gamma} = \omega_\gamma = 0. \tag{2.71}$$

Therefore, Equations (2.68) and (2.69) become the same form as in Equations (2.50) and (2.51)[8]:

$$\boldsymbol{\Sigma}_{\mathbf{xy}} \mathbf{u}_{y_p} - \lambda \boldsymbol{\Sigma}_{\mathbf{xx}} \mathbf{u}_{x_p} = \mathbf{0}, \tag{2.72}$$

$$\boldsymbol{\Sigma}_{\mathbf{yx}} \mathbf{u}_{x_p} - \mu \boldsymbol{\Sigma}_{\mathbf{yy}} \mathbf{u}_{y_p} = \mathbf{0}. \tag{2.73}$$

---

[8]Note that now we have verified the assumptions in Equations (2.61) and (2.64).

Algorithm 2.3 Canonical correlation analysis (CCA)

Input: Two paired datasets with M training samples each: $\{\mathbf{x}_1, \mathbf{x}_2, ..., \mathbf{x}_M\}$, and $\{\mathbf{y}_1, \mathbf{y}_2, ..., \mathbf{y}_M\}$, where $\mathbf{x}_m \in \mathbb{R}^I$ and $\mathbf{y}_m \in \mathbb{R}^J$.

Process:

1: Calculate the sample covariance and cross-covariance matrices $\boldsymbol{\Sigma}_{\mathbf{xx}}$, $\boldsymbol{\Sigma}_{\mathbf{yy}}$, and $\boldsymbol{\Sigma}_{\mathbf{xy}}$ from Equation (2.42), (2.43), and (2.44), respectively.

2: Get the first $P = \min\{I, J\}$ generalized eigenvectors of the generalized eigenvalue problems in Equations (2.56) and (2.55) to form $\tilde{\mathbf{U}}_x$ and $\tilde{\mathbf{U}}_y$, respectively.

Output: A pair of projection matrices $\{\tilde{\mathbf{U}}_x, \tilde{\mathbf{U}}_y\}$.

Considering the zero-correlation constraints, $\tilde{\mathbf{u}}_{y_p}$ corresponds to the generalized eigenvector associated with the pth largest generalized eigenvalue of Equation (2.55). The projection vector $\tilde{\mathbf{u}}_{x_p}$ for $\{\mathbf{x}_m\}$ can be similarly obtained. The maximum number of pairs P is bounded as $P \leq \min\{I, J\}$. The P pairs of $\{\tilde{\mathbf{u}}_{x_p}, \tilde{\mathbf{u}}_{y_p}\}$ form a pair of projection matrices $\{\tilde{\mathbf{U}}_x, \tilde{\mathbf{U}}_y\}$, where $\tilde{\mathbf{U}}_x \in \mathbb{R}^{I \times P}$ and $\tilde{\mathbf{U}}_y \in \mathbb{R}^{J \times P}$.

We would like to make the following remarks regarding CCA:

1. As in PCA, CCA projections decorrelate the data in the output spaces. Furthermore, not only are the individual datasets uncorrelated after projection, but the cross-correlations of the projected data also vanish.

2. Canonical correlations are invariant with respect to any nonsingular linear transformations [Anderson, 2003].

3. CCA is designed for unsupervised learning. However, it can be used in a supervised way, where one of the two paired datasets contains the class labels.

Algorithm 2.3 provides the pseudocode for CCA. It should be noted that as the inverse of $\boldsymbol{\Sigma}_{\mathbf{xx}}$ and/or $\boldsymbol{\Sigma}_{\mathbf{yy}}$ is involved, the generalized eigenvalue problem can be *ill-conditioned* or *poorly conditioned* in practice. Thus, regularization or Cholesky decomposition is often used to make the problem better conditioned [Hardoon et al., 2004].

2.5 Partial Least Squares Analysis

PLS analysis is the last LSL method to be discussed. Similar to CCA, PLS considers the relationships between two sets of data as well. The difference is that it seeks projections that maximize the covariance rather than the correlation between the two datasets. The underlying assumption of PLS is that the observed data are generated by a system or process that is driven by a small number of *latent variables* [Rosipal and Krämer, 2006]. PLS is an attractive alternative to CCA when $\boldsymbol{\Sigma}_{\mathbf{xx}}$ and $\boldsymbol{\Sigma}_{\mathbf{yy}}$ are singular [Marden, 2011].

As in CCA, PLS considers two paired datasets $\{\mathbf{x}_m \in \mathbb{R}^I\}$, $\{\mathbf{y}_m \in \mathbb{R}^J\}$, $m = 1, ..., M$, $I \neq J$ in general. The paired datasets $\{\mathbf{x}_m\}$ and $\{\mathbf{y}_m\}$ form two data matrices: $\mathbf{X} \in \mathbb{R}^{I \times M}$ and $\mathbf{Y} \in \mathbb{R}^{J \times M}$, which are usually assumed to have zero-mean in PLS[9]. PLS seeks P pairs of projection vectors $\{\mathbf{u}_{x_p}, \mathbf{u}_{y_p}\}$, $\mathbf{u}_{x_p} \in \mathbb{R}^I$, $\mathbf{u}_{y_p} \in \mathbb{R}^J$ to successively maximize the sample covariance between $\{\mathbf{w}_p, \mathbf{z}_p\}$, the *coordinate vectors* in the projected subspaces as defined (element-wise) in Equation (2.41). As in the case of PCA, the maximum will not be achieved for finite $\mathbf{u}_{x_p}$ and $\mathbf{u}_{y_p}$. Thus, we need to include normalization constraints as

$$(\tilde{\mathbf{u}}_{x_p}, \tilde{\mathbf{u}}_{y_p}) = \arg \max_{\mathbf{u}_{x_p}, \mathbf{u}_{y_p}} \frac{\mathbf{u}_{x_p}^T \boldsymbol{\Sigma}_{\mathbf{xy}} \mathbf{u}_{y_p}}{\sqrt{(\mathbf{u}_{x_p}^T \mathbf{u}_{x_p})(\mathbf{u}_{y_p}^T \mathbf{u}_{y_p})}}, \qquad (2.74)$$

where $\boldsymbol{\Sigma}_{\mathbf{xy}}$ is defined in Equation (2.44).

PLS model for two datasets: In PLS literature, $\{\mathbf{u}_{x_p}, \mathbf{u}_{y_p}\}$ are called the *weight vectors*, and $\{\mathbf{w}_p, \mathbf{z}_p\}$ are called the *score vectors*[10], which form the *score matrices* $\mathbf{W} \in \mathbb{R}^{M \times P}$ and $\mathbf{Z} \in \mathbb{R}^{M \times P}$. The *loading vectors* $\mathbf{v}_{x_p} \in \mathbb{R}^I$ and $\mathbf{v}_{y_p} \in \mathbb{R}^J$ are defined as the coefficients of regressing $\mathbf{X}$ on $\mathbf{w}_p$ and $\mathbf{Y}$ on $\mathbf{z}_p$, respectively:

$$\mathbf{v}_{x_p} = \mathbf{X}\mathbf{w}_p/(\mathbf{w}_p^T\mathbf{w}_p), \qquad (2.75)$$

$$\mathbf{v}_{y_p} = \mathbf{Y}\mathbf{z}_p/(\mathbf{z}_p^T\mathbf{z}_p). \qquad (2.76)$$

They form the *loading matrices* $\mathbf{V}_{\mathbf{x}} \in \mathbb{R}^{I \times P}$ and $\mathbf{V}_{\mathbf{y}} \in \mathbb{R}^{J \times P}$. Next, let $\mathbf{E}_{\mathbf{x}} \in \mathbb{R}^{I \times M}$ and $\mathbf{E}_{\mathbf{y}} \in \mathbb{R}^{J \times M}$ denote the *residual matrices* (the error terms). The PLS model of the relations between the two datasets is then [Rosipal and Krämer, 2006]

$$\mathbf{X} = \mathbf{V}_{\mathbf{x}}\mathbf{W}^T + \mathbf{E}_{\mathbf{x}}, \qquad (2.77)$$

$$\mathbf{Y} = \mathbf{V}_{\mathbf{y}}\mathbf{Z}^T + \mathbf{E}_{\mathbf{y}}. \qquad (2.78)$$

The NIPALS algorithm: A classical solution to the PLS problem is the nonlinear iterative partial least squares (NIPALS) algorithm [Wold, 1975], as described in Algorithm 2.4. NIPALS has an iterative process. It starts with an initial score vector $\mathbf{z}_p$ and repeats a sequence of calculations until convergence. After the extraction of the score vectors $\{\mathbf{w}_p, \mathbf{z}_p\}$, the matrices $\mathbf{X}$ and $\mathbf{Y}$ are deflated by subtracting their rank-one approximations based on $\{\mathbf{w}_p, \mathbf{z}_p\}$.

---

[9]In most literature on PLS, the data matrices in the PLS model are in the form of $\mathbf{X}^T$ and $\mathbf{Y}^T$ instead, with each sample as a row. Here, we form the data matrix with each sample as a column to be consistent with the presentation of other LSL algorithms, where vector input data are often column vectors rather than row vectors.

[10]They are equivalent to the *canonical variates* in CCA and they are referred to as the *coordinate vectors* in a general context.

Algorithm 2.4 Nonlinear iterative partial least squares (NIPALS)

Input: Two zero-mean paired datasets $\mathbf{X} \in \mathbb{R}^{I \times M}$ and $\mathbf{Y} \in \mathbb{R}^{J \times M}$, a small number η as the stopping criterion.

Process:

1: $p = 0$.
2: **repeat**
3: $\quad p \leftarrow p + 1$.
4: $\quad$ Initialization: set the score vector $\mathbf{z}_p^T$ equal to the first row of $\mathbf{Y}$.
5: $\quad$ **repeat**
6: $\qquad$ Calculate the X-weight vector $\tilde{\mathbf{u}}_{x_p} = \mathbf{X}\mathbf{z}_p / \parallel \mathbf{X}\mathbf{z}_p \parallel$.
7: $\qquad$ Calculate the X-score vector $\mathbf{w}_p = \mathbf{X}^T \tilde{\mathbf{u}}_{x_p}$.
8: $\qquad$ Calculate the Y-weight vector $\tilde{\mathbf{u}}_{y_p} = \mathbf{Y}\mathbf{w}_p / \parallel \mathbf{Y}\mathbf{w}_p \parallel$.
9: $\qquad$ Calculate the Y-score vector $\mathbf{z}_p = \mathbf{Y}^T \tilde{\mathbf{u}}_{y_p}$.
10: $\quad$ **until** $\parallel \mathbf{w}_{p_{old}} - \mathbf{w}_{p_{new}} \parallel / \parallel \mathbf{w}_{p_{new}} \parallel < \eta$.
11: $\quad$ Calculate the X-loading vector $\mathbf{v}_{x_p} = \mathbf{X}\mathbf{w}_p / (\mathbf{w}_p^T \mathbf{w}_p)$.
12: $\quad$ Calculate the Y-loading vector $\mathbf{v}_{y_p} = \mathbf{Y}\mathbf{z}_p / (\mathbf{z}_p^T \mathbf{z}_p)$.
13: $\quad$ Rank-one deflation: $\mathbf{X} \leftarrow \mathbf{X} - \mathbf{v}_{x_p}\mathbf{w}_p^T$ and $\mathbf{Y} \leftarrow \mathbf{Y} - \mathbf{v}_{y_p}\mathbf{z}_p^T$.
14: **until** $\parallel \mathbf{X} \parallel < \eta$.

Output: P pairs of weight vectors $\{\tilde{\mathbf{u}}_{x_p}, \tilde{\mathbf{u}}_{y_p}\}$, $p = 1, ..., P$.

Alternatively, comparing Equation (2.74) with Equation (2.46), we find that they are equivalent if $\boldsymbol{\Sigma}_{\mathbf{xx}}$ and $\boldsymbol{\Sigma}_{\mathbf{yy}}$ are identity matrices. Therefore, the weight vectors $\tilde{\mathbf{u}}_{x_1}$ and $\tilde{\mathbf{u}}_{y_1}$ are the eigenvectors corresponding to the largest eigenvalue of the following eigenvalue problems from Equations (2.56) and (2.55):

$$\boldsymbol{\Sigma}_{\mathbf{xy}}\boldsymbol{\Sigma}_{\mathbf{yx}}\mathbf{u}_{x_1} = \lambda^2 \mathbf{u}_{x_1}, \qquad (2.79)$$

$$\boldsymbol{\Sigma}_{\mathbf{yx}}\boldsymbol{\Sigma}_{\mathbf{xy}}\mathbf{u}_{y_1} = \lambda^2 \mathbf{u}_{y_1}. \qquad (2.80)$$

The subsequent weight vectors are dominant eigenvectors of the deflated versions of the above eigenvalue problems. In [Höskuldsson, 1988], it was pointed out that the NIPALS algorithm actually performs in a similar way to the *power method* of determining the largest eigenvalue for a matrix (see Section A.1.11).

PLS-based regression: The PLS method presented above follows the original design in [Wold, 1975] to model the relations between two datasets where the relationship is symmetric. In practice, PLS is most frequently used for *regression* where the relationship between $\mathbf{X}$ and $\mathbf{Y}$ is asymmetric: $\mathbf{X}$ is the *predictor* while $\mathbf{Y}$ is the *response* to be predicated. PLS regression assumes [Rosipal and Krämer, 2006]

1. The score vectors $\{\mathbf{w}_p\}$ are good predictors of $\mathbf{Y}$.

2. The relation between $\mathbf{W}$ and $\mathbf{Z}$ can be approximated linearly by

$$\mathbf{Z} = \mathbf{W}\mathbf{D} + \mathbf{E}, \qquad (2.81)$$

Algorithm 2.5 PLS1 regression

Input: Two paired datasets $\mathbf{X} \in \mathbb{R}^{I \times M}$ and $\mathbf{y} \in \mathbb{R}^{M \times 1}$, the number of latent factors P.

Process:

1: **for** $p = 1$ to P **do**
2:     Calculate the X-weight vector $\tilde{\mathbf{u}}_{x_p} = \mathbf{X}\mathbf{y}/ \parallel \mathbf{X}\mathbf{y} \parallel$.
3:     Calculate the X-score vector $\mathbf{w}_p = \mathbf{X}^T \tilde{\mathbf{u}}_{x_p}$.
4:     Calculate the regression coefficient $d_p = \mathbf{w}_p^T \mathbf{y}/(\mathbf{w}_p^T \mathbf{w}_p)$.
5:     Calculate the X-loading vector $\mathbf{v}_{x_p} = \mathbf{X}\mathbf{w}_p/(\mathbf{w}_p^T \mathbf{w}_p)$.
6:     Rank-one deflation: $\mathbf{X} \leftarrow \mathbf{X} - \mathbf{v}_{x_p}\mathbf{w}_p^T$ and $\mathbf{y} \leftarrow \mathbf{y} - d_p\mathbf{w}_p$.
7: **end for**

Output: P weight vectors $\{\tilde{\mathbf{u}}_{x_p}\}$, and loading vectors $\{\mathbf{v}_{x_p}\}$, and the regression coefficients $\{d_p\}$, $p = 1, ..., P$.

---

where $\mathbf{E}$ is the matrix of residuals, and $\mathbf{D}$ is a $P \times P$ diagonal matrix with its pth diagonal element

$$d_p = \mathbf{z}_p^T \mathbf{w}_p/(\mathbf{w}_p^T \mathbf{w}_p). \tag{2.82}$$

Here, d_p plays the role of regression coefficients.

PLS1 regression: PLS1 is a PLS regression method where the second dataset is simply a vector $\mathbf{y} \in \mathbb{R}^{M \times 1}$ ($J = 1$). Thus, PLS1 only seeks projection vectors $\{\mathbf{u}_{x_p}\}$ to successively maximize the sample covariance between the projection $\mathbf{w}_p$ and $\mathbf{y}$, which can be seen as a degenerated case. The PLS1 regression algorithm is described in Algorithm 2.5 [Jørgensen and Goegebeur, 2007]. Form matrices $\tilde{\mathbf{U}}_x$ and $\mathbf{V}_x$ with $\{\tilde{\mathbf{u}}_{x_p}\}$ and $\{\mathbf{v}_{x_p}\}$, respectively. Form a regression vector $\mathbf{d}$ with $\{d_p\}$. The linear regression between $\mathbf{X}$ and $\mathbf{y}$ becomes

$$\mathbf{y} \quad = \quad \mathbf{X}^T \tilde{\mathbf{U}}_x (\mathbf{V}_x^T \tilde{\mathbf{U}}_x)^{-1} \mathbf{d}. \tag{2.83}$$

The PLS1 algorithm has a loop with only P iterations. In contrast, the number of iterations is controlled by a stopping criterion for convergence in the NIPALS algorithm.

As in CCA, PLS can also work as a supervised learning method for classification by encoding the class membership in an appropriate *indicator vector/matrix* as the second dataset, which is closely related to Fisher discriminant analysis [Barker and Rayens, 2003].

2.6 Unified View of PCA, LDA, CCA, and PLS

We have covered five LSL algorithms up to now: PCA, ICA, LDA, CCA, and PLS. Except for ICA[11], all four other LSL algorithms aim to maximize a *generalized Rayleigh quotient* of the form

$$\frac{\mathbf{v}^T \mathbf{A} \mathbf{v}}{\mathbf{v}^T \mathbf{B} \mathbf{v}}, \tag{2.84}$$

which can all be solved through generalized eigenvalue problems of the form

$$\mathbf{A} \mathbf{v} = \lambda \mathbf{B} \mathbf{v}. \tag{2.85}$$

Here, we examine the first projection vectors (or pairs) in PCA, LDA, CCA, and PLS. The solution for the first PCA projection vector in Equation (2.9) can be written as

$$\mathbf{S}_T \mathbf{u}_1 = \lambda \mathbf{I} \mathbf{u}_1. \tag{2.86}$$

Equation (2.36) in LDA is already in the form of Equation (2.85). In CCA, Equations (2.50) and (2.51) can be written as a single generalized eigenvalue

---

[11] Nevertheless, the JADE algorithm for ICA maximizes the "sum of squares of the diagonal" measure through solving eigendecomposition problems [Cardoso and Souloumiac, 1993; Cardoso, 1998].

TABLE 2.1: PCA, LDA, CCA and PLS can all be viewed as solving the generalized eigenvalue problem $\mathbf{A}\mathbf{v} = \lambda \mathbf{B}\mathbf{v}$ (adapted from [Bie et al., 2005]).

	Maximize	$\mathbf{A}$	$\mathbf{B}$	$\mathbf{v}$
PCA	Total scatter (variation)	$\mathbf{S}_T$	$\mathbf{I}$	$\mathbf{u}_1$
LDA	Between-class to within-class scatter ratio	$\mathbf{S}_B$	$\mathbf{S}_W$	$\mathbf{u}_1$
CCA	Correlation	$\begin{bmatrix} \mathbf{0} & \mathbf{\Sigma}_{\mathbf{xy}} \\ \mathbf{\Sigma}_{\mathbf{yx}} & \mathbf{0} \end{bmatrix}$	$\begin{bmatrix} \mathbf{\Sigma}_{\mathbf{xx}} & \mathbf{0} \\ \mathbf{0} & \mathbf{\Sigma}_{\mathbf{yy}} \end{bmatrix}$	$\begin{bmatrix} \mathbf{u}_{x_1} \\ \mathbf{u}_{y_1} \end{bmatrix}$
PLS	Covariance	$\begin{bmatrix} \mathbf{0} & \mathbf{\Sigma}_{\mathbf{xy}} \\ \mathbf{\Sigma}_{\mathbf{yx}} & \mathbf{0} \end{bmatrix}$	$\begin{bmatrix} \mathbf{I} & \mathbf{0} \\ \mathbf{0} & \mathbf{I} \end{bmatrix}$	$\begin{bmatrix} \mathbf{u}_{x_1} \\ \mathbf{u}_{y_1} \end{bmatrix}$

problem below:

$$\begin{bmatrix} \mathbf{0} & \boldsymbol{\Sigma}_{\mathbf{xy}} \\ \boldsymbol{\Sigma}_{\mathbf{yx}} & \mathbf{0} \end{bmatrix} \begin{bmatrix} \mathbf{u}_{x_1} \\ \mathbf{u}_{y_1} \end{bmatrix} = \lambda \begin{bmatrix} \boldsymbol{\Sigma}_{\mathbf{xx}} & \mathbf{0} \\ \mathbf{0} & \boldsymbol{\Sigma}_{\mathbf{yy}} \end{bmatrix} \begin{bmatrix} \mathbf{u}_{x_1} \\ \mathbf{u}_{y_1} \end{bmatrix}. \tag{2.87}$$

Similarly, the first projection in PLS can be written as

$$\begin{bmatrix} \mathbf{0} & \boldsymbol{\Sigma}_{\mathbf{xy}} \\ \boldsymbol{\Sigma}_{\mathbf{yx}} & \mathbf{0} \end{bmatrix} \begin{bmatrix} \mathbf{u}_{x_1} \\ \mathbf{u}_{y_1} \end{bmatrix} = \lambda \begin{bmatrix} \mathbf{I} & \mathbf{0} \\ \mathbf{0} & \mathbf{I} \end{bmatrix} \begin{bmatrix} \mathbf{u}_{x_1} \\ \mathbf{u}_{y_1} \end{bmatrix}. \tag{2.88}$$

Thus, we can summarize them in Table 2.1 [Bie et al., 2005]. When $\mathbf{B}$ is nonsingular, the generalized eigenvalue problem in Equation (2.85) is equivalent to the following eigenvalue problem:

$$\mathbf{B}^{-1}\mathbf{A}\mathbf{v} = \lambda\mathbf{v}. \tag{2.89}$$

On the other hand, if $\mathbf{B}$ is singular or close to singular, the generalized eigenvalue problem in Equation (2.85) often becomes ill-conditioned with poor stability, in which case *regularization* is frequently employed to improve the stability.

2.7 Regularization and Model Selection

In machine learning, the *training error* may not be a good estimate of the *test error*. We are more concerned about the **generalization** performance, which refers to the prediction capability of a learning method on independent test data [Friedman et al., 2001]. **Overfitting** refers to the case where the training error is very low (approaching zero) while the test error is high, which means poor generalization. Regularization is a very useful technique employed widely in LSL to tackle *overfitting* and improve *generalization*. There are various motivations for regularization. Here, we discuss the two most relevant ones.

> **Bias, variance, and overfitting:** To understand the overfitting problem, the generalization error can be decomposed into *bias* and *variance* [Friedman et al., 2001]. In the context of machine learning, bias refers to the tendency of a learner to learn the same wrong thing consistently, while variance refers to its tendency to learn random things regardless of the real signal [Domingos, 2012].

2.7.1 Regularizing Covariance Matrix Estimation

As seen in previous sections, LSL algorithms often involve the estimation of covariance (scatter) matrices from finite samples. When the number of available samples is small compared to the number of parameters to be estimated, we have the small sample size (SSS) problem. The estimation problem in such a setting could be poorly posed [Friedman, 1989] and the estimated parameters could be highly unreliable, giving rise to high variance. This problem is particularly severe for LDA and CCA, where the matrix **B** (in Table 2.1) is more likely to be singular or ill-conditioned. A regularization technique has been proved to be effective in tackling the SSS problem.

Regularization was first introduced to tackle the SSS problem for LDA in *regularized discriminant analysis* (RDA) [Friedman, 1989]. It was pointed out that a small number of training samples tends to result in a biased estimation of eigenvalues, that is, the largest ones are biased high while the smallest ones are biased low. On the other hand, sample-based covariance estimates from these poorly posed problems are usually highly variable and unreliable. Two regularization parameters were introduced by Friedman [1989] to account for these undesirable effects. It works by biasing the estimates away from their sample-based values toward more "physically plausible" values, which reduces the variance of the sample-based estimates while tending to increase bias. This *bias-variance trade-off* is controlled by the regularization parameters. A simplified form is a popular regularization scheme that adds a small constant γ to the diagonal of the covariance matrix estimate as

$$\mathbf{B}_r = \mathbf{B} + \gamma \mathbf{I}, \tag{2.90}$$

which is equivalent to adding γ to each eigenvalue.

2.7.2 Regularizing Model Complexity

Generalization performance is closely related to *model complexity*. A model with high complexity tends to have low bias but high variance, while a model with low complexity tends to have high bias but low variance. In general, training error decreases with model complexity, typically dropping to zero if we increase the model complexity enough. However, a model with zero training error overfits to the training data and will typically generalize poorly [Friedman et al., 2001]. Thus, a powerful complicated learner may not necessarily outperform a less powerful simple one because the former usually needs more data to avoid overfitting.

Occam's razor, also spelled Ockham's razor, is a principle stated by William of Ockham that "Plurality should not be posited without necessity" [Domingos, 1999]. It is also called law of economy, or parsimony, succinctness. The principle is interpreted as preferring the simpler one of two competing theories (when the two theories make exactly the same predictions).

Regularization methods are frequently used to control model complexity by penalizing more complex models. Such regularization usually imposes a *penalty* for complexity based on the number of their parameters. Typical examples of regularization for model complexity include ℓ_1 norm and ℓ_2 norm of the parameter vector. The ℓ_1 norm regularization is called least absolute shrinkage and selection operator (Lasso), which leads to sparse representations and has been a hot topic of recent research [Wright et al., 2009].

"the effort to generalize often means that the solution is simple. Often by stopping and saying, 'This is the problem he wants but this is characteristic of so and so. Yes, I can attack the whole class with a far superior method than the particular one because I was earlier embedded in needless detail.' The business of abstraction frequently makes things simple. "

Richard Hamming (1915–1998)
Mathematician, Turing Award Winner

2.7.3   Model Selection

Model selection often refers to a determination of the various parameters in an algorithm, for example, selection of the regularization parameter(s). **Cross-validation** is the simplest and most widely used model selection method. In a K-fold cross-validation, available data are split into K roughly equal-sized partitions. For $k = 1, 2, ..., K$, we hold the kth partition to calculate the classification error of the model trained on the other $K - 1$ partitions. The average over the K rounds is the K-fold cross-validation result. Thus, all data samples are used for both training and validation while each sample is used only once for validation.

Training, validation, and test: When there are enough data, we can divide them into three parts, namely the training set, the validation set, and the test set. We use the training set to train the models. We then use the validation set to estimate classification error to select the model with the smallest error. Finally, we test the selected model on the test set to report its (generalization) error. A typical split suggested in [Friedman et al., 2001] is 50% for training, and 25% each for validation and testing.

Other model selection techniques include the Akaike information criterion (AIC) [Akaike, 1992], the Bayesian information criterion (BIC) [Schwarz, 1978], minimum description length (MDL) [Rissanen, 1983], and structural risk minimization (SRM) [Vapnik and Sterin, 1977].

2.8   Ensemble Learning

Ensemble learning [Zhou, 2012] combines the strengths of multiple simpler *base learners* to achieve better classification performance than that which could be obtained from any of the constituent learners. Base learners with high diversity tend to produce better results by complementing each other. Thus, more random algorithms can produce a stronger ensemble than deliberate algorithms. Ensemble learning usually consists of two tasks: development of base learners from training data, and combination of these learners to form the *composite learner* [Friedman et al., 2001].

It should be noted, however, that ensemble learning typically requires more computation than evaluating a single learner. If online real-time learning is required, fast learners such as *decision trees*, for example, *random forest* [Ho, 1998], are preferred. In the following, we discuss two important ensemble learning methods: bagging and boosting.

2.8.1   Bagging

Bagging [Breiman, 1996] is based on the *bootstrapping* [Hillis and Bull, 1993] and *aggregation* [Tsybakov, 2004] concepts so it is also called *bootstrap aggregating*. It is a model averaging approach aiming to reduce variance and help to avoid overfitting such that stability and classification accuracy can be improved.

Given a training set $\mathbf{X}$ of size M, bagging generates B bootstrap datasets by taking M' samples from $\mathbf{X}$ uniformly and with replacement, where $M' \leq M$. Then, B models are fitted to the B bootstrap datasets. Next, they are combined by averaging the output (for regression) or voting (for classification).

Bagging is effective for high-variance, low-bias procedures such as decision trees where there is great diversity [Friedman et al., 2001]. On the other hand, it is not useful for improving very stable models like k nearest neighbors.

2.8.2   Boosting

Boosting is a general supervised learning method that can be used in conjunction with many other learning algorithms to improve their performance. It is motivated by the question of whether a set of *weak learners*, which only performs slightly better than random guessing, can be boosted into an arbitrarily accurate *strong learner* [Kearns and Valiant, 1994; Freund and Schapire, 1999]. Boosting produces a very accurate predication rule by combining rough and moderately accurate rules of thumb. The motivation is that finding many rough rules of thumb can be much easier than finding a single, highly accurate predication rule.

The boosting algorithm starts with a weak learner that can find the rough

---

Algorithm 2.6 Adaptive boosting (AdaBoost)

Input: A set of M training samples, $(\mathbf{y}_1, c_1), \cdots, (\mathbf{y}_M, c_M)$, $\mathbf{y}_m \in \mathbb{R}^I$, and
$\quad$ $c_m \in \{-1, +1\}$, $m = 1, ..., M$.
Process:
$\quad$ 1: **Initialize** $D_1(m) = \frac{1}{M}$.
$\quad$ 2: **for** $t = 1$ **to** T **do**
$\quad$ 3: $\quad$ Train weak learner using the sample distribution D_t.
$\quad$ 4: $\quad$ Get weak hypothesis $h_t : \mathbb{R}^I \to \{-1, +1\}$.
$\quad$ 5: $\quad$ Calculate ϵ_t according to Equation (2.92). If $\epsilon_t > 0.5$, stop (fail).
$\quad$ 6: $\quad$ Choose α_t according to Equation (2.93).
$\quad$ 7: $\quad$ Update $D_{t+1}(m)$ according to Equation (2.94).
$\quad$ 8: **end for**
Output: The final hypothesis $h(\mathbf{y})$ according to Equation (2.95).

---

rules of thumb. It then repeatedly calls this weak learner by feeding it a different subset of training samples. Thus, each call generates a new weak prediction rule. The boosting algorithm combines these weak rules into a single, (hopefully) very accurate prediction rule [Schapire, 2003]. It has been shown through both theoretical study and empirical testing that boosting is particularly robust in preventing overfitting and reducing the generalization error by increasing the so-called **margins** of the training examples [Breiman, 1998; Schapire et al., 1997]. The *margin* is defined as the minimal distance of an example to the decision surface of classification [Vapnik, 1995]. A larger expected margin of training data generally leads to a lower generalization error.

Among the many boosting algorithms, the adaptive boosting (AdaBoost) formulated in [Freund and Schapire, 1997] is a very popular one with great success [Skurichina and Duin, 2002; Lu et al., 2006e]. The pseudocode for AdaBoost is given in Algorithm 2.6. The algorithm takes a training set of M samples $(\mathbf{y}_1, c_1), ..., (\mathbf{y}_M, c_M)$ as the input, where $\mathbf{y}_m \in \mathbb{R}^I$ and $c_m \in \{-1, +1\}$, $m = 1, ..., M$. It calls a weak learner repeatedly in a series of rounds $t = 1, ..., T$. In each call, a distribution (set of weights) is maintained over the training set. The weight of this distribution on training sample $\mathbf{y}_m$ in round t is denoted by $D_t(m)$. All weights are initialized to be equal $D_1(m) = 1/M$ for $t = 1$.

In the boosting step t, the weak learner produces a weak hypothesis

$$h_t : \mathbb{R}^I \to \{-1, +1\} \tag{2.91}$$

for the distribution D_t. The goodness of h_t is then measured by the error ϵ_t defined as

$$\epsilon_t = \sum_{m: h_t(\mathbf{y}_m) \neq c_m} D_t(m). \tag{2.92}$$

Next, AdaBoost chooses a parameter α_t as

$$\alpha_t = \frac{1}{2} \ln \left(\frac{1 - \epsilon_t}{\epsilon_t} \right),$$ (2.93)

which measures the importance of h_t. The distribution D_t is updated as

$$
\begin{aligned}
D_{t+1}(m) &= \frac{D_t(m)}{Z_t} \times \begin{cases} e^{-\alpha_t} & \text{if } h_t(\mathbf{y}_m) = c_m \\ e^{\alpha_t} & \text{if } h_t(\mathbf{y}_m) \neq c_m \end{cases} \\
&= \frac{D_t(m) \exp(-\alpha_t c_m h_t(\mathbf{y}_m))}{Z_t},
\end{aligned}
$$ (2.94)

where Z_t is a normalization factor to ensure that D_{t+1} is a probability distribution. The update effectively decreases the weights of those samples correctly classified by h_t and increases the weights of those classified incorrectly. In this way, the weak learner is forced to focus on the more difficult training samples in the next round [Freund and Schapire, 1999]. Finally, the final strong hypothesis $h(\mathbf{y})$ is simply the weighted majority vote of all the weak hypothesis $h_t(\mathbf{y})$ as

$$h(\mathbf{y}) = \text{sign} \left(\sum_{t=1}^{T} \alpha_t h_t(\mathbf{y}) \right).$$ (2.95)

2.9 Summary

- Linear subspace learning takes high-dimensional vectors as input to find low-dimensional representations optimizing a certain criterion.

- Principal component analysis produces uncorrelated features capturing as much variation (measured by total scatter) as possible.

- Independent component analysis finds statistically independent features by maximizing the non-Gaussianity or minimization of mutual information.

- Linear discriminant analysis learns features to separate different classes well by maximizing the between-class scatter while minimizing the within-class scatter.

- Canonical correlation analysis finds several pairs of projections for two sets of paired data such that the same pair projections have maximal correlation while they are uncorrelated for other pairs.

- Partial least squares analysis aims to maximize the covariance of projected features for two paired datasets.

- Four of these five algorithms (except ICA) can be formulated as a (generalized) eigenvalue problem.

- In practice, LDA and CCA often need regularization to improve stability and generalization performance.

- Multiple (diverse) learners can be combined to achieve better performance through ensemble learning.

2.10 Further Reading

Linear subspace learning has a long history and it is extensively studied in the literature. LSL algorithms are covered in many textbooks and there are several dedicated books too. Burges [2010] gives a very good overview of these projective methods (except PLS) and covers a broader topic of dimensionality reduction. It is a first-hand reference to understand and dive deeper into the literature of LSL and dimensionality reduction in general from a machine learning perspective.

For a deeper understanding of the background in multivariate mathematics and statistics, Anderson [2003] wrote a classical textbook for reference and a free online source is provided by Marden [2011]. The work by Bie et al. [2005] gives an in-depth analysis of the connections between four LSL algorithms, which is the basis of our discussion in Section 2.6. The book chapter on face recognition in subspaces by Shakhnarovich and Moghaddam [2004] is a good reference for understanding LSL and dimensionality reduction from an application point of view. Recently, De la Torre [2012] has proposed a unifying least-square framework for several LSL algorithms.

PCA is known to be first described by Pearson [1901] so it is now over 100 years old. The approach of derivation in Section 2.1 follows that by Hotelling [1933]. The most comprehensive reference on PCA is the book by Jolliffe [2002]. The eigenface paper by Turk and Pentland [1991] is probably the most influential work that triggered the exploration of appearance-based object recognition via subspace learning.

ICA was introduced in the early 1980s in the neurophysiological setting according to the book by [Hyvärinen et al., 2001], which has extensive coverage of ICA. A more compact introduction to ICA can be found in the tutorial paper by Hyvärinen and Oja [2000]. The application of ICA to face recognition using two architectures was pioneered by Bartlett et al. [2002].

The history of LDA traces back to the work in 1936 by Fisher [1936]. While there is a book on LDA in a wider context [McLachlan, 1992], the discussions on the four LDA criteria by Fukunaga [1990, Chap. 10] are very useful to read. Another good reference is the short introduction on LDA by Duda et al. [2001, Chap. 3] with simple yet intuitive examples.

CCA was introduced by Hotelling [1936]. Anderson [2003] gives an in-depth treatment of this topic while Weenink [2003] covers different computational solutions pretty well.

PLS was first developed by Wold [1975] in the 1970s. It is most widely used in in the field of chemometrics as a standard tool for processing a wide spectrum of chemical data problems. A comprehensive review of this topic is in [Rosipal and Krämer, 2006].

A classic book on statistical machine learning was written by Friedman et al. [2001], covering regularization, model selection and assessment, ensemble learning, and many other important topics. Another classic is the book by Duda et al. [2001]. Zhou [2012] has written a book on ensemble learning in this machine learning and pattern recognition book series.

Chapter 3

Fundamentals of Multilinear Subspace Learning

The previous chapter covered background materials on linear subspace learning. From this chapter on, we shall proceed to multiple dimensions with tensor-level computational thinking. Multilinear algebra is the foundation of multilinear subspace learning (MSL). Thus, we first review the basic notations and operations in multilinear algebra, as well as popular tensor decompositions. In the presentation, we include some discussions of the second-order case (for matrix data) as well, which can be understood in the context of linear algebra. Next, we introduce the important concept of *multilinear projections* for direct mapping of tensors to a lower-dimensional representation, as shown in Figure 3.1. They include elementary multilinear projection (EMP), tensor-to-vector projection (TVP), and tensor-to-tensor projection (TTP), which project an input tensor to a scalar, a vector, and a tensor, respectively. Their relationships are analyzed in detail subsequently. Finally, we extend commonly used vector-based scatter measures to tensors and scalars for optimality criterion construction in MSL.

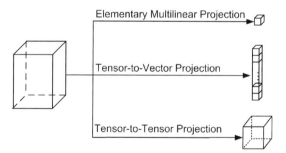

FIGURE 3.1: Multilinear subspace learning finds a lower-dimensional representation by direct mapping of tensors through a multilinear projection.

3.1   Multilinear Algebra Preliminaries

Multilinear algebra, the basis of tensor-based computing, has been studied in mathematics for several decades [Greub, 1967]. A *tensor* is a multidimensional (multiway) array. As pointed out by Kolda and Bader [2009], this notion of tensors is different from the same term referring to *tensor fields* in physics and engineering. In the following, we review the notations and some basic multilinear operations needed in introducing MSL.

3.1.1   Notations and Definitions

In this book, we have tried to remain consistent with the notations and terminologies in applied mathematics, particularly the seminal paper by De Lathauwer et al. [2000b] and the recent SIAM review paper on tensor decomposition by Kolda and Bader [2009].

Vectors are denoted by lowercase boldface letters, for example, $\mathbf{a}$; matrices by uppercase boldface, for example, $\mathbf{A}$; and tensors by calligraphic letters, for example, $\mathcal{A}$. Indices are denoted by lowercase letters and span the range from 1 to the uppercase letter of the index whenever appropriate, for example, $n = 1, 2, ..., N$. Throughout this book, we restrict the discussions to real-valued vectors, matrices, and tensors.

Definition 3.1. *The number of dimensions (ways) of a tensor is its* **order**, *denoted by N. Each dimension (way) is called a* **mode**.

As shown in Figure 3.2, a scalar is a zero-order tensor ($N = 0$), a vector is a first-order tensor ($N = 1$), and a matrix is a second-order tensor ($N = 2$). Tensors of order three or higher are called *higher-order tensors*. An Nth-order tensor is an element in a *tensor space* of degree N, the *tensor product* (*outer product*) of N vector spaces (Section A.1.4) [Lang, 1984].

Mode addressing: There are two popular ways to refer to a *mode*: n-mode or mode-n. In [De Lathauwer et al., 2000b], only n-mode is used. In [De Lathauwer et al., 2000a], n-mode and mode-n are used interchangeably. In [Kolda and Bader, 2009], n-mode and mode-n are used in different contexts. In this book, we prefer to use mode-n to indicate the nth mode for clarity.

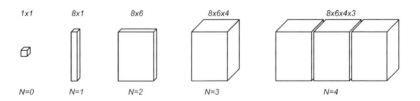

FIGURE 3.2: Illustration of tensors of order $N = 0, 1, 2, 3, 4$.

For example, "1-mode" (referring to the mode index) may be confused with one mode (referring to the number of modes).

An Nth-order tensor has N indices $\{i_n\}$, $n = 1, ..., N$, with each index $i_n(= 1, ..., I_n)$ addressing mode-n of $\mathcal{A}$. Thus, we denote an Nth-order tensor explicitly as $\mathcal{A} \in \mathbb{R}^{I_1 \times I_2 \times ... \times I_N}$.

When we have a set of N vectors or matrices, one for each mode, we denote the nth (i.e., mode-n) vector or matrix using a superscript in parenthesis, for example, as $\mathbf{u}^{(n)}$ or $\mathbf{U}^{(n)}$ and the whole set as $\{\mathbf{u}^{(1)}, \mathbf{u}^{(2)}, ..., \mathbf{u}^{(N)}\}$ or $\{\mathbf{U}^{(1)}, \mathbf{U}^{(2)}, ..., \mathbf{U}^{(N)}\}$, or more compactly as $\{\mathbf{u}^{(n)}\}$ or $\{\mathbf{U}^{(n)}\}$. Figures 3.5, 3.9, and 3.10 provide some illustrations.

Element addressing: For clarity, we adopt the MATLAB style to address elements (entries) of a tensor (including vector and matrix) with indices in parentheses. for example, a single element (entry) is denoted as $\mathbf{a}(2)$, $\mathbf{A}(3, 4)$, or $\mathcal{A}(5, 6, 7)$ in this book, which are denoted as a_2, a_{34}, or a_{567} in conventional notations[1]. To address part of a tensor (a subarray), ":" denotes the full range of the corresponding index and $i : j$ denotes indices ranging from i to j, for example, $\mathbf{a}(2 : 5)$, $\mathbf{A}(3, :)$ (the third row), or $\mathcal{A}(1 : 3, :, 4 : 5)$.

Definition 3.2. *The **mode-n vectors**[2] of $\mathcal{A}$ are defined as the I_n-dimensional vectors obtained from $\mathcal{A}$ by varying the index i_n while keeping all the other indices fixed.*

For example, $\mathcal{A}(:, 2, 3)$ is a mode-1 vector. For second-order tensors (matrices), mode-1 and mode-2 vectors are the column and row vectors, respectively. Figures 3.3(b), 3.3(c), and 3.3(d) give visual illustrations of the mode-1, mode-2, and mode-3 vectors of the third-order tensor $\mathcal{A}$ in Figure 3.3(a), respectively.

Definition 3.3. *The i_nth **mode-n slice** of $\mathcal{A}$ is defined as an $(N-1)$th-order tensor obtained by fixing the mode-n index of $\mathcal{A}$ to be i_n: $\mathcal{A}(:, ..., :, i_n, :, ..., :)$.*

For example, $\mathcal{A}(:, 2, :)$ is a mode-2 slice of $\mathcal{A}$. For second-order tensors (matrices), a mode-1 slice is a mode-2 (row) vector, and a mode-2 slice is a mode-1 (column) vector. Figures 3.4(b), 3.4(c), and 3.4(d) give visual illustrations of the mode-1, mode-2, and mode-3 slices of the third-order tensor $\mathcal{A}$ in Figure 3.4(a), respectively. This definition of a slice is consistent with that in [Bader and Kolda, 2006]; however, it is different from the definition of a slice in [Kolda and Bader, 2009], where a slice is defined as a two-dimensional section of a tensor.

Definition 3.4. *A **rank-one tensor** $\mathcal{A} \in \mathbb{R}^{I_1 \times I_2 \times ... \times I_N}$ equals the outer product[3] of N vectors:*

$$\mathcal{A} = \mathbf{u}^{(1)} \circ \mathbf{u}^{(2)} \circ ... \circ \mathbf{u}^{(N)}, \tag{3.1}$$

---

[1] In Appendix A, we follow the conventional notations of an element.

[2] Mode-n vectors are renamed as mode-n *fibers* in [Kolda and Bader, 2009].

[3] Here, we use a notation 'o' different from the conventional notation '⊗' to better differentiate the outer product of vectors from the Kronecker product of matrices.

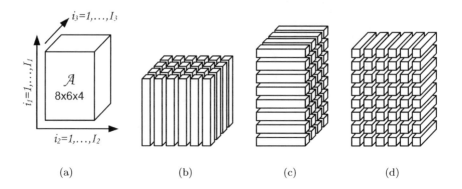

FIGURE 3.3: Illustration of the mode-*n* vectors: (a) a tensor $\mathcal{A} \in \mathbb{R}^{8 \times 6 \times 4}$, (b) the mode-1 vectors, (c) the mode-2 vectors, and (d) the mode-3 vectors.

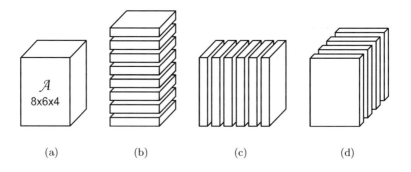

FIGURE 3.4: Illustration of the mode-*n* slices: (a) a tensor $\mathcal{A} \in \mathbb{R}^{8 \times 6 \times 4}$, (b) the mode-1 slices, (c) the mode-2 slices, and (d) the mode-3 slices.

which means that

$$\mathcal{A}(i_1, i_2, ..., i_N) = \mathbf{u}^{(1)}(i_1) \cdot \mathbf{u}^{(2)}(i_2) \cdot ... \cdot \mathbf{u}^{(N)}(i_N) \tag{3.2}$$

for all values of indices.

A rank-one tensor for $N = 2$ (i.e., a second-order rank-one tensor) is a rank-one matrix, with an example shown in Figure 3.5. Some examples of rank-one tensors for $N = 2, 3$ learned from real data are shown in Figures 1.10(c) and 1.11(c) in Chapter 1.

Definition 3.5. *A cubical tensor* $\mathcal{A} \in \mathbb{R}^{I_1 \times I_2 \times ... \times I_N}$ *has the same size for every mode, that is,* $I_n = I$ *for* $n = 1, ..., N$ *[Kolda and Bader, 2009].*

Thus, a square matrix is a second-order cubical tensor by this definition.

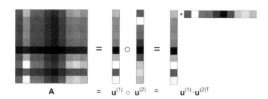

FIGURE 3.5: An example of second-order rank-one tensor (that is, rank-one matrix): $\mathbf{A} = \mathbf{u}^{(1)} \circ \mathbf{u}^{(2)} = \mathbf{u}^{(1)}\mathbf{u}^{(2)^T}$.

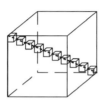

FIGURE 3.6: The diagonal of a third-order cubical tensor.

Definition 3.6. *A **diagonal tensor** $\mathcal{A} \in \mathbb{R}^{I_1 \times I_2 \times \dots \times I_N}$ has non-zero entries, i.e., $\mathcal{A}(i_1, i_2, \dots, i_N) \neq 0$, only for $i_1 = i_2 = \dots = i_N$ [Kolda and Bader, 2009].*

A diagonal tensor of order 2 ($N = 2$) is simply a diagonal matrix. A diagonal tensor of order 3 ($N = 3$) has non-zero entries only along its diagonal as shown in Figure 3.6. A vector $\mathbf{d} \in \mathbb{R}^I$ consisting of the diagonal of a cubical tensor can be defined as

$$\mathbf{d} = diag(\mathcal{A}), \text{ where } \mathbf{d}(i) = \mathcal{A}(i, i, \dots, i). \tag{3.3}$$

3.1.2 Basic Operations

Definition 3.7 (Unfolding: tensor to matrix transformation[4]). *A tensor can be unfolded into a matrix by rearranging its mode-n vectors. The **mode-n unfolding** of $\mathcal{A}$ is denoted by $\mathbf{A}_{(n)} \in \mathbb{R}^{I_n \times (I_1 \times \dots \times I_{n-1} \times I_{n+1} \times \dots \times I_N)}$, where the column vectors of $\mathbf{A}_{(n)}$ are the mode-n vectors of $\mathcal{A}$.*

The (column) order of the mode-n vectors in $\mathbf{A}_{(n)}$ is usually not important as long as it is consistent throughout the computation. For a second-order tensor (matrix) $\mathbf{A}$, its mode-1 unfolding is itself $\mathbf{A}$ and its mode-2 unfolding is its transpose $\mathbf{A}^T$. Figure 3.7 shows the mode-1 unfolding of the third-order tensor $\mathcal{A}$ on the left.

Definition 3.8 (Vectorization: tensor to vector transformation). *Similar to the vectorization of a matrix, the **vectorization** of a tensor is a linear*

---

[4]Unfolding is also known as flattening or matricization [Kolda and Bader, 2009].

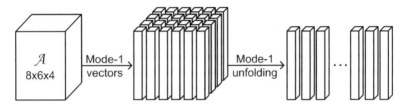

FIGURE 3.7: Visual illustration of the mode-1 unfolding.

transformation that converts the tensor $\mathcal{A} \in \mathbb{R}^{I_1 \times I_2 \times \dots \times I_N}$ *into a column vector* $\mathbf{a} \in \mathbb{R}^{\Pi_{n=1}^{N} I_n}$, *denoted as* $\mathbf{a} = vec(\mathcal{A})$.

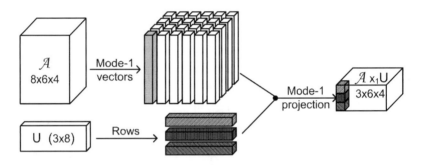

FIGURE 3.8: Visual illustration of the mode-n (mode-1) multiplication.

Definition 3.9 (Mode-n product: tensor matrix multiplication). *The* **mode-n product** *of a tensor* $\mathcal{A} \in \mathbb{R}^{I_1 \times I_2 \times \dots \times I_N}$ *by a matrix* $\mathbf{U} \in \mathbb{R}^{J_n \times I_n}$ *is a tensor* $\mathcal{B} \in \mathbb{R}^{I_1 \times \dots \times I_{n-1} \times J_n \times I_{n+1} \times \dots \times I_N}$, *denoted as*

$$\mathcal{B} = \mathcal{A} \times_n \mathbf{U}, \tag{3.4}$$

where each entry of $\mathcal{B}$ *is defined as the sum of products of corresponding entries in* $\mathcal{A}$ *and* $\mathbf{U}$:

$$\mathcal{B}(i_1, \dots, i_{n-1}, j_n, i_{n+1}, \dots, i_N) = \sum_{i_n} \mathcal{A}(i_1, \dots, i_N) \cdot \mathbf{U}(j_n, i_n). \tag{3.5}$$

This is equivalent to premultiplying each mode-n vector of $\mathcal{A}$ by $\mathbf{U}$. Thus, the mode-n product above can be written using the mode-n unfolding as

$$\mathbf{B}_{(n)} = \mathbf{U}\mathbf{A}_{(n)}. \tag{3.6}$$

For second order tensors (matrices) $\mathbf{A}$ and $\mathbf{U}$ of proper sizes,

$$\mathbf{A} \times_1 \mathbf{U} = \mathbf{U}\mathbf{A}, \quad \mathbf{A} \times_2 \mathbf{U} = \mathbf{A}\mathbf{U}^T. \tag{3.7}$$

Figure 3.8 demonstrates how the mode-1 multiplication $\mathcal{A} \times_1 \mathbf{U}$ is obtained.

The product $\mathcal{A} \times_1 \mathbf{U}$ is computed as the inner products between the mode-1 vectors of $\mathcal{A}$ and the rows of $\mathbf{U}$. In the mode-1 multiplication in Figure 3.8, each mode-1 vector of $\mathcal{A}$ ($\in \mathbb{R}^8$) is projected by $\mathbf{U} \in \mathbb{R}^{3 \times 8}$ to obtain a vector ($\in \mathbb{R}^3$), as the differently shaded vector indicates in the right of the figure.

Tensor matrix multiplication has the following two properties [De Lathauwer et al., 2000b].

Property 3.1. *Given a tensor* $\mathcal{A} \in \mathbb{R}^{I_1 \times I_2 \times \ldots \times I_N}$, *and two matrices* $\mathbf{U} \in \mathbb{R}^{J_n \times I_n}$ *and* $\mathbf{V} \in \mathbb{R}^{J_m \times I_m}$, *where* $m \neq n$, *we have*

$$(\mathcal{A} \times_m \mathbf{U}) \times_n \mathbf{V} = (\mathcal{A} \times_n \mathbf{V}) \times_m \mathbf{U}. \tag{3.8}$$

Property 3.2. *Given a tensor* $\mathcal{A} \in \mathbb{R}^{I_1 \times I_2 \times \ldots \times I_N}$, *and two matrices* $\mathbf{U} \in \mathbb{R}^{J_n \times I_n}$ *and* $\mathbf{V} \in \mathbb{R}^{K_n \times I_n}$, *we have*

$$(\mathcal{A} \times_n \mathbf{U}) \times_n \mathbf{V} = \mathcal{A} \times_n (\mathbf{V} \cdot \mathbf{U}). \tag{3.9}$$

Definition 3.10 (Mode-n product: tensor vector multiplication). *The* **mode-n product** *of a tensor* $\mathcal{A} \in \mathbb{R}^{I_1 \times I_2 \times \ldots \times I_N}$ *by a vector* $\mathbf{u} \in \mathbb{R}^{I_n \times 1}$ *is a tensor* $\mathcal{C} \in \mathbb{R}^{I_1 \times \ldots \times I_{n-1} \times 1 \times I_{n+1} \times \ldots \times I_N}$, *denoted as*

$$\mathcal{C} = \mathcal{A} \times_n \mathbf{u}^T, \tag{3.10}$$

where each entry of $\mathcal{C}$ *is defined as*

$$\mathcal{C}(i_1, \ldots, i_{n-1}, 1, i_{n+1}, \ldots, i_N) = \sum_{i_n} \mathcal{A}(i_1, \ldots, i_N) \cdot \mathbf{u}(i_n). \tag{3.11}$$

Multiplication of a tensor by a vector can be viewed as a special case of tensor matrix multiplication with $J_n = 1$ (so $\mathbf{U} \in \mathbb{R}^{J_n \times I_n} = \mathbf{u}^T$). This product $\mathcal{A} \times_1 \mathbf{u}^T$ can be computed as the inner products between the mode-1 vectors of $\mathcal{A}$ and $\mathbf{u}$. Note that the nth dimension of $\mathcal{C}$ is 1, so effectively the order of $\mathcal{C}$ is reduced to $N - 1$. The *squeeze*() function in MATLAB can remove all modes with dimension equal to one.

Definition 3.11. *The* **scalar product** *(inner product) of two same-sized tensors* $\mathcal{A}, \mathcal{B} \in \mathbb{R}^{I_1 \times I_2 \times \ldots \times I_N}$ *is defined as*

$$< \mathcal{A}, \mathcal{B} > = \sum_{i_1} \sum_{i_2} \ldots \sum_{i_N} \mathcal{A}(i_1, i_2, \ldots, i_N) \cdot \mathcal{B}(i_1, i_2, \ldots, i_N). \tag{3.12}$$

This can be seen as a generalization of the inner product in linear algebra (Section A.1.4).

Definition 3.12. *The* **Frobenius norm** *of* $\mathcal{A}$ *is defined as*

$$\| \mathcal{A} \|_F = \sqrt{< \mathcal{A}, \mathcal{A} >}. \tag{3.13}$$

This is a straightforward extension of the matrix Frobenius norm (Section A.1.5).

3.1.3   Tensor/Matrix Distance Measure

The Frobenius norm can be used to measure the distance between tensors $\mathcal{A}$ and $\mathcal{B}$ as

$$dist(\mathcal{A}, \mathcal{B}) = \parallel \mathcal{A} - \mathcal{B} \parallel_F . \tag{3.14}$$

Although this is a tensor-based measure, it is equivalent to a distance measure of corresponding vector representations denoted as $vec(\mathcal{A})$ and $vec(\mathcal{B})$, as to be shown in the following. We first derive a property regarding the scalar product between two tensors:

Proposition 3.1. $< \mathcal{A}, \mathcal{B} > = < vec(\mathcal{A}), vec(\mathcal{B}) > = [vec(\mathcal{B})]^T vec(\mathcal{A})$.

Proof. Let $\mathbf{a} = vec(\mathcal{A})$ and $\mathbf{b} = vec(\mathcal{B})$ for convenience. From Equation (3.12), $< \mathcal{A}, \mathcal{B} >$ is the summing the products between all corresponding entries in $\mathcal{A}$ and $\mathcal{B}$. We can have the same results by the sum of products between all corresponding entries in $\mathbf{a}$ and $\mathbf{b}$, their vectorizations. Thus, we have

$$
\begin{aligned}
< \mathcal{A}, \mathcal{B} > &= \sum_{i_1=1}^{I_1} \sum_{i_2=1}^{I_2} \cdots \sum_{i_N=1}^{I_N} \mathcal{A}(i_1, i_2, ..., i_N) \cdot \mathcal{B}(i_1, i_2, ..., i_N) \\
&= \sum_{i=1}^{\Pi_{n=1}^{N} I_n} \mathbf{a}(i) \cdot \mathbf{b}(i) \\
&= < \mathbf{a}, \mathbf{b} > \\
&= [\mathbf{b}]^T \mathbf{a}.
\end{aligned}
$$

$\square$

Then, it is straightforward to show the equivalence.

Proposition 3.2. $dist(\mathcal{A}, \mathcal{B}) = \parallel vec(\mathcal{A}) - vec(\mathcal{B}) \parallel_2$.

Proof. From Proposition 3.1,

$$
\begin{aligned}
dist(\mathcal{A}, \mathcal{B}) &= \parallel \mathcal{A} - \mathcal{B} \parallel_F \\
&= \sqrt{< (\mathcal{A} - \mathcal{B}), (\mathcal{A} - \mathcal{B}) >} \\
&= \sqrt{< vec(\mathcal{A}) - vec(\mathcal{B}), vec(\mathcal{A}) - vec(\mathcal{B}) >} \\
&= \parallel vec(\mathcal{A}) - vec(\mathcal{B}) \parallel_2 .
\end{aligned}
$$

$\square$

Proposition 3.2 indicates that the Frobenius norm of the difference between two tensors equals the Euclidean distance between their vectorized representations. The tensor Frobenius norm is a point-based measurement [Lu et al., 2004] without taking the tensor structure into account.

For second-order tensors, that is, matrices, their distance can be measured by the matrix Frobenius norm, which equals the square root of the trace of the difference matrix:

$$dist(\mathbf{A}, \mathbf{B}) = \| \mathbf{A} - \mathbf{B} \|_F = \sqrt{\operatorname{tr}((\mathbf{A} - \mathbf{B})^T(\mathbf{A} - \mathbf{B}))}. \tag{3.15}$$

An alternative for matrix distance is the so-called *volume measure* used in [Meng and Zhang, 2007]. The volume measure between matrices $\mathbf{A}$ and $\mathbf{B}$ is defined as

$$dist(\mathbf{A}, \mathbf{B}) = vol(\mathbf{A} - \mathbf{B}) = \sqrt{|(\mathbf{A} - \mathbf{B})^T(\mathbf{A} - \mathbf{B})|}, \tag{3.16}$$

where $|\cdot|$ denotes the determinant.

The matrix Frobenius norm is further generalized as the *assembled matrix distance* (AMD) in [Zuo et al., 2006] as

$$d_{AMD}(\mathbf{A}, \mathbf{B}) = \left(\sum_{i_1=1}^{I_1} \left(\sum_{i_2=1}^{I_2} (\mathbf{A}(i_1, i_2) - \mathbf{B}(i_1, i_2))^2 \right)^{p/2} \right)^{1/p}, \tag{3.17}$$

where the power p weights the differences between elements. AMD is a variation of the p-norm for vectors and it treats a matrix as a vector effectively. AMD with $p = 2$ is equivalent to the matrix Frobenius norm and the Euclidean norm for vectors. The AMD measure can be generalized to general higher-order tensors, or it can also be modified to take data properties (such as shape and connectivity) into account as in [Porro-Muñoz et al., 2011].

Distance measures are frequently used by classifiers to measure similarity or dissimilarity. Furthermore, it is possible to design classifiers by taking into account the matrix/tensor representation or structure. For example, Wang et al. [2008] proposed a classifier specially designed for matrix representations of patterns and showed that such a classifier has advantages in features extracted from matrix representations.

3.2 Tensor Decompositions

Multilinear subspace learning is based on tensor decompositions. This section reviews two most important works in this area.

3.2.1 CANDECOMP/PARAFAC

Hitchcock [1927b,a] first proposed the idea of expressing a tensor as the sum of rank-one tensors in polyadic form. It became popular in the psychometrics community with the independent introduction of canonical decomposition (CANDECOMP) by Carroll and Chang [1970] and parallel factors (PARAFAC) by Harshman [1970].

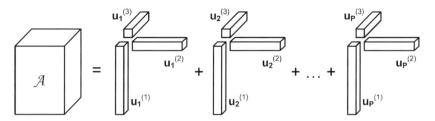

FIGURE 3.9: The CANDECOMP/PARAFAC decomposition of a third-order tensor.

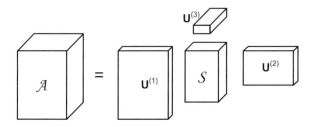

FIGURE 3.10: The Tucker decomposition of a third-order tensor.

With the CANDECOMP/PARAFAC decomposition (CP decomposition), a tensor $\mathcal{A}$ can be factorized into a linear combination of P rank-one tensors:

$$\mathcal{A} = \sum_{p=1}^{P} \lambda_p \mathbf{u}_p^{(1)} \circ \mathbf{u}_p^{(2)} \circ ... \circ \mathbf{u}_p^{(N)}, \tag{3.18}$$

where $P \leq \prod_{n=1}^{N} I_n$. Figure 3.9 illustrates this decomposition.

3.2.2 Tucker Decomposition and HOSVD

The Tucker decomposition [Tucker, 1966] was introduced in the 1960s. It decomposes an Nth-order tensor $\mathcal{A}$ into a core tensor $\mathcal{S}$ multiplied by N matrices, one in each mode:

$$\mathcal{A} = \mathcal{S} \times_1 \mathbf{U}^{(1)} \times_2 \mathbf{U}^{(2)} \times ... \times_N \mathbf{U}^{(N)}, \tag{3.19}$$

where $P_n \leq I_n$ for $n = 1, ..., N$, and $\mathbf{U}^{(n)} = \left[\mathbf{u}_1^{(n)} \mathbf{u}_2^{(n)} ... \mathbf{u}_{P_n}^{(n)} \right]$ is an $I_n \times P_n$ matrix often assumed to have orthonormal column vectors. Figure 3.10 illustrates this decomposition.

The Tucker decomposition was investigated mainly in psychometrics after its initial introduction. It was reintroduced by De Lathauwer et al. [2000b] as the *higher-order singular value decomposition* (HOSVD) to the communities of numerical algebra and signal processing, followed by many other disciplines. When $P_n = I_n$ for $n = 1, ..., N$ and $\{\mathbf{U}^{(n)}, n = 1, ..., N\}$ are all orthogonal

$I_n \times I_n$ matrices, then from Equation (3.19), the core tensor can be written as

$$\mathcal{S} = \mathcal{A} \times_1 \mathbf{U}^{(1)^T} \times_2 \mathbf{U}^{(2)^T} ... \times_N \mathbf{U}^{(N)^T}. \qquad (3.20)$$

Because $\mathbf{U}^{(n)}$ has orthonormal columns, we have [De Lathauwer et al., 2000a]

$$\| \mathcal{A} \|_F^2 = \| \mathcal{S} \|_F^2 . \qquad (3.21)$$

A matrix representation of this decomposition can be obtained by unfolding $\mathcal{A}$ and $\mathcal{S}$ as

$$\mathbf{A}_{(n)} = \mathbf{U}^{(n)} \cdot \mathbf{S}_{(n)} \cdot \left(\mathbf{U}^{(n+1)} \otimes ... \otimes \mathbf{U}^{(N)} \otimes \mathbf{U}^{(1)} \otimes ... \otimes \mathbf{U}^{(n-1)} \right)^T, \qquad (3.22)$$

where $\otimes$ denotes the Kronecker product (Section A.1.4). The decomposition can also be written as

$$\mathcal{A} = \sum_{i_1=1}^{I_1} \sum_{i_2=1}^{I_2} ... \sum_{i_N=1}^{I_N} \mathcal{S}(i_1, i_2, ..., i_N) \mathbf{u}_{i_1}^{(1)} \circ \mathbf{u}_{i_2}^{(2)} \circ ... \circ \mathbf{u}_{i_N}^{(N)}, \qquad (3.23)$$

that is, any tensor $\mathcal{A}$ can be written as a linear combination of $\prod_{n=1}^{N} I_n$ rank-one tensors. Comparison of Equation (3.23) against Equation (3.18) reveals the *equivalence* between Tucker decomposition and CP decomposition.

3.3   Multilinear Projections

A *tensor subspace* is defined through a multilinear projection that maps the input data from a high-dimensional space to a low-dimensional space [He et al., 2005a]. Therefore, multilinear projection is an important concept to grasp before proceeding to multilinear subspace learning.

This section presents three basic multilinear projections named in terms of the input and output representations of the projection: the traditional vector-to-vector projection, the tensor-to-tensor projection, and the tensor-to-vector projection. Furthermore, we investigate the relationships among these projections.

3.3.1   Vector-to-Vector Projection

Linear projection is a standard transformation used widely in various applications [Duda et al., 2001; Moon and Stirling, 2000]. A linear projection takes a vector $\mathbf{x} \in \mathbb{R}^I$ as input and projects it to a vector $\mathbf{y} \in \mathbb{R}^P$ using a projection matrix $\mathbf{U} \in \mathbb{R}^{I \times P}$:

$$\mathbf{y} = \mathbf{U}^T \mathbf{x} = \mathbf{x} \times_1 \mathbf{U}^T. \qquad (3.24)$$

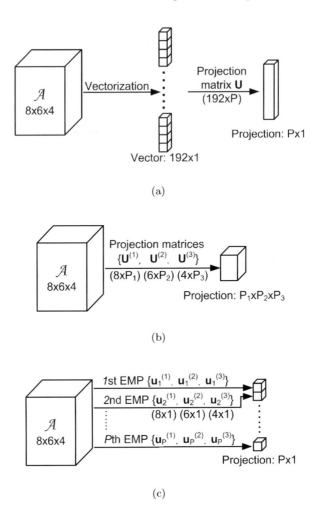

(a)

(b)

(c)

FIGURE 3.11: Illustration of (a) vector-to-vector projection, (b) tensor-to-tensor projection, and (c) tensor-to-vector projection, where EMP stands for elementary multilinear projection.

If we name this projection according to its input and output representations, linear projection is a vector-to-vector projection because it maps a vector to another vector. When the input data is a matrix or a higher-order tensor, it needs to be vectorized (reshaped into a vector) before projection. Figure 3.11(a) illustrates the vector-to-vector projection of a tensor object $\mathcal{A}$.

Denote each column of $\mathbf{U}$ as $\mathbf{u}_p$, so $\mathbf{U} = [\mathbf{u}_1 \ \mathbf{u}_2 \ ... \ \mathbf{u}_P]$. Then each column $\mathbf{u}_p$ projects $\mathbf{x}$ to a scalar $\mathbf{y}(p)$ or y_p:

$$y_p = \mathbf{y}(p) = \mathbf{u}_p^T \mathbf{x}. \qquad (3.25)$$

3.3.2 Tensor-to-Tensor Projection

In addition to traditional vector-to-vector projection, a tensor can also be projected to another tensor (of the same order usually), called the tensor-to-tensor projection (TTP). It is formulated based on the Tucker decomposition.

Consider the second-order case first. A second-order tensor (matrix) $\mathbf{X}$ resides in the *tensor space* denoted as $\mathbb{R}^{I_1} \bigotimes \mathbb{R}^{I_2}$, which is defined as the *tensor product* (*outer product*) of two vector spaces $\mathbb{R}^{I_1}$ and $\mathbb{R}^{I_2}$. For the projection of a matrix $\mathbf{X}$ in a tensor space $\mathbb{R}^{I_1} \bigotimes \mathbb{R}^{I_2}$ to another tensor $\mathbf{Y}$ in a lower-dimensional tensor space $\mathbb{R}^{P_1} \bigotimes \mathbb{R}^{P_2}$, where $P_n \leq I_n$ for $n = 1, 2$, two projection matrices $\mathbf{U}^{(1)} \in \mathbb{R}^{I_1 \times P_1}$ and $\mathbf{U}^{(2)} \in \mathbb{R}^{I_2 \times P_2}$ (usually with orthonormal columns) are used so that [De Lathauwer et al., 2000a]

$$\mathbf{Y} = \mathbf{X} \times_1 \mathbf{U}^{(1)^T} \times_2 \mathbf{U}^{(2)^T} = \mathbf{U}^{(1)^T} \mathbf{X} \mathbf{U}^{(2)}. \tag{3.26}$$

For the general higher-order case, an Nth-order tensor $\mathcal{X}$ resides in the tensor space $\mathbb{R}^{I_1} \bigotimes \mathbb{R}^{I_2} ... \bigotimes \mathbb{R}^{I_N}$ [De Lathauwer et al., 2000b], which is the *tensor product* (*outer product*) of N vector spaces $\mathbb{R}^{I_1}, \mathbb{R}^{I_2}, ..., \mathbb{R}^{I_N}$. For the projection of a tensor $\mathcal{X}$ in a tensor space $\mathbb{R}^{I_1} \bigotimes \mathbb{R}^{I_2} ... \bigotimes \mathbb{R}^{I_N}$ to another tensor $\mathcal{Y}$ in a lower-dimensional tensor space $\mathbb{R}^{P_1} \bigotimes \mathbb{R}^{P_2} ... \bigotimes \mathbb{R}^{P_N}$, where $P_n \leq I_n$ for all n, N projection matrices $\{\mathbf{U}^{(n)} \in \mathbb{R}^{I_n \times P_n}, n = 1, ..., N\}$ (usually with orthonormal columns) are used so that [De Lathauwer et al., 2000a]

$$\mathcal{Y} = \mathcal{X} \times_1 \mathbf{U}^{(1)^T} \times_2 \mathbf{U}^{(2)^T} ... \times_N \mathbf{U}^{(N)^T}. \tag{3.27}$$

These N projection matrices used for TTP can be concisely written as $\{\mathbf{U}^{(n)}\}$. Figure 3.11(b) demonstrates the TTP of a tensor object $\mathcal{A}$ to a smaller tensor of size $P_1 \times P_2 \times P_3$. This multilinear projection can be carried out through N mode-n multiplications, as illustrated in Figure 3.8.

3.3.3 Tensor-to-Vector Projection

The third multilinear projection is from a tensor space to a vector space, and it is called the tensor-to-vector projection (TVP)[5]. It is formulated based on the CANDECOMP/PARAFAC model.

As a vector can be viewed as multiple scalars, the projection from a tensor to a vector can be viewed as multiple projections, each of which projects a tensor to a scalar, as illustrated in Figure 3.11(c). In the figure, the TVP of a tensor $\mathcal{A} \in \mathbb{R}^{8 \times 6 \times 4}$ to a $P \times 1$ vector consists of P projections, each projecting $\mathcal{A}$ to a scalar. Thus, the projection from a tensor to a scalar is the building block for TVP and it is considered first.

A second-order tensor (matrix) $\mathbf{X} \in \mathbb{R}^{I_1 \times I_2}$ can be projected to a scalar y through two unit projection vectors $\mathbf{u}^{(1)}$ and $\mathbf{u}^{(2)}$ as

$$y = \mathbf{X} \times_1 \mathbf{u}^{(1)^T} \times_2 \mathbf{u}^{(2)^T} = \mathbf{u}^{(1)^T} \mathbf{X} \mathbf{u}^{(2)}, \ \| \mathbf{u}^{(1)} \| = \| \mathbf{u}^{(2)} \| = 1, \tag{3.28}$$

---

[5]The tensor-to-vector projection is referred to as the rank-one projections in some works [Wang and Gong, 2006; Tao et al., 2006; Hua et al., 2007].

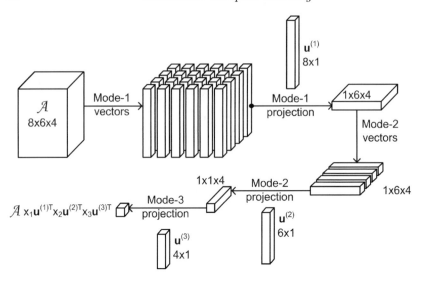

FIGURE 3.12: Illustration of an elementary multilinear projection.

where $\| \cdot \|$ is the Euclidean norm for vectors (see Section A.1.4). It can be written as the inner product between $\mathbf{X}$ and the outer products of $\mathbf{u}^{(1)}$ and $\mathbf{u}^{(2)}$:

$$y = < \mathbf{X}, \mathbf{u}^{(1)}\mathbf{u}^{(2)^{T}} > . \tag{3.29}$$

A general tensor $\mathcal{X} \in \mathbb{R}^{I_1 \times I_2 \times ... \times I_N}$ can be projected to a point y through N unit projection vectors $\{\mathbf{u}^{(1)}, \mathbf{u}^{(2)}, ..., \mathbf{u}^{(N)}\}$, which can also be written as $\{\mathbf{u}^{(n)}, n = 1, ..., N\}$ or simply as $\{\mathbf{u}^{(n)}\}$:

$$y = \mathcal{X} \times_1 \mathbf{u}^{(1)^{T}} \times_2 \mathbf{u}^{(2)^{T}} ... \times_N \mathbf{u}^{(N)^{T}}, \ \| \mathbf{u}^{(n)} \| = 1 \text{ for } n = 1, ..., N. \tag{3.30}$$

It can be written as the scalar product (Equation (3.12)) of $\mathcal{X}$ and the outer product of $\{\mathbf{u}^{(n)}\}$:

$$y = < \mathcal{X}, \mathbf{u}^{(1)} \circ \mathbf{u}^{(2)} \circ ... \circ \mathbf{u}^{(N)} > . \tag{3.31}$$

Denote $\mathcal{U} = \mathbf{u}^{(1)} \circ \mathbf{u}^{(2)} \circ ... \circ \mathbf{u}^{(N)}$, then $y = < \mathcal{X}, \mathcal{U} >$. This multilinear projection through $\{\mathbf{u}^{(1)}, \mathbf{u}^{(2)}, ..., \mathbf{u}^{(N)}\}$ is named as an *elementary multilinear projection* (EMP) [Lu et al., 2009c], which is a projection of a tensor to a scalar[6]. It is a rank-one projection and it consists of one projection vector in each mode. Figure 3.12 illustrates an EMP of a tensor $\mathcal{A} \in \mathbb{R}^{8 \times 6 \times 4}$.

Thus, the TVP of a tensor $\mathcal{X}$ to a vector $\mathbf{y} \in \mathbb{R}^P$ in a P-dimensional vector space consists of P EMPs,

$$\{\mathbf{u}_p^{(1)}, \mathbf{u}_p^{(2)}, ..., \mathbf{u}_p^{(N)}\}, p = 1, ..., P, \tag{3.32}$$

---

[6]We call it EMP rather than tensor-to-scalar projection for two reasons. One is that it is used as a building block in TVP. The other is that we want to emphasize that it is an elementary operation.

which can be written concisely as $\{\mathbf{u}_p^{(n)}, n = 1, ..., N\}_{p=1}^P$ or simply as $\{\mathbf{u}_p^{(n)}\}_N^P$. The TVP from $\mathcal{X}$ to $\mathbf{y}$ is then written as

$$\mathbf{y} = \mathcal{X} \times_{n=1}^N \{\mathbf{u}_p^{(n)}, n = 1, ..., N\}_{p=1}^P = \mathcal{X} \times_{n=1}^N \{\mathbf{u}_p^{(n)}\}_N^P, \qquad (3.33)$$

where the pth entry of $\mathbf{y}$ is obtained from the pth EMP as

$$y_p = \mathbf{y}(p) = \mathcal{X} \times_1 \mathbf{u}_p^{(1)^T} \times_2 \mathbf{u}_p^{(2)^T} ... \times_N \mathbf{u}_p^{(N)^T} = \mathcal{X} \times_{n=1}^N \{\mathbf{u}_p^{(n)}\}. \qquad (3.34)$$

Figure 3.11(c) shows the TVP of a tensor $\mathcal{A}$ to a vector of size $P \times 1$.

3.4 Relationships among Multilinear Projections

With the introduction of the three basic multilinear projections, it is worthwhile to investigate their relationships.

Degenerated conditions: It is easy to verify that the vector-to-vector projection is the special case of the tensor-to-tensor projection and the tensor-to-vector projection for $N = 1$. The elementary multilinear projection is the degenerated version of the tensor-to-tensor projection with $P_n = 1$ for all n.

EMP view of TTP: Each projected element in the tensor-to-tensor projection can be viewed as the projection by an elementary multilinear projection formed by taking one column from each modewise projection matrix. Thus, a projected tensor in the tensor-to-tensor projection is obtained through $\prod_{n=1}^N P_n$ elementary multilinear projections with shared projection vectors (from the projection matrices) in effect, while in the tensor-to-vector projection, the P elementary multilinear projections do not have shared projection vectors.

Equivalence between EMP and VVP: Recall that the projection using an elementary multilinear projection $\{\mathbf{u}^{(1)}, \mathbf{u}^{(2)}, ..., \mathbf{u}^{(N)}\}$ can be written as

$$y =< \mathcal{X}, \mathcal{U} >=< vec(\mathcal{X}), vec(\mathcal{U}) >= [vec(\mathcal{U})]^T vec(\mathcal{X}), \qquad (3.35)$$

by Proposition 3.1. Thus, an elementary multilinear projection is equivalent to a linear projection of $vec(\mathcal{X})$, the vectorized representation of $\mathcal{X}$, by a vector $vec(\mathcal{U})$ as in Equation (3.25). Because $\mathcal{U} = \mathbf{u}^{(1)} \circ \mathbf{u}^{(2)} \circ ... \circ \mathbf{u}^{(N)}$, Equation (3.35) indicates that the elementary multilinear projection is equivalent to a linear projection for $P = 1$ with a constraint on the projection vector such that it is the vectorized representation of a rank-one tensor.

Equivalence between TTP and TVP: Given a TVP $\{\mathbf{u}_p^{(n)}\}_N^P$, we can form N matrices $\{\mathbf{V}^{(n)}\}$, where

$$\mathbf{V}^{(n)} = [\mathbf{u}_1^{(n)}, ..., \mathbf{u}_p^{(n)}, ..., \mathbf{u}_P^{(n)}] \in \mathbb{R}^{I_n \times P}. \qquad (3.36)$$

These matrices can be viewed as a TTP with $P_n = P$ for $n = 1, ..., N$ (equal subspace dimensions in all modes). Thus, the TVP of a tensor by $\{\mathbf{u}_p^{(n)}\}_N^P$ is equivalent to the diagonal of a corresponding TTP of the same tensor by $\{\mathbf{V}^{(n)}\}$ as defined in Equation (3.36):

$$\mathbf{y} = \mathcal{X} \times_{n=1}^N \{\mathbf{u}_p^{(n)}\}_N^P \tag{3.37}$$

$$= diag\left(\mathcal{X} \times_1 \mathbf{V}^{(1)^T} \times_2 \mathbf{V}^{(2)^T} ... \times_N \mathbf{V}^{(N)^T}\right). \tag{3.38}$$

In the second-order case, this equivalence is

$$\mathbf{y} = diag\left(\mathbf{V}^{(1)^T} \mathbf{X} \mathbf{V}^{(2)}\right). \tag{3.39}$$

Number of parameters to estimate: The number of parameters to be estimated in a particular projection indicates model complexity, an important concern in practice. Compared with a projection vector of size $I \times 1$ in a VVP specified by I parameters ($I = \prod_{n=1}^N I_n$ for an Nth-order tensor), an EMP in a TVP is specified by $\sum_{n=1}^N I_n$ parameters. Hence, to project a tensor of size $\prod_{n=1}^N I_n$ to a vector of size $P \times 1$, TVP needs to estimate only $P \cdot \sum_{n=1}^N I_n$ parameters, while VVP needs to estimate $P \cdot \prod_{n=1}^N I_n$ parameters. The implication is that TVP has fewer parameters to estimate while being more constrained on the solutions, and VVP has less constraint on the solutions sought while having more parameters to estimate. In other words, TVP has a simpler model than VVP. For TTP with the same amount of dimensionality reduction $\prod_{n=1}^N P_n = P$, $\sum_{n=1}^N P_n \times I_n$ parameters need to be estimated. Thus, due to shared projection vectors, TTP may need to estimate even fewer parameters and its model can be even simpler.

Table 3.1 contrasts the number of parameters to be estimated by the three projections for the same amount of dimensionality reduction for several cases. Figure 3.13 further illustrates the first three cases, where the numbers of parameters are normalized with respect to that by VVP for better visualization. From the table and figure, we can see that for higher-order tensors, the conventional VVP model becomes extremely complex and parameter estimation becomes extremely difficult. This often leads to the small sample size (SSS) problem in practice when there are limited number of training samples available.

3.5 Scatter Measures for Tensors and Scalars

3.5.1 Tensor-Based Scatters

In analogy to the definitions of scatters in Equations (2.1), (2.27), and (2.25) for vectors used in linear subspace learning, we define tensor-based scatters to be used in multilinear subspace learning (MSL) through TTP.

TABLE 3.1: Number of parameters to be estimated by three multilinear projections.

Input	Output	VVP	TVP	TTP
$\prod_{n=1}^{N} I_n$	P	$P \cdot \prod_{n=1}^{N} I_n$	$P \cdot \sum_{n=1}^{N} I_n$	$\sum_{n=1}^{N} P_n \times I_n$
10×10	4	400	80	40 $(P_n = 2)$
100×100	4	40,000	800	400 $(P_n = 2)$
$100 \times 100 \times 100$	8	8,000,000	2,400	600 $(P_n = 2)$
$\prod_{n=1}^{4} 100$	16	1,600,000,000	6,400	800 $(P_n = 2)$

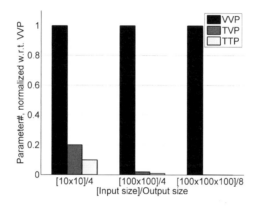

FIGURE 3.13: Comparison of the number of parameters to be estimated by VVP, TVP, and TTP, normalized with respect to the number by VVP for visualization.

Definition 3.13. *Let* $\{\mathcal{A}_m, m = 1, ..., M\}$ *be a set of* M *tensor samples in* $\mathbb{R}^{I_1} \otimes \mathbb{R}^{I_2} ... \otimes \mathbb{R}^{I_N}$. *The total scatter of these tensors is defined as*

$$\Psi_{T_\mathcal{A}} = \sum_{m=1}^{M} \| \mathcal{A}_m - \bar{\mathcal{A}} \|_F^2, \tag{3.40}$$

where $\bar{\mathcal{A}}$ *is the* mean tensor *calculated as*

$$\bar{\mathcal{A}} = \frac{1}{M} \sum_{m=1}^{M} \mathcal{A}_m. \tag{3.41}$$

The mode-n total scatter matrix of these samples is then defined as

$$\mathbf{S}_{T_\mathcal{A}}^{(n)} = \sum_{m=1}^{M} \left(\mathbf{A}_{m(n)} - \bar{\mathbf{A}}_{(n)} \right) \left(\mathbf{A}_{m(n)} - \bar{\mathbf{A}}_{(n)} \right)^T, \tag{3.42}$$

where $\mathbf{A}_{m(n)}$ *and* $\bar{\mathbf{A}}_{(n)}$ *are the mode-n unfolding of* $\mathcal{A}_m$ *and* $\bar{\mathcal{A}}$, *respectively.*

Definition 3.14. *Let* $\{\mathcal{A}_m, m = 1, ..., M\}$ *be a set of M tensor samples in* $\mathbb{R}^{I_1} \otimes \mathbb{R}^{I_2} ... \otimes \mathbb{R}^{I_N}$. *The between-class scatter of these tensors is defined as*

$$\Psi_{B_A} = \sum_{c=1}^{C} M_c \parallel \bar{\mathcal{A}}_c - \bar{\mathcal{A}} \parallel_F^2, \tag{3.43}$$

and the within-class scatter of these tensors is defined as

$$\Psi_{W_A} = \sum_{m=1}^{M} \parallel \mathcal{A}_m - \bar{\mathcal{A}}_{c_m} \parallel_F^2, \tag{3.44}$$

where C is the number of classes, M_c is the number of samples for class c, c_m is the class label for the mth sample $\mathcal{A}_m$, $\bar{\mathcal{A}}$ is the mean tensor, and the class mean tensor is

$$\bar{\mathcal{A}}_c = \frac{1}{M_c} \sum_{m,c_m=c} \mathcal{A}_m. \tag{3.45}$$

Next, the mode-n between-class and within-class scatter matrices are defined accordingly.

Definition 3.15. *The mode-n between-class scatter matrix of these samples is defined as*

$$\mathbf{S}_{B_A}^{(n)} = \sum_{c=1}^{C} M_c \cdot \left(\bar{\mathbf{A}}_{c(n)} - \bar{\mathbf{A}}_{(n)} \right) \left(\bar{\mathbf{A}}_{c(n)} - \bar{\mathbf{A}}_{(n)} \right)^T, \tag{3.46}$$

and the mode-n within-class scatter matrix of these samples is defined as

$$\mathbf{S}_{W_A}^{(n)} = \sum_{m=1}^{M} \left(\mathbf{A}_{m(n)} - \bar{\mathbf{A}}_{c_m(n)} \right) \left(\mathbf{A}_{m(n)} - \bar{\mathbf{A}}_{c_m(n)} \right)^T, \tag{3.47}$$

where $\bar{\mathbf{A}}_{c_m(n)}$ is the mode-n unfolding of $\bar{\mathcal{A}}_{c_m}$.

From the definitions above, the following properties are derived:

Property 3.3. *Because* $\mathrm{tr}(\mathbf{A}\mathbf{A}^T) = \parallel \mathcal{A} \parallel_F^2$ *and* $\parallel \mathcal{A} \parallel_F^2 = \parallel \mathbf{A}_{(n)} \parallel_F^2$,

$$\Psi_{B_A} = \mathrm{tr} \left(\mathbf{S}_{B_A}^{(n)} \right) = \sum_{c=1}^{C} M_c \parallel \bar{\mathbf{A}}_{c(n)} - \bar{\mathbf{A}}_{(n)} \parallel_F^2 \tag{3.48}$$

and

$$\Psi_{W_A} = \mathrm{tr} \left(\mathbf{S}_{W_A}^{(n)} \right) = \sum_{m=1}^{M} \parallel \mathbf{A}_{m(n)} - \bar{\mathbf{A}}_{c_m(n)} \parallel_F^2 \tag{3.49}$$

for all n.

Scatters for matrices: As a special case, when $N = 2$, we have a set of M matrix samples $\{\mathbf{A}_m, m = 1, ..., M\}$ in $\mathbb{R}^{I_1} \otimes \mathbb{R}^{I_2}$. The total scatter of these matrices is defined as

$$\Psi_{T_\mathbf{A}} = \sum_{m=1}^{M} \| \mathbf{A}_m - \bar{\mathbf{A}} \|_F^2, \tag{3.50}$$

where $\bar{\mathbf{A}}$ is the mean matrix calculated as

$$\bar{\mathbf{A}} = \frac{1}{M} \sum_{m=1}^{M} \mathbf{A}_m. \tag{3.51}$$

The between-class scatter of these matrix samples is defined as

$$\Psi_{B_\mathbf{A}} = \sum_{c=1}^{C} M_c \| \bar{\mathbf{A}}_c - \bar{\mathbf{A}} \|_F^2, \tag{3.52}$$

and the within-class scatter of these matrix samples is defined as

$$\Psi_{W_\mathbf{A}} = \sum_{m=1}^{M} \| \mathbf{A}_m - \bar{\mathbf{A}}_{c_m} \|_F^2, \tag{3.53}$$

where $\bar{\mathbf{A}}$ is the mean matrix, and the class mean matrix is

$$\bar{\mathbf{A}}_c = \frac{1}{M_c} \sum_{m,c_m=c} \mathbf{A}_m. \tag{3.54}$$

3.5.2 Scalar-Based Scatters

While the tensor-based scatters defined above are useful for developing MSL algorithms based on TTP, they are not applicable to those based on the TVP/EMP. Therefore, scalar-based scatters need to be defined for MSL through TVP/EMP. They can be viewed as the degenerated versions of the vector-based or tensor-based scatters.

Definition 3.16. *Let $\{a_m, m = 1, ..., M\}$ be a set of M scalar samples. The total scatter of these scalars is defined as*

$$S_{T_a} = \sum_{m=1}^{M} (a_m - \bar{a})^2, \tag{3.55}$$

where $\bar{a}$ is the mean scalar calculated as

$$\bar{a} = \frac{1}{M} \sum_{m=1}^{M} a_m. \tag{3.56}$$

Thus, the total scatter for scalars is simply a scaled version of the sample *variance*.

Definition 3.17. *Let $\{a_m, m = 1, ..., M\}$ be a set of M scalar samples. The between-class scatter of these scalars is defined as*

$$S_{B_a} = \sum_{c=1}^{C} M_c(\bar{a}_c - \bar{a})^2, \qquad (3.57)$$

and the within-class scatter of these scalars is defined as

$$S_{W_a} = \sum_{m=1}^{M} (a_m - \bar{a}_{c_m})^2, \qquad (3.58)$$

where

$$\bar{a}_c = \frac{1}{M_c} \sum_{m, c_m = c} a_m. \qquad (3.59)$$

3.6 Summary

- An Nth-order tensor is an N-dimensional array with N modes.

- Most tensor operations can be viewed as operations on the mode-n vectors. This is key to understanding the connections between tensor operations and matrix/vector operations.

- Linear projection is a vector-to-vector projection. We can project a tensor directly to a tensor or vector through a tensor-to-tensor projection or a tensor-to-vector projection, respectively. Most of the connections among them can be revealed through elementary multilinear projection, which maps a tensor to a scalar.

- For the same amount of dimensionality reduction, TTP and TVP need to estimate many fewer parameters than VVP (linear projection) for higher-order tensors. Thus, TTP and TVP tend to have simpler models and lead to better generalization performance.

- Scatter measures (and potentially other measures/criteria) employed in VVP-based learning can be extended to tensors for TTP-based learning and to scalars for TVP-based learning.

3.7 Further Reading

De Lathauwer et al. [2000b] give a good introduction to multilinear algebra preliminaries for readers with a basic linear algebra background, so it is recommended to those unfamiliar with multilinear algebra. Those interested in HOSVD can find an in-depth treatment in this seminal paper. Its companion paper [De Lathauwer et al., 2000a] focuses on tensor approximation and it is also worth reading.

Kolda and Bader [2009] give a very comprehensive review of tensor decompositions. This paper also covers the preliminaries of multilinear algebra very well, with much additional material. It discusses many variations and various issues related to tensor decompositions. Cichocki et al. [2009] provides another good reference, where Section 1.4 covers the basics of multilinear algebra and Section 1.5 covers tensor decompositions.

We first named the tensor-to-vector projection and the elementary multilinear projection in [Lu et al., 2007b], commonly referred to as rank-one decomposition/projection. We then named the Tucker/HOSVD-style projection as the tensor-to-tensor projection in [Lu et al., 2009c] to suit subspace learning context better. Our survey paper [Lu et al., 2011] further examined the relationships among various projections, formulated the MSL framework, and gave a unifying view of the various scatter measures for MSL algorithms.

For multilinear algebra in a broader context, there are books that are several decades old [Greub, 1967; Lang, 1984], and there is also a book by mathematician Hackbusch [2012] with a modern treatment, which can be a good resource to consult for future development of MSL theories.

Chapter 4

Overview of Multilinear Subspace Learning

Chapter 3 covered the fundamental concepts and operations for multilinear subspace learning (MSL). We learned that a high-dimensional tensor can be projected directly to another tensor or vector of lower dimension through tensor-to-tensor projection (TTP) or tensor-to-vector projection (TVP), respectively. In this chapter, we formulate the general MSL framework and give an overview of this field.

The MSL framework considers the steps involved in designing an MSL algorithm, as shown in Figure 4.1. We compare linear subspace learning (LSL) and MSL in several important aspects. Then, we review various multilinear extensions of principal component analysis (PCA) and linear discriminant analysis (LDA) and categorize them under the MSL framework to reveal their connections. Next, we discuss the historical background and closely related works. Finally, we explore future research directions of MSL in developing new

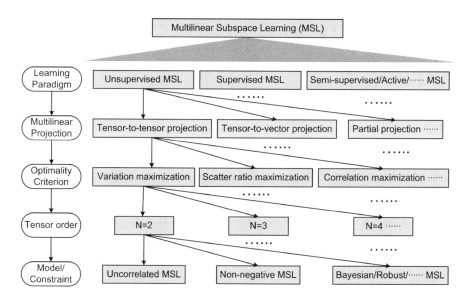

FIGURE 4.1: The multilinear subspace learning (MSL) framework.

MSL algorithms and solving important problems in applications involving big multidimensional data.

4.1   Multilinear Subspace Learning Framework

Definition 4.1. *In MSL, a set of M Nth-order tensor samples $\{\mathcal{X}_1, \mathcal{X}_2,$ ..., $\mathcal{X}_M\}$ are available for training, where each sample $\mathcal{X}_m$ is an $I_1 \times I_2 \times$ ... $\times I_N$ tensor in a tensor space $\mathbb{R}^{I_1 \times I_2 \times ... \times I_N}$, the tensor product (outer product) of N vector spaces $\mathbb{R}^{I_1}$, $\mathbb{R}^{I_2}$, ..., $\mathbb{R}^{I_N}$. For a particular learning paradigm, the MSL objective is to find a multilinear projection such that the extracted features, that is, the sample projections in the subspace, satisfy some optimality criterion, where the subspace dimension is lower than the original input space dimension. For certain desired properties, we may assume some particular model or impose some constraint.*

MSL is an extension of LSL. On the other hand, LSL can be viewed as a special case of MSL when $N = 1$ (first-order tensors). The features extracted by MSL algorithms can be used in a similar way as those extracted by LSL algorithms to perform various tasks. For example, in classification, they can be fed into a classifier such as the nearest neighbor classifier, with similarity calculated according to some distance measure.

Figure 4.1 demonstrates the MSL framework defined above. A particular MSL algorithm can be designed through determining *the learning paradigm to follow, the multilinear projection to employ, the criterion to be optimized, the order of tensor representation,* and *additional model/constraints to be imposed.* On the other hand, we can also analyze existing MSL algorithms under this framework to understand them or their connections better. This formulation is important for evaluating, comparing, and further developing MSL solutions.

Learning paradigm: Two popular learning paradigms are supervised learning and unsupervised learning. *Supervised learning* takes labeled data for training and outputs (predicts) the label of test data. The label usually indicates the desired output, such as class/group labels for (discrete) data. *Unsupervised learning* takes unlabeled data for training, trying to find some hidden structure. It can be used for clustering, or classification/regression with a classifier/regressor. In addition, *semi-supervised learning* makes use of both labeled and unlabeled data in training, while *active learning* (actively) selects samples to get labeled in training.

Multilinear projection: In MSL, the projection for mapping to a low-dimensional subspace can be any of the three types of basic multilinear projections discussed in Section 3.3. Thus, the well-studied linear subspace learning can be viewed as a special (degenerated) case of MSL for $N = 1$, where the

TABLE 4.1: Linear versus multilinear subspace learning.

Comparison	Linear Subspace Learning	Multilinear Subspace Learning
Representation	Reshaped vector representation	Natural tensor representation
Structure (before mapping)	Break natural structure	Preserve natural structure
Parameters (model complexity)	Estimate a large number of parameters	Estimate much fewer parameters
SSS problem	More severe SSS problem	Less SSS problem
Big data	Hardly applicable	Capable & promising
Optimization (in most cases)	Closed-form solution	Suboptimal, iterative solution

projection to be solved is a vector-to-vector projection. This book focuses on MSL based on tensor-to-tensor and tensor-to-vector projections.

Optimality criterion: Optimality criteria in MSL algorithms can be constructed in a similar way as respective LSL algorithms, for example, using the scatter measures defined in Section 3.5. MSL criteria can often be better understood by examining them for the elementary multilinear projections (EMPs) or by examining the lower-order/degenerated case ($N = 1/2$).

Tensor order: As discussed in Section 3.4 (Table 3.1 and Figure 3.13), computational efficiency has close relation with the order of tensor data. Many tensor data have a natural tensor order, for example, 2D/3D/4D data illustrated in Figures. 1.1, 1.2, and 1.3. On the other hand, there might be some flexibility in choosing tensors of different orders as the desired representations. For example, a gray-level image sequence can be naturally represented as a third-order tensor. If a feature vector instead of the gray-level image is used for each frame, it can be represented as a second-order tensor. If a set of Gabor filters turns each frame into a 3D multi-resolution representation, it can be represented as a fourth-order tensor. Different choices of tensor order may be more appropriate in different application scenarios.

Model/Constraint: Practical algorithms often have assumed certain model/constraint for data. For example, PCA derives uncorrelated features, while ICA assumes data are generated from independent sources. More MSL algorithms can be obtained by extending the model/constraint in LSL counterparts to the multilinear case. It is also possible to have some model/constraint that is specific to tensors, for example, in the form of interactions/relationships between different modes.

LSL versus MSL: Table 4.1 summarizes the key differences between LSL and MSL. In the table, *big data* refers to data with its dimensionality beyond the processing power of common computational hardware when linear subspace learning algorithms are used, such as face images with very high resolution, standard gait silhouette sequences, hyperspectral cubes, multichannel EEG signals, or large-scale sensor network data. On the other hand, in *mobile computing,* even small data for desktop computers can be big data for mobile devices due to limited computing power and battery life. MSL has the potential to learn more compact and useful representations than LSL. Nonetheless, MSL solutions are often iterative and we have to pay attention to related issues such as initialization, termination, and convergence.

4.2   PCA-Based MSL Algorithms

The development of PCA-based MSL started with the treatment of images directly as matrices rather than vectors. Figure 4.2 depicts a taxonomy for PCA-based MSL algorithms. They are discussed briefly in the following. More details and related works are presented in Chapter 6.

4.2.1   PCA-Based MSL through TTP

Yang et al. [2004] proposed a two-dimensional PCA (2DPCA) algorithm in 2004. This algorithm solves for a linear transformation that projects an image to a low-dimensional matrix while maximizing the variance measure. It works

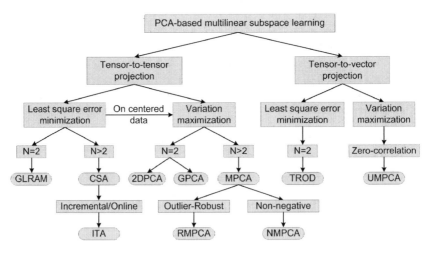

FIGURE 4.2: A taxonomy of PCA-based MSL algorithms.

directly on image matrices but there is only one linear transformation in one mode. Thus, an image is projected in mode-2 (the row mode) only while the projection in mode-1 (the column mode) is ignored, resulting in poor dimensionality reduction. The projection can be viewed as a special case of second-order TTP, where the mode-1 projection matrix is an identity matrix.

Ye [2005a] introduced a more general algorithm called the generalized low rank approximation of matrices (GLRAM). It takes into account the spatial correlation of the image pixels within a localized neighborhood and applies two linear transformations to both the left and right sides of input image matrices. This algorithm solves for two linear transformations that project an image to a low-dimensional matrix while minimizing the least-square (reconstruction) error measure. Thus, projections in both modes are involved and the projection is a second-order TTP ($N = 2$). GLRAM obtained better dimensionality reduction results than 2DPCA in [Ye, 2005a].

Although GLRAM exploits both modes for subspace learning, it is formulated for matrices only. Later, the work in [Wang and Ahuja, 2008] presents tensor approximation methods for third-order tensors using slice projection, while the concurrent subspaces analysis (CSA) is formulated in [Xu et al., 2008] as a generalization of GLRAM for higher-order tensors. The CSA algorithm solves for a TTP minimizing a reconstruction error metric for tensors.

GLRAM and CSA are both formulated with the objective of optimal reconstruction or approximation of tensors. Thus, they did not consider centering or variation maximization as PCA. Centering may not be essential for reconstruction or approximation problem as the (sample) mean is the main focus of attention. For variation maximization, non-centering (mean different from zero) can potentially affect the eigendecomposition in each mode and lead to a solution that captures the variation with respect to the origin rather than capturing the true variation of the data (with respect to the data center).

The generalized PCA (GPCA) proposed in [Ye et al., 2004a] is an extension of PCA to matrices. It is formulated as a variation maximization problem. Thus, GPCA is different from GLRAM as it takes centered data rather than the original non-centered data as input.

The multilinear PCA (MPCA) algorithm proposed in [Lu et al., 2008b] generalizes GPCA to tensors of any order. The objective of MPCA is to find a TTP that captures most of the original tensor input variations. Furthermore, the MPCA work proposed systematic methods for subspace dimension determination, in contrast to the heuristic methods in [Ye, 2005a; Xu et al., 2008]. With the introduction of a discriminative tensor feature selection mechanism, MPCA is further combined with LDA for general tensor object recognition in [Lu et al., 2008b].

In [Inoue et al., 2009], two robust MPCA (RMPCA) algorithms are proposed, where iterative algorithms are derived on the basis of Lagrange multipliers to deal with sample outliers and intra-sample outliers. In [Panagakis et al., 2010], the nonnegative MPCA (NMPCA) extends MPCA to constrain the projection matrices to be nonnegative to preserve the nonnegativity of the original tensor samples when the underlying data factors have physical or psychological interpretation.

In addition, Sun et al. [2008a] proposed an incremental tensor analysis (ITA) framework for summarizing higher-order data streams represented as tensors. The data summary is obtained through TTP and is updated incrementally. Three variants of ITA are introduced in [Sun et al., 2008a]: dynamic tensor analysis (DTA), streaming tensor analysis (STA), and window-based tensor analysis (WTA). The ITA framework focuses on an approximation problem; hence, the objective is to minimize the least square error and it can be considered an incremental version of CSA for streaming data.

4.2.2   PCA-Based MSL through TVP

In comparison to the large number of PCA-based MSL algorithms using TTP, there are much fewer PCA-based MSL algorithms using TVP. The tensor rank-one decomposition (TROD) algorithm introduced in [Shashua and Levin, 2001] is a TVP-based algorithm. It is formulated for image matrices. This algorithm looks for a second-order TVP that projects an image to a low-dimensional vector while minimizing a least-square (reconstruction) error measure. The input data are not centered before learning. The solution of TROD relies on a heuristic procedure of successive *residue calculation* (*deflation*), that is, after obtaining the pth EMP, the input image is replaced by its residue.

PCA derives uncorrelated features, which contain minimum redundancy and ensure linear independence among features. However, none of the above PCA-based MSL algorithms consider the correlations among extracted features and shares this important property with PCA. An uncorrelated MPCA (UMPCA) algorithm is proposed in [Lu et al., 2009d] to extract uncorrelated multilinear features through TVP while capturing most of the variation in the original data input. The UMPCA solution consists of sequential iterative steps for successive variance maximization. A systematic way is derived to determine the maximum number of uncorrelated multilinear features that can be extracted by the method.

4.3   LDA-Based MSL Algorithms

Similar to PCA-based MSL, the development of LDA-based MSL started with 2D extensions of LDA. Figure 4.3 depicts a taxonomy for LDA-based MSL algorithms. They are discussed briefly in the following. More details and related works are to be presented in Chapter 7.

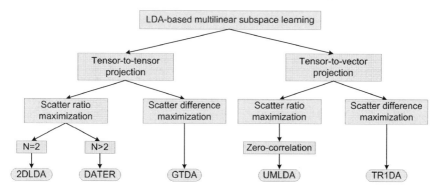

FIGURE 4.3: A taxonomy of LDA-based MSL algorithms.

4.3.1 LDA-Based MSL through TTP

Like GLRAM and GPCA, the 2D LDA (2DLDA) introduced in [Ye et al., 2004b] solves for two linear transformations that project an image to a low-dimensional matrix, but with a different objective criterion. For the input image samples, the between-class and within-class scatter measures are defined for matrix representations. A matrix-based discrimination criterion is then defined as the scatter ratio to be maximized in 2DLDA. Unlike the PCA-based MSL algorithms reviewed above, 2DLDA does not converge over iterations.

Later, the discriminant analysis with tensor representation (DATER) was proposed as a higher-order extension of 2DLDA to perform discriminant analysis on more general tensor inputs [Yan et al., 2005, 2007a]. The DATER algorithm solves for a TTP maximizing the tensor-based scatter ratio. This algorithm does not converge over iterations either.

Tao et al. [2007b] proposed the general tensor discriminant analysis (GTDA) algorithm. This algorithm also solves for a TTP. The difference with DATER is that it maximizes a tensor-based scatter difference criterion, with a tuning parameter involved [Liu et al., 2006]. In contrast with 2DLDA/DATER, this algorithm has good convergence property [Tao et al., 2007b].

4.3.2 LDA-Based MSL through TVP

In this category, the first algorithm is the tensor rank-one discriminant analysis (TR1DA) algorithm proposed in [Wang and Gong, 2006; Tao et al., 2006, 2008a], derived from the TROD algorithm [Shashua and Levin, 2001]. The TR1DA algorithm is formulated for general tensor data. It looks for a TVP that projects a tensor to a low-dimensional vector while maximizing the scalar scatter difference criterion. The method needs to determine a tuning parameter as in GTDA. Furthermore, this algorithm relies on the repeatedly calculated residues used in TROD, which is more suitable for tensor approximation

[Kolda, 2001] than discrimination. Hence, the adoption of this heuristic procedure here lacks theoretical explanation for a discriminative criterion.

The LDA-based MSL algorithms discussed so far do not take the correlations among features into account and they do not derive uncorrelated features in the learned subspace as in the classical LDA [Jin et al., 2001b; Ye et al., 2006]. Thus, an uncorrelated multilinear discriminant analysis (UMLDA) algorithm is formulated in [Lu et al., 2009c] to extract uncorrelated discriminative features directly from tensor data by solving a TVP maximizing a scalar scatter ratio criterion. The solution consists of sequential iterative processes and incorporates an adaptive regularization procedure to enhance the performance in the SSS scenario. Furthermore, an aggregation scheme is adopted to combine differently initialized and regularized UMLDA learners for improved generalization performance while alleviating the regularization parameter selection problem. This further extension is called regularized UMLDA with aggregation (R-UMLDA-A) [Lu et al., 2009c].

4.4 History and Related Works

This section puts MSL in its historical context and looks at several closely related works.

4.4.1 History of Tensor Decompositions

Multilinear algebra, the extension of linear algebra, has been well studied in mathematics around the middle of the 20th century [Greub, 1967; Lang, 1984]. It built on the concept of tensors and developed the theory of tensor spaces. Figure 4.4 gives an overview of the history of tensor decompositions and multilinear subspace learning.

In 1927, Hitchcock [1927a,b] studied the decomposition of a tensor into

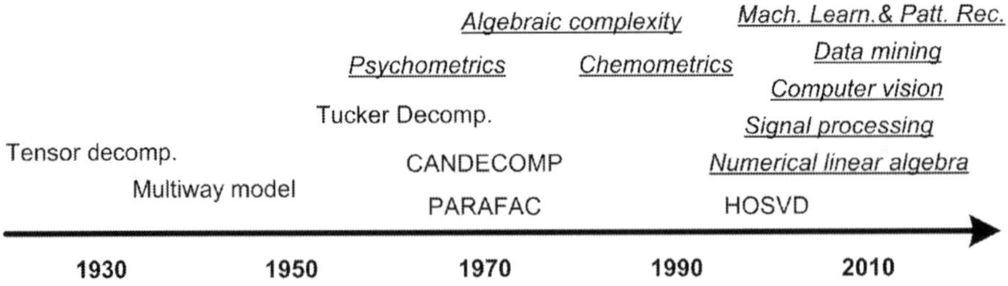

FIGURE 4.4: Overview of the history of tensor decomposition and multilinear subspace learning.

a minimal sum of outer products of vectors. Cattell [1944, 1952] further developed the idea of a multiway model in 1944. Starting in the 1960s and 1970s, this idea became popular in the so-called *multiway analysis* developed in psychometrics and chemometrics for *factor analysis* of multiway datasets[1] [Tucker, 1966; Harshman, 1970; Kroonenberg and Leeuw, 1980]. There are two main types of decomposition methods developed in this field: the Tucker decomposition [Tucker, 1966], and the canonical decomposition (CANDE-COMP) [Carroll and Chang, 1970], which is also known as the parallel factors (PARAFAC) decomposition [Harshman, 1970]. These two decompositions are the most popular ones and they are reviewed in Section 3.2. In the field of algebraic complexity, there are also decompositions of bilinear forms [Kruskal, 1977].

In the 1990s, developments in the field of higher-order statistics of multivariate stochastic variables attracted interest in higher-order tensors from the signal processing community [Comon and Mourrain, 1996; Comon, 1994; De Lathauwer, 1997; Kofidis and Regalia, 2001]. De Lathauwer et al. [2000b] reintroduced the Tucker decomposition and further developed it as the higher-order singular value decomposition (HOSVD) solution, an extension of the SVD to higher-order tensors. Its computation leads to the calculation of N different matrix SVDs of matrix unfoldings. De Lathauwer et al. [2000a] studied the alternating least squares (ALS) algorithm for the best Rank-$(R_1, R_2, ..., R_N)$ approximation of higher-order tensors, where tensor data were projected into a lower-dimensional tensor space iteratively. The application of HOSVD truncation and the best Rank-$(R_1, R_2, ..., R_N)$ approximation to dimensionality reduction in independent component analysis (ICA) was discussed in [De Lathauwer and Vandewalle, 2004]. In addition, Kofidis and Regalia [2002] studied the best rank-1 approximation of higher-order super-symmetric tensors.

In addition to signal processing, tensor decompositions have been further developed in numerical linear algebra [Kolda, 2001; Zhang and Golub, 2001; Kolda, 2003], data mining [Sun et al., 2006; Faloutsos et al., 2007; Liu et al., 2005], and computer vision [Vasilescu and Terzopoulos, 2002b; Vasilescu, 2002; Vasilescu and Terzopoulos, 2002a, 2003].

4.4.2 Nonnegative Matrix and Tensor Factorizations

Paatero and Tapper [1994] proposed the *nonnegative matrix factorization* (NMF), which became popular following the development by Lee and Seung [1999]. NMF imposes nonnegativity constraints in factorization of nonnegative data, such as images and environmental models. In NMF, a nonnegative

---

[1]Multiway (multivariate) datasets are higher-order tensors characterized by several sets of categorical variables that are measured in a crossed fashion [Faber et al., 2003; Kroonenberg, 1983].

matrix $\mathbf{X}$ is factorized into two nonnegative matrices $\mathbf{W}$ and $\mathbf{H}$:

$$\mathbf{X} = \mathbf{WH}, \qquad (4.1)$$

that is, all elements in these matrices must be equal to or greater than zero. The motivation is to retain the nonnegative characteristics of the original data for easier interpretation.

NMF has been extended to *nonnegative tensor factorizations* (NTF) by imposing the nonnegativity constraint on tensor decompositions. NTF methods based on the CANDECOMP/PARAFAC decomposition have been proposed in [Bro and De Jong, 1997; Friedlander and Hatz, 2008; Paatero, 1997; Welling and Weber, 2001; Shashua and Hazan, 2005; Zafeiriou, 2009b; Zafeiriou and Petrou, 2011], while NTF methods based on Tucker decomposition have been proposed in [Bro and De Jong, 1997; Mørup et al., 2008].

4.4.3 Tensor Multiple Factor Analysis and Multilinear Graph-Embedding

The HOSVD [De Lathauwer et al., 2000b,a] has led to the development of new multilinear algorithms and the exploration of new application areas. Multilinear analysis of biometric data was pioneered by the TensorFace method [Vasilescu and Terzopoulos, 2002b,a, 2005, 2003], which employs the multilinear algorithms proposed in [De Lathauwer et al., 2000b,a] to analyze the factors involved in the formation of facial images. Similar analysis has also been done for motion signatures [Vasilescu, 2002] and gait sequences [Lee and Elgammal, 2005].

However, in these *multiple factor analysis* works, input data such as images or video sequences are still represented as vectors. These vectors are arranged into a tensor according to the multiple factors involved in their formation for subsequent analysis. Such tensor formation needs a large number of training samples captured under various conditions, which may be impractical and may have the missing-data problem. They also require the forming factors to be known for each data sample; that is, they require data to be labeled and hence they are **supervised learning** methods though many of them are extensions of unsupervised learning methods. Furthermore, the size of formed tensors is usually huge, leading to high memory and computational demands.

Finally, in addition to multilinear extensions of the linear subspace learning algorithms, multilinear extensions of linear graph-embedding algorithms have also been introduced in [He et al., 2005a; Dai and Yeung, 2006; Yan et al., 2007b; Xu et al., 2007; Hua et al., 2007; Chen et al., 2010]. Some of them will be discussed in Chapters 6 and 7.

4.5 Future Research on MSL

As an emerging dimensionality reduction approach for direct feature learning from tensor data, MSL research is still in its infancy with many problems to be explored. This section outlines several research topics worth further investigation, which can be summarized in two main directions: one is toward the development of new MSL solutions, while the other is toward the exploration of new MSL applications.

4.5.1 MSL Algorithm Development

Following existing multilinear extensions of classical linear subspace learning algorithms, more multilinear extensions can be developed for rich ideas and algorithms in vector-based computational methods. In future research, existing algorithms can be further enhanced and new algorithms can be investigated along the following directions:

1. **Probabilistic MSL:** Real-world problems often have a lot of uncertainties and noise. Probabilistic models are very useful in quantifying the likelihood of unknown variables and the noise in measurement [Szeliski, 2010]. They can infer the best estimates of desired quantities and analyze their resulting uncertainties.

   There have been several probabilistic extensions of multilinear PCA, which will be reviewed in more details in Section 6.6.8. The Bayesian PCA [Bishop, 1999] has been extended to the multilinear case using TTP in [Tao et al., 2008b], and the probabilistic PCA in [Tipping and Bishop, 1999b,a] has been extended to 2D in [Zhao et al., 2012]. However, this direction is not well studied and these algorithms are not widely adopted or applied yet. Further research in probabilistic MSL is important in achieving robustness when solving real-world problems.

2. **Online/incremental MSL**: An online algorithm processes its input piece-by-piece in serial fashion as the input is fed to the algorithm[2]. It is typically used when the entire input is not available from the beginning. *Online learning* [Mairal et al., 2010; Kivinen et al., 2004; Lee and Kriegman, 2005] is very important for real-time processing and big data applications. For big data, even if the entire dataset is available, there may not be enough computational power or resources to deal with it, and online or *incremental learning* may be the only viable option.

   There are a few incremental extensions of MPCA for data stream mining [Sun et al., 2008a, 2006] and object tracking [Wang et al., 2011a], which will be discussed in Section 6.6.7. Some incremental variations of

---

[2]In contrast, an *offline algorithm* has the entire problem data from the beginning.

multilinear discriminant analysis extensions are reported in [Wen et al., 2010] for color-based tracking and in [Jia et al., 2011] for action recognition. Most of these methods are studied in a limited context on a limited number of problems. Further research in online or incremental MSL is important in achieving efficiency when solving many real-world problems, especially real-time and/or big data problems.

3. **Nonlinear extension of MSL via kernel/graph-embedding:** MSL is a linear modeling technique so MSL algorithms do not capture nonlinear relationships in tensor data. There are two popular dimensionality reduction approaches that take nonlinearity into account. One is *kerned-based learning*, the other is *graph-embedding-based learning*. Several multilinear extensions of these two approaches have appeared in the literature. However, users usually need to select more parameters (compared with MSL) so it may be tricky to make them work well. The MSL framework could be useful in understanding and further developing these methods.

In kernel-based learning [Müller et al., 2001; Cristianini and Shawe-Taylor, 2000; Schölkopf et al., 1998], the input data are mapped to an even-higher dimensional space for possible better separation. The most popular kernel-based algorithm is probably the *support vector machines* (SVM) [Steinwart and Christmann, 2008; Chang and Lin, 2011; Burges, 1998]. Several multilinear extensions of SVM, often called *support tensor machines* analogously, have been developed in [Tao et al., 2007a; Kotsia and Patras, 2010a,b; Guo et al., 2012; Hao et al., 2013].

Graph-embedding algorithms [Yan et al., 2007b] are traditionally known as *manifold learning* algorithms. They are usually designed to capture local structures or relationships rather than global ones. Popular algorithms in this category include Isomap [Tenenbaum et al., 2000], locally linear embedding [Roweis and Saul, 2000], and locality preserving projections [He et al., 2005b; Cai et al., 2006]. Many of them have been extended to tensors in [He et al., 2005a; Dai and Yeung, 2006; Yan et al., 2007b; Xu et al., 2007; Hua et al., 2007].

4. **Multilinear Gaussian processes:** Gaussian processes [MacKay, 1998; Williams, 1998] are stochastic processes widely used in probabilistic classification and regression. Realizations of Gaussian processes consist of random values associated with every point in a time/space range and each random variable has a normal distribution. Hence, any finite collection of these random variables has a multivariate normal distribution.

A Gaussian process can be considered an infinite-dimensional generalization of the multivariate normal distribution [Rasmussen and Williams, 2006]. Thus, it is also connected with kernel machines. In addition, PCA can also be formulated as a particular Gaussian process prior on a mapping from a latent space to the observed data space, which leads to the

Gaussian process latent variable models (GPLVM) [Lawrence, 2004]. Therefore, it is possible to develop a further generalization to the multilinear case.

5. **Combination of MSL with other methods:** There have been several approaches combining MSL algorithms directly with classical learning algorithms such as MPCA+LDA [Lu et al., 2006b, 2008b] and MPCA-based boosting [Lu et al., 2007a, 2009a], with promising results obtained. It is also possible to combine ideas from MSL and other learning paradigms, such as *semi-supervised learning* [Chapelle et al., 2006; Belkin and Niyogi, 2004], *active learning* [Tong and Koller, 2002; Tong and Chang, 2001], *transfer learning* [Pan and Yang, 2010; Raina et al., 2007], *reinforcement learning* [Sutton and Barto, 1998; Kaelbling et al., 1996], *sparse coding or sparse representation* [Wright et al., 2009; Mairal et al., 2010], and *deep learning* [Hinton and Salakhutdinov, 2006; Hinton et al., 2006]. It will be interesting to study what combinations are effective and what are not, for example, to find out whether simple cascading is good enough or special combination schemes need to be developed. Furthermore, we can study the combination of different MSL algorithms such as MPCA with DATER or GTDA. The solutions developed in some MSL algorithms, such as the regularization and aggregation scheme for UMLDA, could be useful for other algorithms as well.

6. **Partial multilinear projection and tensor reorganization:** Most research works in MSL compare multilinear solutions using natural tensor representation against linear solutions using vector representation. It could be an interesting topic to study hybrid approaches for higher-order tensors where a selected number of modes are vectorized to result in tensors with order greater than one but less than N (e.g., $(N - d)$th order tensors), from which features are learned. The CCA extension introduced in [Zhang et al., 2011] is an interesting work along this direction.

   On the other hand, we may also consider reorganizing the input Nth-order tensor into even higher order, for example, an $(N + d)$th-order tensor to learn more compact features and for more efficient processing.

7. **Optimization in MSL:** Last but not least, in MSL, there are still many unsolved problems remaining, such as the optimal initialization, the optimal projection order, and the optimal stopping criterion. Some attempts have been made in solving some of these problems in the literature [Lu et al., 2008b]. It will be beneficial if further research can lead to a deeper understanding of these issues. Alternative optimization strategies can be explored, such as the enhanced line search [Rajih et al., 2008] and other methods discussed in [Comon et al., 2009]. In addition, the performance of many MSL algorithms depends on the values of several hyper-parameters. Though aggregation has been proposed in [Lu et al.,

2009c] to partly address this problem, new ways can be investigated to determine or at least guide the optimal parameter setting automatically or semi-automatically. These issues are more fundamental in nature and their progress could have a profound impact on MSL and tensor-based learning in general.

4.5.2 MSL Application Exploration

Many real-world data are naturally tensor objects. Most big data can be represented as tensors too. Thus, MSL and tensor-based learning in general can benefit a wide range of traditional as well as new big data applications. In particular, many classical applications of linear subspace learning algorithms can be explored for their multilinear counterparts. Although MSL algorithms have been applied in various applications (to be discussed in detail in Chapter 9), we believe they will become more and more important as data grow bigger and bigger. In the following, we point out several directions to explore in applying MSL to real-world problems:

1. **Computer/mobile vision:** Machine learning algorithms are very useful in computer vision. Many MSL algorithms have been applied to vision problems such as face recognition [Yan et al., 2007a] and gait recognition [Lu et al., 2008b]. In face recognition, in addition to the traditional 2-D image-based approach, high-resolution and three-dimensional face detection and recognition have also emerged as important research directions [Bowyer et al., 2006; Li et al., 2005; Colombo et al., 2006; Phillips et al., 2005; Liu, 2006]. Other computer vision tasks where tensor-based processing can be useful include generic image or 3D object recognition tasks [Sahambi and Khorasani, 2003]; medical image analysis, image clustering and categorization [Xu and Wunsch II, 2005]; content-based image/video retrieval [He, 2004]; space-time analysis of video sequences for gesture recognition [Nolker and Ritter, 2002]; activity recognition [Green and Guan, 2004]; and space-time superresolution [Shechtman et al., 2005] for digital cameras with limited spatial and temporal resolution.

   In addition to the areas above, it will be interesting to investigate whether MSL can contribute to two important areas in computer vision: local feature extraction [Lowe, 2004] and fast approximate nearest neighbor search for feature matching [Muja and Lowe, 2009], though a related work has been reported in [Han et al., 2012]. MSL can also play a role in research on large-scale (structured) datasets [Deng et al., 2009; Torralba et al., 2008] where tensor-based analysis could greatly improve the efficiency. Finally, as we are moving to *mobile Internet*, more and more computation will be done on mobile devices and the requirement on efficiency is becoming even higher due to limited computing power and battery life. Learning-based approaches became popular in mobile vision recently. For example, the Features from Accelerated Segment Test (FAST) [Rosten et al., 2010] corner detector is probably the fastest corner detector that can achieve real-time performance on mobile phones. Thus, MSL may find applications in mobile visual computing where efficient processing is crucial for success.

2. **Data mining:** As shown in Figures 1.2(d), 1.2(e), 1.2(f), and 1.3(b), streaming data and mining data are frequently organized as higher-order tensors [Faloutsos et al., 2007; Sun et al., 2006, 2008b]. Data in environmental sensor monitoring are often organized in three modes: time, location, and type [Faloutsos et al., 2007]. Data in network forensics are often organized in three modes of time, source, and destination, and data in web graph mining are commonly organized in three modes of source, destination, and text [Sun et al., 2006]. Sun et al. [2008a] have developed several methods for analysis of these big data. Nonetheless, there are many other big data that can be mined using MSL or more general tensor-based algorithms [Han and Kamber, 2006; Witten and Frank, 2005]. Such data can often be organized in much higher order than natural objects (so they are good candidates for studying tensor reorganization). Because we expect the advantage of tensor-based learning to be bigger with higher-order tensors, MSL-based methods could be a very useful tool in mining such data.

3. **Audio and speech processing:** While audio and speech processing [Gold et al., 2011], especially natural language processing [Manning and Schütze, 1999], are very important areas with increasingly bigger data as well, there are only a few MSL algorithms applied in this area, for example, the work on music genre recognition in [Panagakis et al., 2010]. To explore opportunities of MSL in this field, it will be useful to examine the application of linear subspace learning (e.g, PCA, LDA) and other linear methods in this field first. Then, respective higher-order extensions can be studied or developed based on MSL for feature extraction from audio or speech signals.

4. **Neurotechnology and bioinformatics:** Neurotechnology [Oweiss, 2010; Ayers et al., 2002] studies the brain while bioinformatics [Saeys et al., 2007; Librado and Rozas, 2009] studies biological data. Electroencephalography (EEG) [Delorme and Makeig, 2004] and functional magnetic resonance imaging or functional MRI (fMRI) [Forman et al., 1995] are two popular brain imaging technologies. With high sampling rate, EEG data can be very big. fMRI scan sequences are very big data as well. Some MSL algorithms have been applied to both types of data in [Li et al., 2009a; Barnathan, 2010]. It will be worthwhile to further explore the usage of MSL on EEG and fMRI for various study purposes. In bioinformatics, large-scale DNA sequencing can result in very big data and MPCA has been extended to analyze such biological sequences in

[Mažgut et al., 2010]. In the future, other MSL algorithms could be employed or developed in DNA sequence analysis.

4.6   Summary

- Given a set of training tensor samples in a particular *learning paradigm*, MSL aims to find a *multilinear projection* to a lower dimensional space, satisfying some *optimality criterion*. Additional *models/constraints* can be enforced to obtain certain desired properties.

- MSL is based on the development in tensor decompositions, including Tucker decomposition, CANDECOMP/PARAFAC, and HOSVD.

- New MSL algorithms can be developed as multilinear counterparts of LSL algorithms. Issues to be further explored include combinations of MSL with other learning algorithms and paradigms, partial multilinear projections and tensor reorganization, and optimization problems in MSL. In addition, MSL can be useful in many applications such as computer/mobile vision, data mining, audio/speech processing, neurotechnology, and bioinformatics.

4.7   Further Reading

For other tensor data analysis developments, readers can refer to the following references. Qi et al. [2007] reviewed numerical multilinear algebra and its applications. Muti and Bourennane [2007] summarized new tensor-based filtering methods for multicomponent data modeled as tensors in noise reduction for color images and multicomponent seismic data. Acar and Yener [2009] surveyed unsupervised multiway data analysis for discovering structures in higher-order datasets in applications such as chemistry, neuroscience, and social network analysis. Kolda and Bader [2009] provided an overview of higher-order tensor decompositions and their applications in psychometrics, chemometrics, and signal processing. These works primarily focus on unsupervised tensor data analysis through factor decomposition.

In addition, Zafeiriou [2009a] provided an overview of both unsupervised and supervised nonnegative tensor factorization (NTF) [Hazan et al., 2005; Shashua and Hazan, 2005] with NTF algorithms and their applications in visual representation and recognition discussed. The book by Cichocki et al. [2009] provides comprehensive coverage of NMF and NTF. There are some other interesting decompositions not mentioned in the overview, such as the Tensor-CUR decomposition [Mahoney et al., 2008], compact matrix decomposition [Sun et al., 2008b], and random tensor decomposition [Tsourakakis, 2009].

For those interested in computer vision, Freeman [2011] has written a very good article on where machine learning can help computer vision problems. It can inspire the application of MSL or more general tensor-based methods to important computer vision problems.

The NSF workshop report on the future of *tensor-based computation and modeling* [NSF, 2009] is a very good reference that gives a big picture of tensor-based computing in general.

Finally, for real impact on real-world applications, [Wagstaff, 2012] pointed out several other important factors to consider in addition to developing learning algorithms, such as problem formulation, data collection, data preprocessing, system evaluation, interpretations, and promotion. It is a particularly inspiring paper to read.

Chapter 5

Algorithmic and Computational Aspects

The previous chapters presented the fundamentals as well as an overview of multilinear subspace learning (MSL). In this final chapter of Part I, we discuss the algorithmic and computational aspects of MSL. This chapter is mainly for those who are interested in developing, implementing, and testing MSL algorithms. We first examine a typical iterative solution, as shown in Figure 5.1. We then discuss various issues, including initialization, projection order, termination, and convergence. Next, we consider the generation of a synthetic tensor dataset with varying characteristics for MSL algorithm analysis. Finally, we deal with feature selection strategies and various computational aspects.

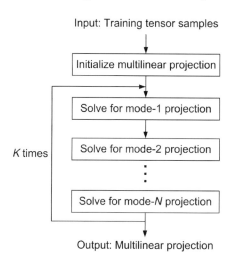

FIGURE 5.1: Typical flow of a multilinear subspace learning algorithm.

5.1   Alternating Partial Projections for MSL

While a linear (vector-to-vector) projection in linear subspace learning (LSL) usually has closed-form solutions, this is often not the case for tensor-to-tensor projection (TTP) and tensor-to-vector projection (TVP) in MSL. These two tensor-based projections have N sets of parameters for N projection matrices/vectors to be solved, one in each mode. In most cases, we are not able to solve for these N sets simultaneously, except when $N = 1$. The solution to one projection matrix/vector often depends on the other projection matrices/vectors (except when $N = 1$, the linear case), making their simultaneous estimation extremely difficult, if not impossible.

Therefore, *alternating partial projections* (APP), a suboptimal, iterative procedure originated from the *alternating least squares* (ALS) algorithm [Kroonenberg and Leeuw, 1980], is commonly employed to solve for the tensor-based projections. Consequently, issues due to the iterative nature of the solution, such as initialization, the order of solving the projections, termination, and convergence, need to be addressed in MSL.

> **Alternating least squares**: The ALS algorithm was first developed in 1970 [Harshman, 1970; Carroll and Chang, 1970] to solve a similar problem in the *three-way factor analysis* [Faber et al., 2003] where parameters in three modes need to be estimated. The principle behind ALS is to reduce the (least square) optimization problem into smaller conditional subproblems that can be solved through simple established methods employed in the linear case. Thus, the parameters for each mode are estimated in turn separately and are conditional on the parameter values for the other modes. At each step, by fixing the parameters in all the modes but one mode, a new objective function depending only on the mode left free to vary is optimized and this conditional subproblem is linear and much simpler. The parameter estimations for each mode are obtained in this way sequentially and iteratively until convergence.

The APP method reduces a multilinear optimization problem into N smaller conditional subproblems that can be solved based on established methods developed for the linear case. It alternates between solving one set of parameters (in one mode) at a time, as shown in Figure 5.1. The mode-n projection matrix/vector is solved one by one separately, conditioned on the projection matrices/vectors in all the other modes ($j \neq n$). In each mode n, it solves a conditional subproblem through mode-n *partial multilinear projections* using the projection matrices/vectors in all the other modes ($j \neq n$). The mode-n partial multilinear projection of a tensor $\mathcal{X}$ in TTP using $\{\mathbf{U}^{(j)}, j \neq n\}$ is written as

$$\hat{y}^{(n)} = \mathcal{X} \times_1 \mathbf{U}^{(1)^T} \times_2 \mathbf{U}^{(2)^T} \cdots \times_{(n-1)} \mathbf{U}^{(n-1)^T} \times_{(n+1)} \mathbf{U}^{(n+1)^T} \cdots \times_N \mathbf{U}^{(N)^T}. \tag{5.1}$$

The mode-n partial multilinear projection of a tensor $\mathcal{X}$ in the pth elementary multilinear projection (EMP) of a TVP using $\{\mathbf{u}_p^{(j)}, j \neq n\}$ is written as

$$\hat{\mathbf{y}}_p^{(n)} = \mathcal{X} \times_1 \mathbf{u}_p^{(1)^T} \times_2 \mathbf{u}_p^{(2)^T} \cdots \times_{(n-1)} \mathbf{u}_p^{(n-1)^T} \times_{(n+1)} \mathbf{u}_p^{(n+1)^T} \cdots \times_N \mathbf{u}_p^{(N)^T}. \tag{5.2}$$

A new objective function depending only on the mode-n projection matrix/vector can be formulated assuming fixed projection matrices/vectors in all other modes $(j \neq n)$. This conditional subproblem is linear and much simpler. Thus, the parameter estimations for each mode can be obtained in this way iteratively until a stopping criterion is met.

In addition, for TTP-based MSL, it is often costly to exhaustively test all the possible combinations of the N values, $P_1, P_2, ..., P_N$, for a desired amount of dimensionality reduction. Thus, a mechanism is often needed to determine the desired subspace dimensions $\{P_1, P_2, ..., P_N\}$ for this approach. In contrast, only one value P needs to be tested for TVP-based MSL.

Typical algorithmic procedures for TTP-based and TVP-based MSL algorithms are shown in Algorithms 5.1 and 5.2, respectively. The iterations discussed above correspond to the loop indexed by k in Algorithms 5.1 and 5.2. In each iteration k, the loop indexed by n in Algorithms 5.1 and 5.2 consists of the N conditional subproblems. Note that in TVP-based MSL, the P

Algorithm 5.1 A typical TTP-based multilinear subspace learning algorithm

Input: A set of tensor samples $\{\mathcal{X}_m \in \mathbb{R}^{I_1 \times I_2 \times \cdots \times I_N}, m = 1, ..., M\}$, the desired tensor subspace dimensions $\{P_1, ..., P_N\}$, and the maximum number of iterations K.

Process:

1: Initialize the tensor-to-tensor projection $\{\mathbf{U}^{(n)}, n = 1, ..., N\}$.

2: Local optimization:

3: **for** $k = 1$ to K **do**

4:     **for** $n = 1$ to N **do**

5:         Obtain the mode-n partial multilinear projection of input samples using the projection matrices in all the other modes as in Equation (5.1).

6:         Solve for the mode-n multilinear projection $\mathbf{U}^{(n)}$ as a linear problem obtained through unfolding the mode-n partially projected tensors in the previous step to matrices.

7:     **end for**

8:     If the algorithm converges, break and output the current TTP.

9: **end for**

Output: The (suboptimal) TTP $\{\tilde{\mathbf{U}}^{(n)}, n = 1, ..., N\}$.

Algorithm 5.2 A typical TVP-based multilinear subspace learning algorithm

Input: A set of tensor samples $\{\mathcal{X}_m \in \mathbb{R}^{I_1 \times I_2 \times \dots \times I_N}, m = 1, \dots, M\}$, the desired tensor subspace dimension P, and the maximum number of iterations K.

Process:

1: **for** $p = 1$ **to** P **do**
2:     Initialize the pth EMP $\{\mathbf{u}_p^{(n)}, n = 1, \dots, N\}$.
3:     Local optimization:
4:     **for** $k = 1$ **to** K **do**
5:         **for** $n = 1$ **to** N **do**
6:             Obtain the mode-n partial multilinear projection of input samples using the projection vectors in all the other modes as in Equation (5.2).
7:             Solve for the pth mode-n multilinear projection $\mathbf{u}_p^{(n)}$ as a linear problem obtained through the vectors produced by the mode-n partial multilinear projections in the previous step.
8:         **end for**
9:         If the algorithm converges, break and output the current EMP.
10:     **end for**
11: **end for**

Output: The (suboptimal) TVP $\{\tilde{\mathbf{u}}_p^{(n)}, n = 1, \dots, N\}_{p=1}^{P}$.

EMPs of a TVP are obtained one by one sequentially (from $p = 1$ to $p = P$), as indexed by p in Algorithm 5.2.

5.2   Initialization

Due to the iterative nature of MSL solutions, the projection matrices or vectors to be solved need to be initialized at the beginning of the iterative process. This section summarizes several popular initialization methods and examines one of them in detail.

5.2.1   Popular Initialization Methods

Popular initialization methods for TTP are

- **Random initialization**: Each element of the mode-n projection matrix is drawn from a zero-mean uniform distribution[1] between $[-0.5, 0.5]$ and these random projection matrices are normalized to have unit norm.

- **Pseudo-identity matrices**: The mode-n projection matrix is initialized as a truncated identity matrix by taking the first P_n columns of an identity matrix of size $I_n \times I_n$.

- **Full projection truncation (FPT)**: The mode-n projection matrix is initialized by truncation of the mode-n full projection matrix, which will be discussed in detail in the next section.

Popular initialization methods for TVP are:

- **Random initialization**: Each element of a mode-n projection vector is drawn from a zero-mean uniform distribution[2] between $[-0.5, 0.5]$ and the initialized projection vector is normalized to have unit length.

- **Uniform initialization**: Each mode-n projection vector is initialized to have unit length and the same value for all its entries, which is equivalent to the all ones vector $\mathbf{1}$ with proper normalization.

Initialization method selection: When computational cost is not a major concern (e.g., in offline training) and there is a well-defined optimization criterion, different initializations can be tested to select the one giving the best result. For example, the empirical studies in [Lu et al., 2009c] indicate that the results of the regularized uncorrelated multilinear discriminant analysis (R-UMLDA) are affected by initialization, and the uniform initialization gives better results. In another study [Lu et al., 2008b], the effects of different initialization methods have been examined for the multilinear principal component analysis (MPCA) algorithm. It was found that different initialization methods do not have a significant impact on the performance of MPCA in practical applications. However, it can affect the speed of convergence for the iterative solution and the FPT method results in much faster convergence.

Among the initialization methods listed above, only the FPT method depends on the data. Therefore, the FPT method for TTP initialization is discussed and analyzed in detail below.

5.2.2 Full Projection Truncation

FPT is a more principled initialization method based on the truncation of the *full projection* in MPCA [Lu et al., 2008b], with $P_n = I_n$ for $n = 1, ..., N$. For each mode n, we have the mode-n total scatter matrix given by Equation (3.42) as

$$\breve{\mathbf{S}}_{T\mathcal{X}}^{(n)} = \sum_{m=1}^{M} \left(\mathbf{X}_{m(n)} - \bar{\mathbf{X}}_{(n)} \right) \left(\mathbf{X}_{m(n)} - \bar{\mathbf{X}}_{(n)} \right)^T, \tag{5.3}$$

---

[1]A Gaussian distribution may be used as well but the uniform distribution is more popular.

[2]Similarly, a Gaussian distribution may be used.

where $\mathbf{X}_{m(n)}$ is the mode-n unfolding of tensor sample $\mathcal{X}_m$, and $\bar{\mathbf{X}}_{(n)}$ is the mode-n unfolding of the mean tensor $\bar{\mathcal{X}}$ as defined in Equation (3.41). $\breve{\mathbf{S}}_{T_{\mathcal{X}}}^{(n)}$ is determined by the input tensor samples only. The full projection matrices $\{\breve{\mathbf{U}}^{(n)} \in \mathbb{R}^{I_n \times I_n}\}$ ($P_n = I_n$ for all n) project $\mathcal{X}_m$ to $\breve{\mathcal{Y}}_m$ as

$$\breve{\mathcal{Y}}_m = \mathcal{X}_m \times_1 \breve{\mathbf{U}}^{(1)^T} ... \times_N \breve{\mathbf{U}}^{(N)^T}, \tag{5.4}$$

where $\breve{\mathbf{U}}^{(n)}$ is comprised of the eigenvectors of $\breve{\mathbf{S}}_{T_{\mathcal{X}}}^{(n)}$. The full projection fully captures the total scatter $\Psi_{T_{\mathcal{X}}}$ as defined in Equation (3.40) in the original data.

In FPT initialization, the first P_n columns of the full projection matrix $\breve{\mathbf{U}}^{(n)}$ are kept to give an initial projection matrix $\mathbf{U}^{(n)}$. The corresponding total scatter is denoted as $\Psi_{T_{y_0}}$. This initialization is equivalent to the higher-order singular value decomposition (HOSVD) truncation [De Lathauwer and Vandewalle, 2004; Lu et al., 2006d].

5.2.3 Interpretation of Mode-n Eigenvalues

Before we analyze the FPT method, we try to illustrate and interpret the meanings of the mode-n eigenvalues, the eigenvalues of the mode-n total scatter matrix. We introduce the *total scatter tensor* $\breve{\mathcal{Y}}_{ST} \in \mathbb{R}^{I_1 \times I_2 \times ... \times I_N}$ of the full projection as another extension of the total scatter matrix in LSL. Each entry of the tensor $\breve{\mathcal{Y}}_{ST}$ is defined as below:

$$\breve{\mathcal{Y}}_{ST}(i_1, i_2, ..., i_N) = \sum_{m=1}^{M} \left[(\breve{\mathcal{Y}}_m - \bar{\breve{\mathcal{Y}}})(i_1, i_2, ..., i_N) \right]^2, \tag{5.5}$$

where $\breve{\mathcal{Y}}_m$ is the full projection of $\mathcal{X}_m$ in Equation (5.4), and

$$\bar{\breve{\mathcal{Y}}} = \frac{1}{M} \sum_{m=1}^{M} \breve{\mathcal{Y}}_m. \tag{5.6}$$

Using the above definition, for the full projection ($P_n = I_n$ for all n), the i_nth mode-n eigenvalue $\breve{\lambda}_{i_n}^{(n)}$ of $\breve{\mathbf{S}}_{T_{\mathcal{X}}}^{(n)}$ is the sum of all the entries of the i_nth mode-n slice of $\breve{\mathcal{Y}}_{ST}$:

$$\breve{\lambda}_{i_n}^{(n)} = \sum_{i_1=1}^{I_1} ... \sum_{i_{n-1}=1}^{I_{n-1}} \sum_{i_{n+1}=1}^{I_{n+1}} ... \sum_{i_N=1}^{I_N} \breve{\mathcal{Y}}_{ST}(i_1, ..., i_{n-1}, i_n, i_{n+1}, ..., i_N). \tag{5.7}$$

Figure 5.2 is a visualization of the mode-n eigenvalues[3] for interpretation. In this figure, a number of third-order tensors, for example, short sequences (three frames) of 5×4 images, are projected to a tensor space of size $5 \times 4 \times 3$ (full projection). A total scatter tensor $\breve{\mathcal{Y}}_{ST} \in \mathbb{R}^{5 \times 4 \times 3}$ is obtained. Each mode-n eigenvalue corresponds to a mode-n slice of $\breve{\mathcal{Y}}_{ST}$, as indicated in the figure.

---

[3]In this book, eigenvalues are all arranged in descending order.

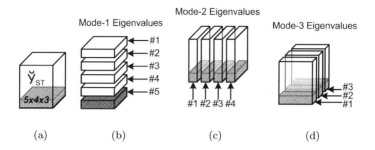

FIGURE 5.2: Visual interpretation of (a) the total scatter tensor, (b) the mode-1 eigenvalues, (c) the mode-2 eigenvalues, and (d) the mode-3 eigenvalues of the respective mode-n total scatter matrix for input samples. A mode-n eigenvalue is equal to the sum of all the entries of the indicated mode-n slice.

Using Equation (3.23), each input tensor $\mathcal{X}_m$ can be written as a linear combination of $P_1 \times P_2 \times ... \times P_N$ rank-one tensors:

$$\breve{\mathcal{U}}_{p_1 p_2 ... p_N} = \breve{\mathbf{u}}_{p_1}^{(1)} \circ \breve{\mathbf{u}}_{p_2}^{(2)} \circ ... \circ \breve{\mathbf{u}}_{p_N}^{(N)}, \tag{5.8}$$

where $\breve{\mathbf{u}}_{p_n}^{(n)}$ is the pth column of $\breve{\mathbf{U}}^{(n)}$. These rank-one tensors can be considered as eigentensors, in analogy to eigenvectors. Each eigentensor is associated with an EMP. The projected tensor $\breve{\mathcal{Y}}_m$ can be viewed as the projection of $\mathcal{X}_m$ onto these eigentensors/EMPs, with each entry of $\breve{\mathcal{Y}}_m$ corresponding to one eigentensor/EMP.

5.2.4 Analysis of Full Projection Truncation

In FPT, when a non-zero eigenvalue is truncated in one mode, the eigenvalues in all the other modes tend to decrease in magnitude and the corresponding eigenvectors change accordingly. Thus, eigendecompositions need to be recomputed in all the other modes, and the projection matrices in all the other modes need to be updated. Because the computations of all the projection matrices in MSL are interdependent, the update of a mode-n projection matrix updates the other projection matrices (in modes $j \neq n$) as well, which are no longer consisting of the eigenvectors of the corresponding (updated) mode-n total scatter matrix and they need to be updated. This update continues until the termination criterion is satisfied. A formal proof is available in [Lu et al., 2008b]

Figure 5.2 provides a visual illustration of the above analysis. Truncation of a column in the mode-n full projection matrix $\breve{\mathbf{U}}^{(n)}$ results in eliminating a mode-n slice of $\breve{\mathcal{Y}}_{ST}$. In Figure 5.2, if the last non-zero (fifth) mode-1 eigenvalue is discarded (shaded in Figure 5.2(b)) and the corresponding (fifth) column in $\breve{\mathbf{U}}^{(1)}$ is truncated, the corresponding (fifth) mode-1 slice of $\breve{\mathcal{Y}}_{ST}$

is removed (shaded in Figure 5.2(a)), resulting in a truncated total scatter tensor $\tilde{\mathcal{Y}}_{ST} \in \mathbb{R}^{4 \times 4 \times 3}$. Discarding this slice will affect all eigenvalues in the remaining modes, whose corresponding slices have a non-empty overlap with the discarded mode-1 slice. In Figures 5.2(c) and 5.2(d), the shaded parts indicate the removed mode-1 slice corresponding to the discarded eigenvalue.

The upper bound Ψ_U and lower bound Ψ_L for the loss of variation due to FPT (measured by the total scatter) are:

$$\Psi_L = \max_n \sum_{i_n = P_n + 1}^{I_n} \check{\lambda}_{i_n}^{(n)}, \quad \Psi_U = \sum_{n=1}^{N} \sum_{i_n = P_n + 1}^{I_n} \check{\lambda}_{i_n}^{(n)}. \tag{5.9}$$

A formal proof of this result is in [Lu et al., 2008b].

From Equation (5.9), we can see that the tightness of the bounds is determined by the eigenvalues in each mode. The bounds can be observed in Figure 5.2. For instance, truncation of the last eigenvector in each of the three modes results in another truncated total scatter tensor $\mathring{\mathcal{Y}}_{ST} \in \mathbb{R}^{4 \times 3 \times 2}$. Thus, the loss of variation (the sum of entries removed from $\tilde{\mathcal{Y}}_{ST}$ to result in $\mathring{\mathcal{Y}}_{ST}$) is upper-bounded by the total of the sums of all the entries in each truncated slice and lower-bounded by the maximum of sums of all the entries in each truncated slice. For FPT, the gap between the actual loss of variation and the upper bound is due to multiple counts of the overlaps between the discarded slice in one mode and the discarded slices in the other modes of $\check{\mathcal{Y}}_{ST}$.

The tightness of the bounds Ψ_U and Ψ_L depends on the order N, the eigenvalue characteristics (distribution) such as the number of zero-valued eigenvalues, and the degree of truncation P_n. For example, for $N = 1$, the case of PCA, $\Psi_L = \Psi_U$ and the full projection truncation is the optimal solution so no iterations are necessary. A larger N results in more terms in the upper bound and tends to have looser bounds, and vice versa. In addition, if all the truncated eigenvectors correspond to zero-valued eigenvalues, $\Psi_L = \Psi_U = 0$, and the FPT results in the optimal solution.

5.3   Projection Order, Termination, and Convergence

Projection order: Being iterative, an MSL algorithm needs to compute N projection matrices or vectors (one in each mode) in a certain order. This order may affect the obtained solution. In practice, a sequential order from 1 to N is usually taken. Similar to the initialization selection, when computational cost is not a major concern and when there is a well-defined optimality criterion, the effects of several/all projection orders can be studied for the best performance. In several studies in the literature [Lu et al., 2008b, 2009c,d], altering the ordering of the projection matrix/vector computation does not result in a significant performance difference in practical situations.

Termination: There are two commonly used termination criteria. One way is by examining the convergence of the objective criterion or the projection matrices/vectors to be solved. A user-specified hyper-parameter is usually used as a threshold. The iteration stops if there is little improvement in the resulting objective criterion or if the difference between subsequent update of the projection matrix/vector is smaller than the threshold. However, it should be noted that not all MSL algorithms will converge, so one should be careful when using this termination criterion to avoid infinite loops. Even if an MSL algorithm has provable convergence behavior, due to numerical approximating, it is still possible that the convergence performance is different from what is expected.

In practice, for computational consideration, another easy way to terminate the iteration is to set the maximum number of iterations allowed to K. The value of K should be set based on experimental study of the algorithm behavior and affordable computational cost.

Convergence: The convergence of an MSL algorithm can be either formally proved or empirically studied. Formal proof is often based on the observation that each iteration step optimizes the current optimality criterion. In this case, it is important to ensure that the definition of the criterion to be optimized is not changing over iterations.

Remark 5.1. *Although it may be very difficult to determine the optimal initialization and the optimal projection order, the aggregation scheme suggested in [Lu et al., 2009c] can reduce the need for their optimal determination.*

5.4 Synthetic Data for Analysis of MSL Algorithms

Due to the high dimensionality of tensor space, specific tensor data can hardly be representative of all possible tensor data. Therefore, one should be careful in generalizing conclusions drawn on a specific tensor dataset. In this section, we discuss synthetic datasets generated with different characteristics that are useful in examining the properties of MSL algorithms.

The core computation of many MSL algorithms is the modewise eigendecomposition, so the distribution of eigenvalues is expected to affect the performance of MSL algorithms. Three synthetic datasets with modewise eigenvalues spanning different magnitude ranges are generated in [Lu et al., 2008b] to study the properties of MSL algorithms on data of different characteristics. Specifically, M third-order tensor samples $\mathcal{A}_m \in \mathbb{R}^{I_1 \times I_2 \times I_3}$ are generated per set according to

$$\mathcal{A}_m = \mathcal{B}_m \times_1 \mathbf{C}^{(1)} \times_2 \mathbf{C}^{(2)} \times_3 \mathbf{C}^{(3)} + \mathcal{D}_m, \qquad (5.10)$$

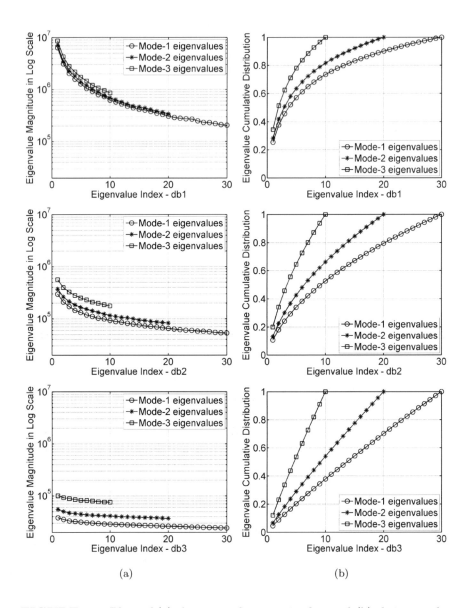

FIGURE 5.3: Plots of (a) the eigenvalue magnitudes, and (b) their cumulative distributions for synthetic datasets db1, db2, and db3.

using a core tensor $\mathcal{B}_m \in \mathbb{R}^{I_1 \times I_2 \times I_3}$, mode-$n$ projection matrix $\mathbf{C}^{(n)} \in \mathbb{R}^{I_n \times I_n}$ $(n = 1, 2, 3)$, and a "noise" tensor $\mathcal{D}_m \in \mathbb{R}^{I_1 \times I_2 \times I_3}$. All entries in $\mathcal{B}_m$ are drawn from a zero-mean unit-variance Gaussian distribution and each entry $\mathcal{B}_m(i_1, i_2, i_3)$ is multiplied by $\left(\frac{I_1 \cdot I_2 \cdot I_3}{i_1 \cdot i_2 \cdot i_3} \right)^f$. In this data generation procedure, f

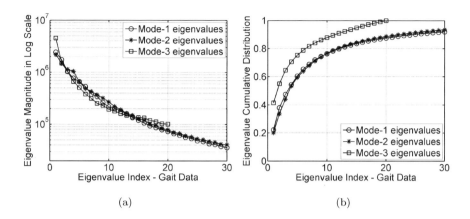

(a)   (b)

FIGURE 5.4: Plots of (a) the eigenvalue magnitudes and (b) their cumulative distributions for the gallery set of the USF Gait database V.1.7. (up to thirty eigenvalues).

controls the eigenvalue distributions, so that the datasets created have eigenvalue magnitudes in different ranges. Smaller f results in a narrower range of eigenvalue spread, and vice versa. The matrices $\mathbf{C}^{(n)}$ ($n = 1, 2, 3$) are orthogonal matrices obtained by applying SVD on random matrices with entries drawn from zero-mean, unit-variance Gaussian distribution. All entries of $\mathcal{D}_m$ are drawn from a zero-mean Gaussian distribution with variance 0.01.

Three synthetic datasets, db1, db2, and db3, were created in [Lu et al., 2008b] with $f = 1/2, 1/4$, and $1/16$, respectively; $M = 100$; and $I_1 = 30$, $I_2 = 20$, $I_3 = 10$. Figure 5.3(a) depicts the spread of eigenvalue magnitudes and Figure 5.3(b) depicts the cumulative distributions of eigenvalues [Dorogovtsev et al., 2002; Jolliffe, 2002] for these three datasets. The eigenvalues of real-world data usually spread a wide range in each mode, which is closer to db1. As an example, Figure 5.4 shows a plot of up to thirty eigenvalues in three modes and their cumulative distributions obtained from the gallery set of the USF gait database V.1.7 [Sarkar et al., 2005], where each data sample is of size $128 \times 88 \times 20$. Several studies on these datasets were reported in [Lu et al., 2008b].

5.5   Feature Selection for TTP-Based MSL

As TTP-based MSL produces features in tensor representations while many existing tools (e.g., classifiers) take vector input, it is common to convert tensor-valued features to vector-valued features through feature selection. This section discusses two popular feature selection strategies.

5.5.1   Supervised Feature Selection

In TTP-based MSL, each entry of a projected tensor feature can be viewed as a (scalar) feature corresponding to a particular EMP (or eigentensor). For classification tasks where class labels are available in training, we can rank and select discriminative entries from tensor-valued features by considering within-class variation and between-class variation. A larger between-class variation relative to the within-class variation indicates good class separability, while a smaller between-class variation relative to the within-class variation indicates poor class separability. Hence, we can select discriminative EMPs (features) according to their class discrimination power [Belhumeur et al., 1997; Wang et al., 2005], measured by the ratio of the between-class scatter over the within-class scatter:

Definition 5.1. *The class discriminability* $\Theta_{p_1 p_2 \ldots p_N}$ *for an EMP* $\{\mathbf{u}_{p_1}^{(1)}, \mathbf{u}_{p_2}^{(2)},$ *...,* $\mathbf{u}_{p_N}^{(N)}\}$ *(or eigentensor* $\tilde{\mathcal{U}}_{p_1 p_2 \ldots p_N}$ *) is defined as*

$$
\Theta_{p_1 p_2 \ldots p_N} = \frac{\sum_{c=1}^{C} M_c \cdot \left[\bar{\mathcal{Y}}_c(p_1, p_2, \ldots, p_N) - \bar{\mathcal{Y}}(p_1, p_2, \ldots, p_N) \right]^2}{\sum_{m=1}^{M} \left[\mathcal{Y}_m(p_1, p_2, \ldots, p_N) - \bar{\mathcal{Y}}_{c_m}(p_1, p_2, \ldots, p_N) \right]^2}, \tag{5.11}
$$

where C *is the number of classes,* M *is the number of training samples,* M_c *is the number of samples for class* c, *and* c_m *is the class label for the* mth *training sample* $\mathcal{X}_m$. $\mathcal{Y}_m$ *is the feature tensor of* $\mathcal{X}_m$ *in the TTP-projected tensor subspace so* $\mathcal{Y}_m(p_1, p_2, \ldots, p_N)$ *is the projection of* $\mathcal{X}_m$ *by the EMP* $\{\mathbf{u}_{p_1}^{(1)}, \mathbf{u}_{p_2}^{(2)}, \ldots, \mathbf{u}_{p_N}^{(N)}\}$. *The mean feature tensor is*

$$
\bar{\mathcal{Y}} = \frac{1}{M} \sum_m \mathcal{Y}_m \tag{5.12}
$$

and the class mean feature tensor is

$$
\bar{\mathcal{Y}}_c = \frac{1}{M_c} \sum_{m, c_m = c} \mathcal{Y}_m. \tag{5.13}
$$

For supervised feature selection, the entries in the projected tensor $\mathcal{Y}_m$ are arranged into a feature vector $\mathbf{y}_m$, ordered according to $\Theta_{p_1 p_2 \ldots p_N}$ in descending order, and only the first $H_{\mathbf{y}}$ most discriminative components of $\mathbf{y}_m$ are kept for classification, with $H_{\mathbf{y}}$ determined empirically, user specified, or by cross-validation (Section 2.7.3). This selection strategy results in a more discriminative subspace, especially for unsupervised MSL algorithms. These selected features can be fed into conventional classifiers such as LDA or a support vector machine (SVM) for classification tasks.

5.5.2 Unsupervised Feature Selection

For unsupervised feature selection where class labels are not available or not used, we define an importance score below:

Definition 5.2. *The importance score* $\Upsilon_{p_1 p_2 ... p_N}$ *for an EMP* $\{\mathbf{u}_{p_1}^{(1)}, \mathbf{u}_{p_2}^{(2)},$
..., $\mathbf{u}_{p_N}^{(N)}\}$ *(or eigentensor* $\tilde{\mathcal{U}}_{p_1 p_2 ... p_N}$*) is defined as*

$$\Upsilon_{p_1 p_2 ... p_N} = \sum_{m=1}^{M} \left[\mathcal{Y}_m(p_1, p_2, ..., p_N) - \bar{\mathcal{Y}}(p_1, p_2, ..., p_N) \right]^2, \qquad (5.14)$$

where $\mathcal{Y}_m$ *is the projection of* $\mathcal{X}_m$ *in the tensor subspace, and* $\bar{\mathcal{Y}}$ *is the mean feature tensor computed as in Equation (5.12).*

For feature selection, the entries in $\mathcal{Y}_m$ are arranged into a feature vector $\mathbf{y}_m$ according to $\Upsilon_{p_1 p_2 ... p_N}$ in descending order. Only the first $H_{\mathbf{y}}$ entries of $\mathbf{y}_m$ are kept for subsequent analysis.

5.6 Computational Aspects

This section considers the memory requirements, storage needs, and computational complexity of MSL algorithms. These factors affect the required computing power and processing (execution) time so they are relative measures of the practicality and usefulness. In addition, we also offer some MAT-LAB implementation tips for large datasets. For simplicity, we assume that $I_1 = I_2 = ... = I_N = \left(\prod_{n=1}^{N} I_n \right)^{\frac{1}{N}} = I$ in the discussions below.

5.6.1 Memory Requirements and Storage Needs

An MSL algorithm can often compute its solution without requiring all data samples in the memory, as respective computations can be done incrementally by reading $\mathcal{X}_m$ sequentially without loss of information. Hence, the memory needed for input data can be as low as $O(I^N)$, although sequential reading may lead to higher I/O costs. This is a major advantage that MSL algorithms enjoy over HOSVD-based tensor decomposition solutions [Lee and Elgammal, 2005; Lu et al., 2006d] that require the formation of an $(N + 1)$th-order tensor when the input tensor samples are of Nth-order, resulting in memory requirement of $O(M \cdot I^N)$ for input data. This is of considerable importance in applications with large datasets as the size of the input data may lead to a significant increase in complexity and memory storage requirement. Furthermore, the training phase can often be done offline so the additional I/O cost is not considered a big disadvantage. In addition, as shown in Figure 1.8, the

TABLE 5.1: Order of computational complexity of eigendecomposition for multilinear subspace learning (MSL) and linear subspace learning (LSL).

K	N	I	$O(K \cdot N \cdot I^3)$ (MSL)	$O(I^{3N})$ (LSL)
1	2	10	$O(2 \times 10^3)$	$O(10^6)$
10	2	10	$O(2 \times 10^4)$	$O(10^6)$
100	2	10	$O(2 \times 10^4)$	$O(10^6)$
10	3	10	$O(3 \times 10^4)$	$O(10^9)$
10	3	100	$O(3 \times 10^7)$	$O(10^{18})$
10	4	100	$O(4 \times 10^7)$	$O(10^{24})$

memory required for intermediate computations for MSL can be several orders of magnitude lower than that for linear subspace learning (LSL), especially for high-dimensional and higher-order tensors.

As the number of parameters in MSL is typically much lower than that in LSL (see Table 3.1), the storage size of TTP or TVP in MSL is typically much smaller than that of linear projection in LSL. Thus, MSL results in more compact projections than LSL.

5.6.2   Computational Complexity

Computational complexity is analyzed here for training and testing phases separately.

In training, as different algorithms involve different computations, we compare only the computational cost of eigendecomposition in MSL and LSL. For the same set of training data with M Nth-order tensors with $I_n = I$ for all n, the computations needed for eigendecomposition in MSL and LSL are on the order of $O(K \cdot N \cdot I^3)$ and $O(I^{3N}) = O(I^{3(N-1)} \cdot I^3)$, respectively. In most applications, $I^{3(N-1)} \gg K \cdot N$, as shown in Table 5.1. Therefore, the computational cost of eigendecomposition in MSL is typically much lower than that in LSL.

In testing phase of MSL, the extraction or projection of features from a test sample is a linear operation. Thus, feature extraction by MSL has the same order of computational complexity as feature extraction by LSL.

5.6.3   MATLAB® Implementation Tips for Large Datasets

When processing large datasets in MATLAB, it is common to see the "out of memory" error, especially on 32-bit MATLAB. Here, we list some tips for avoiding/reducing such errors when working with MSL algorithms on tensor data.

1. Use the 64-bit version MATLAB on a 64-bit machine, which has a much larger addressing space. In addition, a large RAM size is also necessary because the required memory cannot exceed the available RAM size. Thus, even with 64-bit MATLAB, it is still useful to consider the following tips in saving memory space required.

2. Clear variables not in use to save memory.

3. Use the most compact data types, for example, 8-bit integer instead of 32-bit double-precision floating point for gray-level image/video data input.

4. Save variables not to be used immediately to mat files and load them only when needed.

5. For a large-size variable, instead of passing it in function calls, save it as a mat file, clear the variable, and load the variable in the calling function. This (as well as Tip 4) reduces the memory required but increases I/O overhead so the running time will be longer but it will not crash due to being out of memory. Effectively, we deal with data residing on the hard disk, using the main memory as a cache.

6. MSL algorithms usually do not require all data to be in the memory at the same time. Thus, we can read training data samples in sequentially rather than in one shot. In this way, we trade higher I/O processing for lower memory demand again.

5.7 Summary

- A typical MSL solution follows the alternating partial projections method to convert the optimization of N projection matrices/vectors into N conditional optimization problems on partial multilinear projections, solving one projection matrix/vector at a time. It is iterative and alternates between modes.

- For an iterative MSL algorithm, we need to choose a good initialization method, a good mode order, and a suitable termination criterion, if possible, for the best performance. MSL algorithms may not always converge (even to a local optimum).

- For rigorous study of the properties of an MSL algorithm, it is useful to utilize synthetic datasets with different characteristics, as conclusions on specific datasets may not be generalized to other datasets.

- We can employ supervised/unsupervised feature selection strategies to convert feature tensors generated by TTP-based MSL algorithms to feature vectors for potentially better performance and/or for combination with conventional classifiers.

- In practice, we need to consider the memory and storage requirement, and computation complexity of MSL algorithms for efficient processing. In training, a single iteration of an MSL algorithm is usually much more efficient than its linear counterpart, especially for higher-order tensors. Thus, for efficient training, it is important to keep the maximum number of iterations low. In testing, feature extraction by MSL is a linear operation with the same order of computation complexity as that by LSL.

5.8 Further Reading

The alternating least squares (ALS) method is several decades old and widely used due to its programming simplicity. It is useful to refer to the relevant discussion by De Lathauwer et al. [2000b]. Recently, Comon et al. [2009] provided a detailed study on tensor decomposition using ALS and pointed out that ALS may not converge in general. Their paper is a very good source for further reading on this topic.

The issues of initialization, projection order, termination, and convergence are often overlooked when developing MSL algorithms. Our work in [Lu et al., 2008b] is probably the first to examine these issues comprehensively, and this work also studied other issues including synthetic data generation, subspace dimension determination, feature selection, and computational complexity. Our other work [Lu et al., 2009c,d] also gave a thorough treatment of many of these issues.

Appendix C of this book includes more discussions on the software issues in a wider and more general scope.

Part II

Algorithms and Applications

Chapter 6

Multilinear Principal Component Analysis

Part II covers specific multilinear subspace learning (MSL) algorithms and applications. This chapter starts with multilinear extensions of principal component analysis (PCA), as shown in Figure 6.1 under the MSL framework.

We describe five algorithms in more detail and cover other algorithms briefly. We begin with the generalized PCA (GPCA) [Ye et al., 2004a], a two-dimensional (2D) extension of PCA. GPCA involves only linear algebra so it is easier to understand. Then we discuss a further higher-order generalization: the multilinear PCA (MPCA) algorithm [Lu et al., 2008b]. GPCA and MPCA are based on the tensor-to-tensor projection (TTP). Next, we study two methods that are based on the tensor-to-vector projection (TVP): the tensor rank-one decomposition (TROD) [Shashua and Levin, 2001] and the uncorrelated MPCA (UMPCA) algorithm [Lu et al., 2009d]. In addition, we examine the combination of MPCA with boosting, in order to give a flavor of fusing MSL algorithms with other state-of-the-art learning methods. Finally, we summarize several other multilinear PCA extensions under the MSL framework.

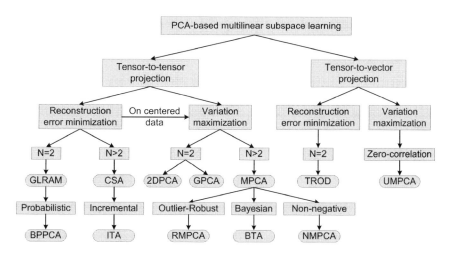

FIGURE 6.1: Multilinear PCA algorithms under the MSL framework.

Remark 6.1. *The multiple factor analysis algorithm in [Vasilescu and Ter-zopoulos, 2002b] is also referred to as MPCA. However, as pointed out in Section 4.4.3, this method requires training data to be labeled with forming factors such as people, views, and illuminations, so it is a **supervised learning** method by design. In contrast, the multilinear extensions of PCA covered in this chapter are unsupervised learning algorithms for unlabeled training data.*

6.1   Generalized PCA[1]

We start the presentation of multilinear PCA algorithms with a simple second-order case of $N = 2$, that is, a 2D extension for matrix data. In this case, the extension mainly involves linear algebra and matrix notations, which are easier for readers to follow.

6.1.1   GPCA Problem Formulation

The GPCA proposed in [Ye et al., 2004a] is an extension of PCA for matrix data, that is, second-order tensors. It is an unsupervised algorithm targeted at maximizing the captured variation in the projected subspace as in the classical PCA. From Definition 3.13 and the special case in Equation (3.50), the GPCA problem is defined as follows.

The GPCA Problem: A set of M matrix data samples $\{\mathbf{X}_1, \mathbf{X}_2, ..., \mathbf{X}_M\}$ is available for training. Each sample $\mathbf{X}_m \in \mathbb{R}^{I_1 \times I_2}$ $(m = 1, ..., M)$ assumes values in the *tensor space* $\mathbb{R}^{I_1} \bigotimes \mathbb{R}^{I_2}$, which is defined as the *tensor product* (*outer product*) of two vector spaces $\mathbb{R}^{I_1}$ and $\mathbb{R}^{I_2}$. The GPCA objective is to find two linear transformations (projection matrices) $\mathbf{U}^{(1)} \in \mathbb{R}^{I_1 \times P_1}$ $(P_1 \leq I_1)$ and $\mathbf{U}^{(2)} \in \mathbb{R}^{I_2 \times P_2}$ $(P_2 \leq I_2)$, with orthonormal columns, that project a sample $\mathbf{X}_m$ to

$$\mathbf{Y}_m = \mathbf{U}^{(1)^T} \mathbf{X}_m \mathbf{U}^{(2)} = \mathbf{X}_m \times_1 \mathbf{U}^{(1)^T} \times_2 \mathbf{U}^{(2)^T} \in \mathbb{R}^{P_1 \times P_2} \qquad (6.1)$$

such that $\{\mathbf{Y}_m \in \mathbb{R}^{P_1} \bigotimes \mathbb{R}^{P_2}, m = 1, ..., M\}$ captures most of the variation observed in the original samples, assuming that the variation is measured by the total scatter $\Psi_{T_{\mathbf{Y}}}$ defined for matrices in Equation (3.50).

We can write the objective function for GPCA as

$$(\tilde{\mathbf{U}}^{(1)}, \tilde{\mathbf{U}}^{(2)}) = \arg \max_{\mathbf{U}^{(1)}, \mathbf{U}^{(2)}} \Psi_{T_{\mathbf{Y}}} = \arg \max_{\mathbf{U}^{(1)}, \mathbf{U}^{(2)}} \sum_{m=1}^{M} \parallel \mathbf{Y}_m - \bar{\mathbf{Y}} \parallel_F^2, \qquad (6.2)$$

---

[1]Another popular PCA extension has the same name generalized PCA [Vidal et al., 2005], which learns multiple subspaces instead of only one subspace.

where

$$\bar{\mathbf{Y}} = \frac{1}{M} \sum_{m=1}^{M} \mathbf{Y}_m. \tag{6.3}$$

Note that GPCA projects an input matrix sample to another matrix of smaller size, instead of a low-dimensional vector in PCA. This projection scheme belongs to the TTP. The subspace dimensions P_1 and P_2 need to be specified.

6.1.2 GPCA Algorithm Derivation

As in PCA, GPCA centers the training samples as

$$\tilde{\mathbf{X}}_m = \mathbf{X}_m - \bar{\mathbf{X}}, \tag{6.4}$$

where $\bar{\mathbf{X}}$ is the mean of the training samples defined as

$$\bar{\mathbf{X}} = \frac{1}{M} \sum_{m=1}^{M} \mathbf{X}_m. \tag{6.5}$$

Thus, from Equations (6.1) and (6.4), the objective function Equation (6.2) becomes

$$
\begin{aligned}
(\tilde{\mathbf{U}}^{(1)}, \tilde{\mathbf{U}}^{(2)}) &= \arg\max_{\mathbf{U}^{(1)}, \mathbf{U}^{(2)}} \sum_{m=1}^{M} \| \mathbf{Y}_m - \bar{\mathbf{Y}} \|_F^2 \\
&= \arg\max_{\mathbf{U}^{(1)}, \mathbf{U}^{(2)}} \sum_{m=1}^{M} \| \mathbf{U}^{(1)^T} \tilde{\mathbf{X}}_m \mathbf{U}^{(2)} \|_F^2 .
\end{aligned}
\tag{6.6}
$$

As discussed in Section 5.1, there is no closed-form solution to this maximization problem because we need to determine two interdependent projection matrices. Thus, the GPCA solution follows the alternating partial projections (APP) method (Section 5.1) originating from the alternating least squares (ALS) [Kroonenberg and Leeuw, 1980] to solve for the two projection matrices one at a time through partial multilinear projections iteratively. In each iteration step, we assume one projection matrix is given to find the other one that maximizes the captured variation, and then vice versa. The solution is stated in the following theorem [Ye et al., 2004a]:

Theorem 6.1. *Let $\tilde{\mathbf{U}}^{(1)}$ and $\tilde{\mathbf{U}}^{(2)}$ be the matrices maximizing the total scatter $\Psi_{T_\mathbf{Y}}$ as in Equation (6.2). Then,*

1. *For a given $\mathbf{U}^{(2)}$, the optimal $\tilde{\mathbf{U}}^{(1)}$ consists of the P_1 eigenvectors of the conditioned mode-1 scatter matrix*

$$\Psi_{T_{\mathbf{Y}1}} = \sum_{m=1}^{M} \tilde{\mathbf{X}}_m \mathbf{U}^{(2)} \mathbf{U}^{(2)^T} \tilde{\mathbf{X}}_m^T \tag{6.7}$$

associated with the P_1 largest eigenvalues.

2. For a given $\mathbf{U}^{(1)}$, the optimal $\tilde{\mathbf{U}}^{(2)}$ consists of the P_2 eigenvectors of the conditioned mode-2 scatter matrix

$$\boldsymbol{\Psi}_{T_{\mathbf{Y}2}} = \sum_{m=1}^{M} \tilde{\mathbf{X}}_m^T \mathbf{U}^{(1)} \mathbf{U}^{(1)^T} \tilde{\mathbf{X}}_m \tag{6.8}$$

associated with the P_2 largest eigenvalues.

Proof. For a given $\mathbf{U}^{(2)}$, we can write Equation (6.6) further as

$$\begin{aligned}
\tilde{\mathbf{U}}^{(1)} &= \arg\max_{\mathbf{U}^{(1)}} \sum_{m=1}^{M} \| \mathbf{U}^{(1)^T} \tilde{\mathbf{X}}_m \mathbf{U}^{(2)} \|_F^2 \\
&= \arg\max_{\mathbf{U}^{(1)}} \sum_{m=1}^{M} \text{tr}(\mathbf{U}^{(1)^T} \tilde{\mathbf{X}}_m \mathbf{U}^{(2)} \mathbf{U}^{(2)^T} \tilde{\mathbf{X}}_m^T \mathbf{U}^{(1)}) \\
&= \arg\max_{\mathbf{U}^{(1)}} \text{tr}(\mathbf{U}^{(1)^T} \boldsymbol{\Psi}_{T_{\mathbf{Y}1}} \mathbf{U}^{(1)}). \tag{6.9}
\end{aligned}$$

Because $\mathbf{U}^{(1)}$ has orthonormal columns, the trace above is maximized only if $\tilde{\mathbf{U}}^{(1)}$ consists of the P_1 eigenvectors of the matrix $\boldsymbol{\Psi}_{T_{\mathbf{Y}1}}$ associated with the P_1 largest eigenvalues [Kokiopoulou et al., 2011].

Similarly, for a given $\mathbf{U}^{(1)}$, we can write Equation (6.6) further as

$$\begin{aligned}
\tilde{\mathbf{U}}^{(2)} &= \arg\max_{\mathbf{U}^{(2)}} \sum_{m=1}^{M} \text{tr}(\mathbf{U}^{(2)^T} \tilde{\mathbf{X}}_m^T \mathbf{U}^{(1)} \mathbf{U}^{(1)^T} \tilde{\mathbf{X}}_m \mathbf{U}^{(2)}) \\
&= \arg\max_{\mathbf{U}^{(2)}} \text{tr}(\mathbf{U}^{(2)^T} \boldsymbol{\Psi}_{T_{\mathbf{Y}2}} \mathbf{U}^{(2)}). \tag{6.10}
\end{aligned}$$

Because $\mathbf{U}^{(2)}$ has orthonormal columns, the trace above is maximized only if $\tilde{\mathbf{U}}^{(2)}$ consists of the P_2 eigenvectors of the matrix $\boldsymbol{\Psi}_{T_{\mathbf{Y}2}}$ associated with the P_2 largest eigenvalues. □

Thus, based on Theorem 6.1, the GPCA problem can be solved using an iterative procedure as summarized in Algorithm 6.1. Note that the GPCA algorithm is initialized using the pseudo-identity matrices (see Section 5.2) in [Ye et al., 2004a], while other initialization methods can also be used.

Remark 6.2. $\boldsymbol{\Psi}_{T_{\mathbf{Y}1}}$ *can be seen as the total scatter measure of the mode-1 partial projections of the input samples* $\{\tilde{\mathbf{X}}_m \mathbf{U}^{(2)}\}$. $\boldsymbol{\Psi}_{T_{\mathbf{Y}2}}$ *can be seen as the total scatter measure of the mode-2 partial projections of the input samples* $\{\tilde{\mathbf{X}}_m^T \mathbf{U}^{(1)}\}$.

6.1.3   Discussions on GPCA

Convergence and termination: The derivation of Theorem 6.1 implies that per iteration, the total scatter $\boldsymbol{\Psi}_{T_{\mathbf{Y}}}$ is a non-decreasing function (it either

---

Algorithm 6.1 Generalized PCA (GPCA)

Input: A set of matrix samples $\{\mathbf{X}_m \in \mathbb{R}^{I_1 \times I_2}, m = 1, ..., M\}$, the desired subspace dimensions P_1 and P_2 (or the percentage of variation kept in each mode Q to determine P_1 and P_2 by Equations (6.12) and (6.13)), and the maximum number of iterations K.

Process:

1: **Centering:** Center the input samples to get $\{\tilde{\mathbf{X}}_m\}$ as in Equation (6.4).
2: **Initialization:** Initialize $\mathbf{U}^{(1)}$ and $\mathbf{U}^{(2)}$ to be truncated identity matrices of size $I_1 \times P_1$ and $I_2 \times P_2$, respectively (initialization of $\mathbf{U}^{(1)}$ can be skipped).
3: [**Local optimization:**]
4: **for** $k = 1$ to K **do**
5:    [**Mode 1**]
6:    Calculate $\mathbf{\Psi}_{T_{\mathbf{Y}1}}$ according to Equation (6.7).
7:    Set the matrix $\mathbf{U}^{(1)}$ to be consisting of the P_1 eigenvectors of $\mathbf{\Psi}_{T_{\mathbf{Y}1}}$, corresponding to the largest P_1 eigenvalues.
8:    [**Mode 2**]
9:    Calculate $\mathbf{\Psi}_{T_{\mathbf{Y}2}}$ according to Equation (6.8).
10:   Set the matrix $\mathbf{U}^{(2)}$ to be consisting of the P_2 eigenvectors of $\mathbf{\Psi}_{T_{\mathbf{Y}2}}$, corresponding to the largest P_2 eigenvalues.
11: **end for**

Output: The two projection matrices $\tilde{\mathbf{U}}^{(1)}$ and $\tilde{\mathbf{U}}^{(2)}$ that maximize the total scatter in the projected space.

---

remains the same or increases) because each update of a projection matrix maximizes $\mathbf{\Psi}_{T_{\mathbf{Y}}}$ conditioned on the other projection matrix. On the other hand, $\mathbf{\Psi}_{T_{\mathbf{Y}}}$ is upper-bounded by $\mathbf{\Psi}_{T_{\mathbf{X}}}$ (the total scatter in the original samples) because the projection matrices have orthonormal columns. Therefore, GPCA is expected to converge over iterations. Nonetheless, in practice, we often terminate the iterations by specifying a maximum number of iterations K (e.g., $K = 10$) for efficiency, as in Algorithm 6.1.

Subspace dimension determination: In [Ye et al., 2004a], the mode-1 and mode-2 subspace dimensions are set to be equal for simplicity assuming similar variations in two modes. Thus, this simple method did not consider the case of different variations in mode-1 and mode-2. A more principled method is the Q-based method in [Lu et al., 2008b], a multilinear extension of the traditional dimension selection strategy in PCA. We consider the conditioned mode-1 and mode-2 scatter matrices for the full projection case (Section 5.2.2), where $P_1 = I_1$ and $P_2 = I_2$ so $\tilde{\mathbf{U}}^{(1)}$ and $\tilde{\mathbf{U}}^{(2)}$ become orthogonal and Equations (6.7) and (6.8) become

$$\breve{\mathbf{\Psi}}_{\mathbf{Y}1} = \sum_{m=1}^{M} \tilde{\mathbf{X}}_m \tilde{\mathbf{X}}_m^T, \qquad \text{and} \qquad \breve{\mathbf{\Psi}}_{\mathbf{Y}2} = \sum_{m=1}^{M} \tilde{\mathbf{X}}_m^T \tilde{\mathbf{X}}_m. \qquad (6.11)$$

Denote the eigenvalues of $\breve{\boldsymbol{\Psi}}_{\mathbf{Y1}}$ and $\breve{\boldsymbol{\Psi}}_{\mathbf{Y2}}$ as $\{\breve{\lambda}_{i_1}^{(1)}, i_1 = 1, ..., I_1\}$ and $\{\breve{\lambda}_{i_2}^{(2)}, i_2 = 1, ..., I_2\}$ in descending order, respectively. Define the ratio

$$Q^{(1)} = \frac{\sum_{i_1=1}^{P_1} \breve{\lambda}_{i_1}^{(1)}}{\sum_{i_1=1}^{I_1} \breve{\lambda}_{i_1}^{(1)}} \qquad \text{and} \qquad Q^{(2)} = \frac{\sum_{i_2=1}^{P_2} \breve{\lambda}_{i_2}^{(2)}}{\sum_{i_2=1}^{I_2} \breve{\lambda}_{i_2}^{(2)}} \tag{6.12}$$

to be the remaining portion of the total scatter when keeping P_1 and P_2 eigenvectors, respectively. In the Q-based method, the first P_1 and P_2 eigenvectors are kept for mode-1 and mode-2, respectively, so that

$$Q^{(1)} = Q^{(2)} = Q. \tag{6.13}$$

The equality usually can hold only approximately in practice so we find the smallest values of P_1 and P_2 satisfying $Q^{(1)} \geq Q$ and $Q^{(2)} \geq Q$, respectively.

6.1.4 Reconstruction Error Minimization

As in PCA, there is an interesting relationship between variation maximization and reconstruction (approximation) error minimization in GPCA for centered samples. For a pair of projection matrices $\mathbf{U}^{(1)} \in \mathbb{R}^{I_1 \times P_1}$ and $\mathbf{U}^{(2)} \in \mathbb{R}^{I_2 \times P_2}$, we have the corresponding projections $\{\tilde{\mathbf{Y}}_m \in \mathbb{R}^{P_1 \times P_2}, m = 1, ..., M\}$ of the centered samples $\{\tilde{\mathbf{X}}_m\}$:

$$\tilde{\mathbf{Y}}_m = \mathbf{U}^{(1)^T} \tilde{\mathbf{X}}_m \mathbf{U}^{(2)}. \tag{6.14}$$

We can use them to reconstruct the centered samples $\{\tilde{\mathbf{X}}_m \in \mathbb{R}^{I_1 \times I_2}, m = 1, ..., M\}$ as[2]

$$\hat{\tilde{\mathbf{X}}}_m = \mathbf{U}^{(1)} \tilde{\mathbf{Y}}_m \mathbf{U}^{(2)^T}. \tag{6.15}$$

Based on the following trace properties (Section A.1.6)

$$\text{tr}(\mathbf{A}\mathbf{A}^T) = \| \mathbf{A} \|_F^2 \quad \text{and} \quad \text{tr}(\mathbf{A}\mathbf{B}\mathbf{C}\mathbf{D}) = \text{tr}(\mathbf{D}\mathbf{A}\mathbf{B}\mathbf{C}), \tag{6.16}$$

---

[2]Subsequently, the original samples can be reconstructed as $\hat{\mathbf{X}}_m = \tilde{\mathbf{X}}_m + \bar{\mathbf{X}}$.

the squared reconstruction error can be calculated as

$$\sum_{m=1}^{M} \| \tilde{\mathbf{X}}_m - \mathbf{U}^{(1)} \tilde{\mathbf{Y}}_m \mathbf{U}^{(2)^T} \|_F^2$$

$$= \sum_{m=1}^{M} \operatorname{tr} \left((\tilde{\mathbf{X}}_m - \mathbf{U}^{(1)} \tilde{\mathbf{Y}}_m \mathbf{U}^{(2)^T})(\tilde{\mathbf{X}}_m - \mathbf{U}^{(1)} \tilde{\mathbf{Y}}_m \mathbf{U}^{(2)^T})^T \right)$$

$$= \sum_{m=1}^{M} \left(\operatorname{tr}(\tilde{\mathbf{X}}_m \tilde{\mathbf{X}}_m^T) - 2\operatorname{tr}(\tilde{\mathbf{X}}_m \mathbf{U}^{(2)} \tilde{\mathbf{Y}}_m^T \mathbf{U}^{(1)^T}) + \operatorname{tr}(\tilde{\mathbf{Y}}_m \tilde{\mathbf{Y}}_m^T) \right)$$

$$= \sum_{m=1}^{M} \left(\operatorname{tr}(\tilde{\mathbf{X}}_m \tilde{\mathbf{X}}_m^T) - 2\operatorname{tr}(\mathbf{U}^{(1)^T} \tilde{\mathbf{X}}_m \mathbf{U}^{(2)} \tilde{\mathbf{Y}}_m^T) + \operatorname{tr}(\tilde{\mathbf{Y}}_m \tilde{\mathbf{Y}}_m^T) \right)$$

$$= \sum_{m=1}^{M} \left(\operatorname{tr}(\tilde{\mathbf{X}}_m \tilde{\mathbf{X}}_m^T) - 2\operatorname{tr}(\tilde{\mathbf{Y}}_m \tilde{\mathbf{Y}}_m^T) + \operatorname{tr}(\tilde{\mathbf{Y}}_m \tilde{\mathbf{Y}}_m^T) \right)$$

$$= \sum_{m=1}^{M} \left(\operatorname{tr}(\tilde{\mathbf{X}}_m \tilde{\mathbf{X}}_m^T) - \operatorname{tr}(\tilde{\mathbf{Y}}_m \tilde{\mathbf{Y}}_m^T) \right)$$

$$= \sum_{m=1}^{M} (\| \tilde{\mathbf{X}}_m \|_F^2 - \| \tilde{\mathbf{Y}}_m \|_F^2)$$

$$= \sum_{m=1}^{M} \| \tilde{\mathbf{X}}_m \|_F^2 - \sum_{m=1}^{M} \| \tilde{\mathbf{Y}}_m \|_F^2 . \tag{6.17}$$

Because for a learning problem, the training data $\{\tilde{\mathbf{X}}_m\}$ is fixed, $\| \tilde{\mathbf{X}}_m \|_F^2$ becomes a constant. Therefore, the maximization of the variation (scatter) $\sum_{m=1}^{M} \| \tilde{\mathbf{Y}}_m \|_F^2$ is equivalent to the minimization of the reconstruction error $\sum_{m=1}^{M} \| \tilde{\mathbf{X}}_m - \mathbf{U}^{(1)} \tilde{\mathbf{Y}}_m \mathbf{U}^{(2)^T} \|_F^2$ from the above result. Note that $\mathbf{U}^{(2)^T} \mathbf{U}^{(2)}$ and $\mathbf{U}^{(1)^T} \mathbf{U}^{(1)}$ are identity matrices in the derivations above because $\mathbf{U}^{(1)}$ and $\mathbf{U}^{(2)}$ have orthonormal columns.

6.2 Multilinear PCA

MPCA [Lu et al., 2008b] is an unsupervised MSL algorithm for general tensors targeting variation maximization as in GPCA and PCA. It solves for a TTP that allows projected tensors to capture most of the variation present in the original tensors. It can be considered a higher-order extension of GPCA [Ye et al., 2004a].

6.2.1   MPCA Problem Formulation

Based on Definition 3.13, the MPCA problem is defined as follows.

The MPCA Problem: A set of M tensor data samples $\{\mathcal{X}_1, \mathcal{X}_2, ..., \mathcal{X}_M\}$ are available for training. Each sample $\mathcal{X}_m \in \mathbb{R}^{I_1 \times I_2 \times ... \times I_N}$ assumes values in a tensor space $\mathbb{R}^{I_1} \bigotimes \mathbb{R}^{I_2} ... \bigotimes \mathbb{R}^{I_N}$, which is the *tensor product (outer product)* of N vector spaces $\mathbb{R}^{I_1}, \mathbb{R}^{I_2}, ..., \mathbb{R}^{I_N}$. The MPCA objective is to find a TTP $\{\mathbf{U}^{(n)} \in \mathbb{R}^{I_n \times P_n}, n = 1, ..., N\}$ that maps the original tensor space $\mathbb{R}^{I_1} \bigotimes \mathbb{R}^{I_2} ... \bigotimes \mathbb{R}^{I_N}$ into a tensor subspace $\mathbb{R}^{P_1} \bigotimes \mathbb{R}^{P_2} ... \bigotimes \mathbb{R}^{P_N}$ (with $P_n \leq I_n$, for $n = 1, ..., N$) as

$$\mathcal{Y}_m = \mathcal{X}_m \times_1 \mathbf{U}^{(1)^T} \times_2 \mathbf{U}^{(2)^T} ... \times_N \mathbf{U}^{(N)^T} \in \mathbb{R}^{P_1 \times P_2 \times ... \times P_N}, m = 1, ..., M, \tag{6.18}$$

such that $\{\mathcal{Y}_m\}$ captures most of the variation observed in the original tensor samples. The variation is measured by the total scatter defined for tensors in Definition 3.13, and $\mathbf{U}^{(n)}$ is the mode-n projection matrix of a TTP, with orthonormal columns as in GPCA.

In other words, the MPCA objective is the determination of the N projection matrices that maximize the total tensor scatter Ψ_{T_y}:

$$\{\tilde{\mathbf{U}}^{(n)}\} = \arg \max_{\{\mathbf{U}^{(n)}\}} \Psi_{T_y} = \arg \max_{\{\mathbf{U}^{(n)}\}} \sum_{m=1}^{M} \| \mathcal{Y}_m - \bar{\mathcal{Y}} \|_F^2, \tag{6.19}$$

where the mean of projected tensor features

$$\bar{\mathcal{Y}} = \frac{1}{M} \sum_{m=1}^{M} \mathcal{Y}_m. \tag{6.20}$$

Here, the subspace dimension for each mode, P_n, is assumed to be known or predetermined first. Section 6.2.4 will discuss adaptive determination of P_n, when it is not known in advance.

6.2.2   MPCA Algorithm Derivation

As in GPCA and PCA, we center the input tensor samples first as

$$\tilde{\mathcal{X}}_m = \mathcal{X}_m - \bar{\mathcal{X}}, m = 1, ..., M, \tag{6.21}$$

where the sample mean is

$$\bar{\mathcal{X}} = \frac{1}{M} \sum_{m=1}^{M} \mathcal{X}_m. \tag{6.22}$$

The projection of the centered sample $\tilde{\mathcal{X}}_m$ is

$$\begin{aligned} \tilde{\mathcal{Y}}_m &= \tilde{\mathcal{X}}_m \times_1 \mathbf{U}^{(1)^T} \times_2 \mathbf{U}^{(2)^T} ... \times_N \mathbf{U}^{(N)^T} & (6.23) \\ &= \mathcal{Y}_m - \bar{\mathcal{Y}}. & (6.24) \end{aligned}$$

As mentioned in Section 5.1, the N projection matrices need the determination of N sets of parameters that are interdependent (except for $N = 1$ or $P_n = I_n$ for all n). Thus, we follow the APP method again. Because the projection to an Nth-order tensor subspace consists of N projections, we solve N optimization subproblems by finding the mode-n projection matrix $\mathbf{U}^{(n)}$ that maximizes the mode-n total scatter conditioned on the projection matrices in all the other modes. The following theorem gives the solution for such a subproblem.

Theorem 6.2. *Let* $\{\tilde{\mathbf{U}}^{(n)}, n = 1, ..., N\}$ *be the solution to Equation (6.19). Then, given all the other projection matrices* $\{\mathbf{U}^{(1)}, ..., \mathbf{U}^{(n-1)}, \mathbf{U}^{(n+1)}, ..., \mathbf{U}^{(N)}\}$, *the projection matrix* $\tilde{\mathbf{U}}^{(n)}$ *consists of the* P_n *eigenvectors corresponding to the largest* P_n *eigenvalues of the matrix*

$$\mathbf{\Phi}^{(n)} = \sum_{m=1}^{M} \tilde{\mathbf{X}}_{m(n)} \mathbf{U}_{\mathbf{\Phi}^{(n)}} \mathbf{U}_{\mathbf{\Phi}^{(n)}}^{T} \tilde{\mathbf{X}}_{m(n)}^{T}, \qquad (6.25)$$

where $\tilde{\mathbf{X}}_{m(n)}$ *is the mode-n unfolding of* $\tilde{\mathcal{X}}_m$ *and*

$$\mathbf{U}_{\mathbf{\Phi}^{(n)}} = \left(\mathbf{U}^{(n+1)} \otimes \mathbf{U}^{(n+2)} \otimes ... \otimes \mathbf{U}^{(N)} \otimes \mathbf{U}^{(1)} \otimes \mathbf{U}^{(2)} \otimes ... \mathbf{U}^{(n-1)} \right). \quad (6.26)$$

The symbol "$\otimes$" denotes the Kronecker product (Section A.1.4).

Proof. Write the objective function Equation (6.19) in terms of the input tensor samples as

$$
\begin{aligned}
\Psi_{T_{\mathcal{Y}}} &= \sum_{m=1}^{M} \parallel \mathcal{Y}_m - \bar{\mathcal{y}} \parallel_F^2 = \sum_{m=1}^{M} \parallel \tilde{\mathcal{Y}}_m \parallel_F^2 \\
&= \sum_{m=1}^{M} \parallel \tilde{\mathcal{X}}_m \times_1 \mathbf{U}^{(1)^T} \times_2 \mathbf{U}^{(2)^T} ... \times_N \mathbf{U}^{(N)^T} \parallel_F^2 . \quad (6.27)
\end{aligned}
$$

From the definition of the Frobenius norm for a tensor and that for a matrix, we have

$$\parallel \mathcal{A} \parallel_F = \parallel \mathbf{A}_{(n)} \parallel_F . \qquad (6.28)$$

From Equation (3.22), we can express $\Psi_{T_{\mathcal{Y}}}$ using the equivalent matrix representation with mode-n unfolding as follows:

$$\Psi_{T_{\mathcal{Y}}} = \sum_{m=1}^{M} \parallel \mathbf{U}^{(n)^T} \tilde{\mathbf{X}}_{m(n)} \mathbf{U}_{\mathbf{\Phi}^{(n)}} \parallel_F^2, \qquad (6.29)$$

where $\mathbf{U}_{\mathbf{\Phi}^{(n)}}$ is defined in Equation (6.26). Because $\parallel \mathbf{A} \parallel_F^2 = \mathrm{tr}(\mathbf{A}\mathbf{A}^T)$, we can write $\Psi_{T_{\mathcal{Y}}}$ in terms of $\mathbf{\Phi}^{(n)}$ defined in Equation (6.25) as

$$
\begin{aligned}
\Psi_{T_{\mathcal{Y}}} &= \sum_{m=1}^{M} \mathrm{tr} \left(\mathbf{U}^{(n)^T} \tilde{\mathbf{X}}_{m(n)} \mathbf{U}_{\mathbf{\Phi}^{(n)}} \mathbf{U}_{\mathbf{\Phi}^{(n)}}^{T} \tilde{\mathbf{X}}_{m(n)}^{T} \mathbf{U}^{(n)} \right) \\
&= \mathrm{tr} \left(\mathbf{U}^{(n)^T} \mathbf{\Phi}^{(n)} \mathbf{U}^{(n)} \right) . \quad (6.30)
\end{aligned}
$$

Equation (6.30) is similar to Equations (6.9) and (6.10). Therefore, for given $\{\mathbf{U}^{(1)}, ..., \mathbf{U}^{(n-1)}, \mathbf{U}^{(n+1)}, ..., \mathbf{U}^{(N)}\}$, Ψ_{T_y} is maximized if and only if $\mathrm{tr}\left(\mathbf{U}^{(n)^T}\mathbf{\Phi}^{(n)}\mathbf{U}^{(n)}\right)$ is maximized. The maximum is obtained if $\tilde{\mathbf{U}}^{(n)}$ consists of the P_n eigenvectors of the matrix $\mathbf{\Phi}^{(n)}$ corresponding to the largest P_n eigenvalues. □

It can be verified that Theorem 6.1 is a special case ($N = 2$) of Theorem 6.2.

Note that the matrix $\mathbf{\Phi}^{(n)}$ in Equation (6.25) above can be written as the mode-n total scatter matrix defined in Equation (3.42) for partial multilinear projections. We can first obtain the mode-n partial multilinear projection of $\tilde{\mathcal{X}}_m$ using the projection matrices in all the other modes as:

$$\hat{\mathcal{Y}}_m^{(n)} = \tilde{\mathcal{X}}_m \times_1 \mathbf{U}^{(1)^T} ... \times_{(n-1)} \mathbf{U}^{(n-1)^T} \times_{(n+1)} \mathbf{U}^{(n+1)^T} ... \times_N \mathbf{U}^{(N)^T}. \quad (6.31)$$

Then, we can get the mode-n total scatter matrix from $\hat{\mathbf{Y}}_{m(n)}$, the mode-n unfolding of $\hat{\mathcal{Y}}_m^{(n)}$, as

$$\mathbf{S}_{T_{\hat{y}}}^{(n)} = \sum_{m=1}^{M} \left(\hat{\mathbf{Y}}_{m(n)} - \bar{\hat{\mathbf{Y}}}_{(n)}\right)\left(\hat{\mathbf{Y}}_{m(n)} - \bar{\hat{\mathbf{Y}}}_{(n)}\right)^T, \quad (6.32)$$

where $\bar{\hat{\mathbf{Y}}}_{(n)}$ is the mean of $\{\hat{\mathbf{Y}}_{m(n)}\}$ or the mode-n unfolding of $\bar{\hat{\mathcal{Y}}}^{(n)}$, the mean of $\{\hat{\mathcal{Y}}_m^{(n)}\}$. Note that $\bar{\hat{\mathcal{Y}}}^{(n)}$ is zero because $\{\tilde{\mathcal{X}}_m\}$ are centered. We can verify that $\mathbf{\Phi}^{(n)} = \mathbf{S}_{T_{\hat{y}}}^{(n)}$.

As seen from Theorem 6.2, the product $\mathbf{U}_{\mathbf{\Phi}^{(n)}}\mathbf{U}_{\mathbf{\Phi}^{(n)}}^T$ depends on $\{\mathbf{U}^{(1)}, ..., \mathbf{U}^{(n-1)}, \mathbf{U}^{(n+1)}, ..., \mathbf{U}^{(N)}\}$, indicating that the optimization of $\mathbf{U}^{(n)}$ depends on the projections in all the other modes except when $N = 1$ or when $P_n = I_n$ for all n (see Lemma 6.1 below). Therefore, there is no closed-form solution to this maximization problem. Instead, we use the APP method to solve Equation (6.19). The pseudocode is summarized in Algorithm 6.2, where the projection matrices are then initialized through the full projection truncation (FPT) described in Section 5.2.2. In the local optimization step, the projection matrices are updated one by one (the "n loop") with all the others fixed. The local optimization procedure is repeated (the "k loop") until the result converges or a maximum number of iterations K is reached.

6.2.3 Discussions on MPCA

Full projection: In the case of *full projection*, where $P_n = I_n$ for $n = 1, ..., N$, we have the following lemma on $\mathbf{U}_{\mathbf{\Phi}^{(n)}}\mathbf{U}_{\mathbf{\Phi}^{(n)}}^T$.

Lemma 6.1. *When* $P_n = I_n$ *for* $n = 1, ..., N$, $\mathbf{U}_{\mathbf{\Phi}^{(n)}}\mathbf{U}_{\mathbf{\Phi}^{(n)}}^T$ *is an identity matrix.*

Algorithm 6.2 Multilinear principal component analysis (MPCA)

Input: A set of tensor samples $\{\mathcal{X}_m \in \mathbb{R}^{I_1 \times I_2 \times \ldots \times I_N}, m = 1, \ldots, M\}$, the desired tensor subspace dimensions $\{P_n, n = 1, 2, \ldots, N\}$ (or the percentage of variation kept in each mode Q to determine $\{P_n\}$ by Equations (6.38) and (6.39)), and the maximum number of iterations K.

Process:

1: **Centering:** Center the input samples to get $\{\tilde{\mathcal{X}}_m\}$ according to Equation (6.21).

2: **Initialization:** Calculate the eigendecomposition of the mode-n total scatter matrix in full projection (Section 5.2.2) $\breve{\mathbf{S}}_{T_{\mathcal{X}}}^{(n)} = \sum_{m=1}^{M} \tilde{\mathbf{X}}_{m(n)} \cdot \tilde{\mathbf{X}}_{m(n)}^T$ and set $\mathbf{U}^{(n)}$ to consist of the eigenvectors corresponding to the most significant P_n eigenvalues, for $n = 1, \ldots, N$.

3: [**Local optimization:**]

4: **for** $k = 1$ **to** K **do**

5:     **for** $n = 1$ **to** N **do**

6:         Calculate the mode-n partial multilinear projections $\{\hat{\mathcal{Y}}_m^{(n)}, m = 1, \ldots, M\}$ according to Equation (6.31).

7:         Calculate $\mathbf{S}_{T_{\hat{y}}}^{(n)}$ according to Equation (6.32).

8:         Set $\mathbf{U}^{(n)}$ to consist of the P_n eigenvectors of $\mathbf{S}_{T_{\hat{y}}}^{(n)}$ corresponding to the largest P_n eigenvalues.

9:     **end for**

10: **end for**

Output: The N projection matrices $\{\tilde{\mathbf{U}}^{(n)} \in \mathbb{R}^{I_n \times P_n}, n = 1, 2, \ldots, N\}$ that maximize the total scatter in the projected space.

Proof. By successive application of the transpose property of the Kronecker product $(\mathbf{A} \otimes \mathbf{B})^T = \mathbf{A}^T \otimes \mathbf{B}^T$ (Section A.1.4):

$$\mathbf{U}_{\Phi^{(n)}}^T = \left(\mathbf{U}^{(n+1)^T} \otimes \mathbf{U}^{(n+2)^T} \otimes \ldots \otimes \mathbf{U}^{(N)^T} \otimes \mathbf{U}^{(1)^T} \otimes \mathbf{U}^{(2)^T} \otimes \ldots \mathbf{U}^{(n-1)^T} \right).$$
(6.33)

By the Kronecker product theorem $(\mathbf{A} \otimes \mathbf{B})(\mathbf{C} \otimes \mathbf{D}) = (\mathbf{AC} \otimes \mathbf{BD})$ (Section A.1.4),

$$\mathbf{U}_{\Phi^{(n)}} \mathbf{U}_{\Phi^{(n)}}^T$$
$$= \left(\mathbf{U}^{(n+1)} \mathbf{U}^{(n+1)^T} \otimes \ldots \otimes \mathbf{U}^{(N)} \mathbf{U}^{(N)^T} \otimes \mathbf{U}^{(1)} \mathbf{U}^{(1)^T} \otimes \ldots \mathbf{U}^{(n-1)} \mathbf{U}^{(n-1)^T} \right). (6.34)$$

When $P_n = I_n$ for all n, $\mathbf{U}^{(n)}$ becomes an orthogonal matrix and $\mathbf{U}^{(n)^T} \mathbf{U}^{(n)} = \mathbf{I}_{I_n}$, where $\mathbf{I}_{I_n}$ is an $I_n \times I_n$ identity matrix. Then, $\mathbf{U}^{(n)^{-1}} = \mathbf{U}^{(n)^T}$ and $\mathbf{U}^{(n)} \mathbf{U}^{(n)^T} = \mathbf{I}_{I_n}$. Thus, $\mathbf{U}_{\Phi^{(n)}} \mathbf{U}_{\Phi^{(n)}}^T = \mathbf{I}_{I_1 \times I_2 \times \ldots \times I_{n-1} \times I_{n+1} \times \ldots \times I_N}$. □

Convergence: The derivation of Theorem 6.2 implies that per iteration, the total scatter $\Psi_{T_{\mathcal{Y}}}$ is a non-decreasing function (it either remains the same or increases) because each update of the mode-n projection matrix $\mathbf{U}^{(n)}$ maximizes Ψ_{T_y}. On the other hand, Ψ_{T_y} is upper-bounded by Ψ_{T_X} (the variation in the original samples) because the projection matrices $\{\mathbf{U}^{(n)}\}$ consist of orthonormal columns. Therefore, MPCA is expected to have good convergence property. Empirical studies [Lu et al., 2008b] show that the MPCA algorithm converges very fast (within five iterations) for typical tensor data. Furthermore, when modewise eigenvalues are all distinct (with multiplicity 1), the projection matrices $\{\mathbf{U}^{(n)}\}$, which maximize Ψ_{T_y}, are expected to converge as well. The convergence of $\{\mathbf{U}^{(n)}\}$ is under the condition that the sign for the first component of each mode-n eigenvector is fixed because eigenvectors are unique up to the sign. Experimental studies in [Lu et al., 2008b] show that the projection matrices $\{\mathbf{U}^{(n)}\}$ do converge within a small number of iterations.

Feature extraction: The projection matrices $\{\tilde{\mathbf{U}}^{(n)}, n = 1, ..., N\}$ obtained by TTP-based MSL can be used to extract tensor features from a set of training tensor samples $\{\mathcal{X}_m, m = 1, ..., M\}$. In testing, a test tensor sample $\mathcal{X}$ is centered by subtracting the mean $\bar{\mathcal{X}}$ (obtained from the training samples) and then projected to the MPCA feature $\mathcal{Y}$:

$$\mathcal{Y} = (\mathcal{X} - \bar{\mathcal{X}}) \times_1 \tilde{\mathbf{U}}^{(1)^T} \times_2 \tilde{\mathbf{U}}^{(2)^T} ... \times_N \tilde{\mathbf{U}}^{(N)^T}. \tag{6.35}$$

In classification tasks, feature selection strategies (Section 5.5) can be applied to get vector-valued features for input to conventional classifiers.

6.2.4   Subspace Dimension Determination

Before solving for the MPCA projection, we need to determine the targeted subspace dimensions $\{P_n, n = 1, ..., N\}$. For TTP-based MSL, their determination could be a difficult problem because there could be $\prod_{n=1}^{N} I_n$ possible subspace dimensions. For example, there are 225,280 possible subspace dimensions for the standard gait recognition problem considered in [Lu et al., 2008b], where exhaustive testing becomes infeasible.

To investigate principled ways of subspace dimension determination, the objective function Equation (6.19) needs to be revised to include a constraint on the desired dimensionality reduction. The revised objective function is as follows:

$$\{\tilde{\mathbf{U}}^{(n)}, \tilde{P}_n\} = \arg \max_{\{\mathbf{U}^{(n)}, P_n\}} \Psi_{T_y} \quad \texttt{subject to} \quad \frac{\prod_{n=1}^{N} P_n}{\prod_{n=1}^{N} I_n} < \Delta, \tag{6.36}$$

where the ratio between the targeted (reduced) subspace dimensionality and the original tensor space dimensionality is utilized to measure the amount of dimensionality reduction, and Δ is a threshold parameter to be specified by the user or determined based on empirical studies. We look at two methods below.

6.2.4.1 Sequential Mode Truncation

The first subspace dimension determination solution is called sequential mode truncation (SMT) [Lu et al., 2008b]. Starting with $P_n = I_n$ for all n at step $\tau = 0$, at each subsequent step $\tau = \tau + 1$, the SMT method truncates, in a selected mode n, the P_nth mode-n eigenvector of the reconstructed input tensors. The truncation could be interpreted as the elimination of the corresponding P_nth mode-n slice (Definition 3.3) of the total scatter tensor (see Figure 5.2 for reference). For mode selection, the scatter loss rate $\vartheta_\tau^{(n)}$ due to truncation of the P_nth eigenvector is calculated for each mode. $\vartheta_\tau^{(n)}$ is defined as follows:

$$\vartheta_\tau^{(n)} = \frac{\Psi_{\mathcal{Y}_{(\tau)}} - \Psi_{\mathcal{Y}_{(\tau-1)}}}{P_n \cdot \prod_{j=1, j\neq n}^{N} P_j - (P_n - 1) \cdot \prod_{j=1, j\neq n}^{N} P_j} = \frac{\tilde{\lambda}_{P_n}^{(n)}}{\prod_{j=1, j\neq n}^{N} P_j}, \quad (6.37)$$

where $\Psi_{\mathcal{Y}_{(\tau)}}$ is the total scatter obtained at step τ; $\prod_{j=1, j\neq n}^{N} P_j$ is the amount of dimensionality reduction achieved; and $\tilde{\lambda}_{P_n}^{(n)}$, the corresponding P_nth mode-n eigenvalue, is the loss of variation due to truncating the P_nth mode-n eigenvector. The mode with the smallest $\vartheta_\tau^{(n)}$ is selected for step-τ truncation. For the selected mode n, P_n is decreased by 1: $P_n = P_n - 1$ and $\frac{\prod_{n=1}^{N} P_n}{\prod_{n=1}^{N} I_n} < \Delta$ is tested. The truncation stops when $\frac{\prod_{n=1}^{N} P_n}{\prod_{n=1}^{N} I_n} < \Delta$ is satisfied. Otherwise, the input tensors are reconstructed according to Equation (3.19) using the current truncated projection matrices and they are used to recompute the mode-n eigenvalues and eigenvectors corresponding to full projection. Because eigenvalues in the other modes are affected by the eigenvector truncation in a given mode (see Section 5.2.4), it is expected that the SMT, which takes into account this effect, constitutes a reasonably good choice for determining P_n in the sense of Equation (6.36).

6.2.4.2 Q-Based Method

The second method is the Q-based method [Lu et al., 2008b], a suboptimal, simplified dimension determination procedure that requires no iteration and is more practical.

Define the ratio

$$Q^{(n)} = \frac{\sum_{i_n=1}^{P_n} \check{\lambda}_{i_n}^{(n)}}{\sum_{i_n=1}^{I_n} \check{\lambda}_{i_n}^{(n)}} \quad (6.38)$$

to be the remaining portion of the mode-n total scatter after truncation of the mode-n eigenvectors beyond the P_nth, where $\check{\lambda}_{i_n}^{(n)}$ is the i_nth full-projection mode-n eigenvalue (Section 5.2.3). In the Q-based method, the first P_n eigenvectors are kept in mode n (for each n) so that

$$Q^{(1)} = Q^{(2)} = \ldots = Q^{(N)} = Q. \quad (6.39)$$

The equality can hold only approximately in practice because one is unlikely to find P_n that gives the exact equality. Thus, we find the smallest values of P_n satisfying $Q^{(n)} \geq Q$ instead. It should be noted that $\sum_{i_n=1}^{I_n} \check{\lambda}_{i_n}^{(n)} = \Psi_{T\mathcal{X}}$ for all n because from Theorem 6.2, the total scatter for the full projection was given as

$$\check{\Psi}_{T_y} = \Psi_{T\mathcal{X}} = \sum_{m=1}^{M} \| \mathbf{Y}_{m(n)} - \bar{\mathbf{Y}}_{(n)} \|_F^2 = \sum_{i_n=1}^{I_n} \check{\lambda}_{i_n}^{(n)}, n = 1, ..., N. \qquad (6.40)$$

This method is an extension of the traditional dimension selection strategy in PCA to the multilinear case. The reason behind this choice is that loss of variation is (approximately) proportional to the sum of the corresponding eigenvalues of the discarded eigenvectors (Section 5.2.4). By discarding the least significant eigenvectors in each mode, the variation loss can be contained and a tighter lower bound for the captured variation is obtained (see Equation (5.9)). The empirical study reported in [Lu et al., 2008b] indicates that the Q-based method provides results similar to those obtained by SMT (as measured in the total scatter captured). Thus, it is preferred over the more computationally expensive SMT method in practice.

6.3   Tensor Rank-One Decomposition

The TROD algorithm was introduced in [Shashua and Levin, 2001]. Although it was originally formulated only for image matrices, it can be easily extended to higher-order tensors. It is an unsupervised TVP-based MSL algorithm. TROD solves for a TVP, that is, P elementary multilinear projections (EMPs) to minimize the least-square reconstruction error.

6.3.1   TROD Problem Formulation

Different from TTP-based MSL approach, TVP-based MSL approach extracts features one at a time. In the TVP setting, given M training tensor samples $\{\mathcal{X}_m\}$, the pth EMP, denoted as $\{\mathbf{u}_p^{(n)}, n = 1, ..., N\}$ or simply as $\{\mathbf{u}_p^{(n)}\}$, projects the mth sample $\mathcal{X}_m$ to a scalar feature y_{m_p} as

$$y_{m_p} = \mathcal{X}_m \times_1 \mathbf{u}_p^{(1)^T} \times_2 \mathbf{u}_p^{(2)^T} ... \times_N \mathbf{u}_p^{(N)^T} = \mathcal{X}_m \times_{n=1}^{N} \{\mathbf{u}_p^{(n)}\}. \qquad (6.41)$$

The TVP, consisting of P such EMPs $\{\mathbf{u}_p^{(n)}\}_N^P$, projects $\mathcal{X}_m$ to a vector $\mathbf{y}_m \in \mathbb{R}^{P \times 1}$ as

$$\mathbf{y}_m = \mathcal{X}_m \times_{n=1}^{N} \{\mathbf{u}_p^{(n)}\}_N^P. \qquad (6.42)$$

Note that $\mathbf{y}_m(p) = y_{m_p}$. TROD aims to minimize the least-square reconstruction error ϵ_{TROD} between the original tensors and the reconstruction (or approximation) from a linear combination of a set of rank-one tensors as

$$\epsilon_{TROD} = \sum_{m=1}^{M} \left\| \mathcal{X}_m - \sum_{p=1}^{P} (y_{m_p} \cdot \mathbf{u}_p^{(1)} \circ \mathbf{u}_p^{(2)} \circ ... \circ \mathbf{u}_p^{(N)}) \right\|_F^2, \qquad (6.43)$$

where each rank-one tensor is the outer product of the projection vectors from an EMP.

Second-order case $N = 2$: The corresponding projections and reconstruction error for matrix input data $\{\mathbf{X}_m\}$ are

$$y_{m_p} = \mathbf{u}_p^{(1)^T} \mathbf{X}_m \mathbf{u}_p^{(2)} \qquad (6.44)$$

and

$$\epsilon_{TROD} = \sum_{m=1}^{M} \left\| \mathbf{X}_m - \sum_{p=1}^{P} (y_{m_p} \cdot \mathbf{u}_p^{(1)} \mathbf{u}_p^{(2)^T}) \right\|_F^2, \qquad (6.45)$$

respectively.

The TROD Problem: A set of M tensor data samples $\{\mathcal{X}_1, \mathcal{X}_2, ..., \mathcal{X}_M\}$ are available for training. Each tensor sample $\mathcal{X}_m \in \mathbb{R}^{I_1 \times I_2 ... \times I_N}$ assumes values in the tensor space $\mathbb{R}^{I_1} \otimes \mathbb{R}^{I_2} ... \otimes \mathbb{R}^{I_N}$. The objective of TROD is to find a TVP, which consists of P EMPs, $\{\mathbf{u}_p^{(n)} \in \mathbb{R}^{I_n \times 1}, n = 1, ..., N\}_{p=1}^{P}$ that maps the original tensor space $\mathbb{R}^{I_1} \otimes \mathbb{R}^{I_2} ... \otimes \mathbb{R}^{I_N}$ into a vector subspace $\mathbb{R}^P$ (with $P < \prod_{n=1}^{N} I_n$), that is, projecting $\{\mathcal{X}_m, m = 1, ..., M\}$ to $\{\mathbf{y}_m, m = 1, ..., M\}$ as in Equation (6.42), such that the reconstruction error in the projected subspace, measured by ϵ_{TROD} in Equation (6.43), is minimized.

6.3.2   Greedy Approach for TROD

Shashua and Levin [2001] proposed a closed-form solution for a special case of $P = M$ (when reduced dimension equals the number of training samples) and a greedy solution for the general case. Because $P = M$ is seldom the case in practice, here we discuss the latter solution. This greedy approach employs a heuristic procedure of successive *residue calculation*, which is similar to the deflation procedure in the partial least squares (PLS) analysis algorithms (Section 2.5) or the power method (Section A.1.11). Algorithm 6.3 is the pseudocode for this algorithm.

There are P steps. In each step p, we solve for the pth EMP $\{\mathbf{u}_p^{(n)}\}$ such that the reconstruction error ϵ_{TROD} is minimized:

$$\{\tilde{\mathbf{u}}_p^{(n)}\} = \arg \min_{\{\mathbf{u}_p^{(n)}\}} \sum_{m=1}^{M} \left\| \mathcal{X}_m - (y_{m_p} \cdot \mathbf{u}_p^{(1)} \circ \mathbf{u}_p^{(2)} \circ ... \circ \mathbf{u}_p^{(N)}) \right\|_F^2. \qquad (6.46)$$

First, following a similar derivation as in Equation (6.17), the minimization problem above is converted to the following maximization problem [Shashua and Levin, 2001]:

$$\{\tilde{\mathbf{u}}_p^{(n)}\} = \arg \max_{\{\mathbf{u}_p^{(n)}\}} \sum_{m=1}^{M} (y_{m_p})^2. \tag{6.47}$$

After obtaining the pth EMP $\{\tilde{\mathbf{u}}_p^{(n)}\}$, each input sample $\mathcal{X}_m$ is replaced by its residue as

$$\mathcal{X}_m \leftarrow \mathcal{X}_m - y_{m_p} \cdot (\tilde{\mathbf{u}}_p^{(1)} \circ \tilde{\mathbf{u}}_p^{(2)} \circ \dots \circ \tilde{\mathbf{u}}_p^{(N)}), \tag{6.48}$$

and the next EMP is obtained by solving Equation (6.47) again, where $\mathcal{X}_m$ has been replaced by its residue as in Equation (6.48).

The second order version ($N = 2$) of residue calculation is

$$\mathbf{X}_m \leftarrow \mathbf{X}_m - y_{m_p} \cdot \tilde{\mathbf{u}}_p^{(1)} \tilde{\mathbf{u}}_p^{(2)^T}. \tag{6.49}$$

6.3.3 Solving for the pth EMP

In order to solve for the pth EMP $\{\mathbf{u}_p^{(n)}\}$, there are N sets of parameters corresponding to the N projection vectors to be determined, $\mathbf{u}_p^{(1)}, \mathbf{u}_p^{(2)}, \dots \mathbf{u}_p^{(N)}$, one in each mode. We employ the APP method again to solve for them one at a time conditioned on all the other projection vectors. The parameters for each mode are estimated in this way sequentially and iteratively until a stopping criterion is met. The iteration for determining each EMP corresponds to the loop indexed by k in Algorithm 6.3, and in each iteration k, the loop indexed by n in Algorithm 6.3 consists of the N conditional subproblems.

To solve for the mode-n projection vector $\mathbf{u}_p^{(n)}$, the mode-n partial multilinear projection of the tensor samples are first obtained using $\{\mathbf{u}_p^{(j)}, j \neq n\}$ as

$$\acute{\mathbf{y}}_{m_p}^{(n)} = \mathcal{X}_m \times_1 \mathbf{u}_p^{(1)^T} \dots \times_{n-1} \mathbf{u}_p^{(n-1)^T} \times_{n+1} \mathbf{u}_p^{(n+1)^T} \dots \times_N \mathbf{u}_p^{(N)^T}, \tag{6.50}$$

where $\acute{\mathbf{y}}_{m_p}^{(n)} \in \mathbb{R}^{I_n}$. The conditional subproblem is then constructed as to determine $\mathbf{u}_p^{(n)}$ that maximizes the quantity $\Psi_{\acute{\mathbf{y}}}$ defined as

$$\Psi_{\acute{\mathbf{y}}} = \sum_{m=1}^{M} (\mathbf{u}_p^{(n)^T} \acute{\mathbf{y}}_{m_p}^{(n)})^2. \tag{6.51}$$

The solution $\tilde{\mathbf{u}}_p^{(n)}$ is obtained as the eigenvector associated with the largest

---

Algorithm 6.3 Tensor rank-one decomposition (TROD)

Input: A set of tensor samples $\{\mathcal{X}_m \in \mathbb{R}^{I_1 \times I_2 \times \dots \times I_N}, m = 1, \dots, M\}$, the desired feature vector length P, the maximum number of iterations K (or a small number η for testing convergence).

Process:

1: **for** $p = 1$ **to** P **do**

2:     [Step p: determine the pth EMP]

3:     Initialize $\mathbf{u}_{p_{(0)}}^{(n)} \in \mathbb{R}^{I_n}$ for $n = 1, \dots, N$.

4:     [**Local optimization:**]

5:     **for** $k = 1$ **to** K **do**

6:       **for** $n = 1$ **to** N **do**

7:         Calculate the partial multilinear projection $\acute{\mathbf{y}}_{m_p}^{(n)}$ according to Equation (6.50), for $m = 1, \dots, M$.

8:         Calculate $\acute{\mathbf{S}}_{T_p}^{(n)}$ according to Equation (6.52).

9:         Set $\mathbf{u}_{p_{(k)}}^{(n)}$ to be the eigenvector of $\acute{\mathbf{S}}_{T_p}^{(n)}$ associated with the largest eigenvalue.

10:      **end for**

11:     **end for**

12:     Set $\tilde{\mathbf{u}}_p^{(n)} = \mathbf{u}_{p_{(K)}}^{(n)}$.

13:     Calculate the projection y_{m_p} according to Equation (6.41) for $m = 1, \dots, M$.

14:     Replace $\mathcal{X}_m$ with its residue according to Equation (6.48).

15: **end for**

Output: The TVP $\{\tilde{\mathbf{u}}_p^{(n)}\}_N^P$ that minimizes the reconstruction error in the projected space.

---

eigenvalue of the following matrix [Shashua and Levin, 2001]:

$$\acute{\mathbf{S}}_{T_p}^{(n)} = \sum_{m=1}^{M} \acute{\mathbf{y}}_{m_p}^{(n)} \acute{\mathbf{y}}_{m_p}^{(n)^T}. \tag{6.52}$$

Feature extraction: Despite working directly on tensor data, TROD feature extraction produces feature vectors (rather than feature tensors) like conventional linear algorithms, due to the nature of TVP. For a test tensor sample $\mathcal{X}$, the feature vector $\mathbf{y}$ is obtained through the TVP obtained by TROD as

$$\mathbf{y} = \mathcal{X} \times_{n=1}^{N} \{\tilde{\mathbf{u}}_p^{(n)}\}_N^P. \tag{6.53}$$

6.4   Uncorrelated Multilinear PCA

UMPCA is a TVP-based unsupervised MSL algorithm developed in [Lu et al., 2008c, 2009d]. It shares an important property with classical PCA, that is, extracting uncorrelated features, that contain minimum redundancy and ensure linear independence among features under the Gaussian assumption. The derivation of UMPCA follows the classical PCA derivation of successive variance maximization [Jolliffe, 2002]. It solves for a number of EMPs one by one to maximize the captured variance, so it shares similarity with TROD. The difference is that it enforces the zero-correlation constraint while TROD depends on a greedy residue calculation procedure. Nevertheless, UMPCA has a limitation in the number of uncorrelated features that can be extracted.

6.4.1   UMPCA Problem Formulation

Following the standard derivation of PCA in [Jolliffe, 2002] as briefly reviewed in Section 2.1, the variance of principal components is considered one by one. The pth principal components are $\{y_{m_p}, m = 1, ..., M\}$, the projections of $\{\mathcal{X}_m\}$ by the pth EMP $\{\mathbf{u}_p^{(n)}\}$ as in Equation (6.41).

Accordingly, from Definition 3.16, the variance is measured by their total scatter $S_{T_{y_p}}$:

$$S_{T_{y_p}} = \sum_{m=1}^{M} (y_{m_p} - \bar{y}_p)^2, \tag{6.54}$$

where

$$\bar{y}_p = \frac{1}{M} \sum_{m=1}^{M} y_{m_p}. \tag{6.55}$$

Next, let $\mathbf{g}_p \in \mathbb{R}^M$ denote the pth *coordinate vector*, with its mth component $\mathbf{g}_p(m) = g_{p_m} = y_{m_p}$ (note that the mth projected sample $\mathbf{y}_m \in \mathbb{R}^P$). We view $\mathbf{g}_p$ as a vector of M observations of the variable g_p. We denote the mean of $\mathbf{g}_p$ as

$$\bar{g}_p = \frac{1}{M} \sum_{m=1}^{M} g_{p_m} = \frac{1}{M} \sum_{m=1}^{M} y_{m_p} = \bar{y}_p. \tag{6.56}$$

The *sample Pearson correlation* between two coordinate vectors $\mathbf{g}_p$ and $\mathbf{g}_q$ is defined as

$$\rho_{pq} = \frac{\sum_{m=1}^{M} [(g_{p_m} - \bar{g}_p)(g_{q_m} - \bar{g}_q)]}{\sqrt{\sum_{m=1}^{M} (g_{p_m} - \bar{g}_p)^2} \sqrt{\sum_{m=1}^{M} (g_{q_m} - \bar{g}_q)^2}} = \frac{(\mathbf{g}_p - \bar{g}_p)^T (\mathbf{g}_q - \bar{g}_q)}{\| \mathbf{g}_p - \bar{g}_p \| \| \mathbf{g}_q - \bar{g}_q \|}. \tag{6.57}$$

A formal definition of the UMPCA problem is then given in the following.

The UMPCA Problem: A set of M tensor data samples $\{\mathcal{X}_1, \mathcal{X}_2, ...,$

$\mathcal{X}_M\}$ are available for training. Each sample $\mathcal{X}_m \in \mathbb{R}^{I_1 \times I_2 \times \ldots \times I_N}$ assumes values in the tensor space $\mathbb{R}^{I_1} \bigotimes \mathbb{R}^{I_2} \ldots \bigotimes \mathbb{R}^{I_N}$. The objective of UMPCA is to find a TVP, which consists of P EMPs, $\{\mathbf{u}_p^{(n)} \in \mathbb{R}^{I_n}, n = 1, \ldots, N\}_{p=1}^P$, that maps the original tensor space $\mathbb{R}^{I_1} \bigotimes \mathbb{R}^{I_2} \ldots \bigotimes \mathbb{R}^{I_N}$ into a vector subspace $\mathbb{R}^P$ (with $P < \prod_{n=1}^N I_n$), that is, projecting $\{\mathcal{X}_m, m = 1, \ldots, M\}$ to $\{\mathbf{y}_m, m = 1, \ldots, M\}$ as in Equation (6.42) such that the variance of the projected samples, measured by $S_{T_{yp}}$ in Equation (6.54), is maximized in each EMP direction, subject to the constraint that the P coordinate vectors $\{\mathbf{g}_p \in \mathbb{R}^M, p = 1, \ldots, P\}$ are all uncorrelated.

Centering: As in MPCA, we would like to center the input data first to obtain $\{\tilde{\mathcal{X}}_m\}$ as in Equation (6.21). The corresponding projections become $\{\tilde{\mathbf{y}}_m\}$, and its element becomes $\{\tilde{y}_{m_p}\}$ with corresponding mean $\bar{\tilde{y}}_p = 0$. Thus, the scatter in Equation (6.54) becomes

$$S_{T_{yp}} = \sum_{m=1}^M \tilde{y}_{m_p}^2. \tag{6.58}$$

The corresponding coordinate vectors become $\{\tilde{\mathbf{g}}_p\}$ and each vector $\tilde{\mathbf{g}}_p$ has zero-mean, so the constraint of uncorrelated features becomes equivalent to that of orthogonal features with the correlation defined in Equation (6.57) becoming

$$\rho_{pq} = \frac{\tilde{\mathbf{g}}_p^T \tilde{\mathbf{g}}_q}{\| \tilde{\mathbf{g}}_p \| \| \tilde{\mathbf{g}}_q \|}. \tag{6.59}$$

Now we can write the UMPCA objective function for the pth EMP as

$$\{\tilde{\mathbf{u}}_p^{(n)}\} = \arg \max_{\{\mathbf{u}_p^{(n)}\}} \sum_{m=1}^M \tilde{y}_{m_p}^2, \tag{6.60}$$

$$\text{subject to } \mathbf{u}_p^{(n)^T} \mathbf{u}_p^{(n)} = 1 \quad \text{and} \quad \frac{\tilde{\mathbf{g}}_p^T \tilde{\mathbf{g}}_q}{\| \tilde{\mathbf{g}}_p \| \| \tilde{\mathbf{g}}_q \|} = \delta_{pq}, \ p, q = 1, \ldots, P,$$

where the projection vectors are constrained to unit length and δ_{pq} is the *Kronecker delta* defined as

$$\delta_{pq} = \begin{cases} 1 & \text{if } p = q \\ 0 & \text{Otherwise.} \end{cases} \tag{6.61}$$

Feature extraction: Similar to TROD, as a TVP-based algorithm, UMPCA produces feature vectors like traditional linear algorithms. The UMPCA feature vector $\mathbf{y}$ for a test sample $\mathcal{X}$ is obtained through the TVP learned by UMPCA as in Equation (6.53).

6.4.2 UMPCA Algorithm Derivation

The solution to the UMPCA problem in Equation (6.60) follows the successive variance maximization approach in [Jolliffe, 2002], as reviewed in Section 2.1.

Similar to TROD, the P EMPs $\{\mathbf{u}_p^{(n)}\}_N^P$ are determined sequentially (one by one) in P steps, with the pth step obtaining the pth EMP. This stepwise process proceeds as follows:

Step 1: Determine the first EMP $\{\mathbf{u}_1^{(n)}\}$ by maximizing $S_{T_{y_1}}$ without any constraint.

Step 2: Determine the second EMP $\{\mathbf{u}_2^{(n)}\}$ by maximizing $S_{T_{y_2}}$ subject to the constraint that $\tilde{\mathbf{g}}_2^T \tilde{\mathbf{g}}_1 = 0$.

Step 3: Determine the third EMP $\{\mathbf{u}_3^{(n)}\}$ by maximizing $S_{T_{y_3}}$ subject to the constraint that $\tilde{\mathbf{g}}_3^T \tilde{\mathbf{g}}_1 = 0$ and $\tilde{\mathbf{g}}_3^T \tilde{\mathbf{g}}_2 = 0$.

Step p ($p = 4, ..., P$): Determine the pth EMP $\{\mathbf{u}_p^{(n)}\}$ by maximizing $S_{T_{y_p}}$ subject to the constraint that $\tilde{\mathbf{g}}_p^T \tilde{\mathbf{g}}_q = 0$ for $q = 1, ..., p-1$.

The following presents the algorithm to compute these EMPs in detail, as summarized in the pseudocode in Algorithm 6.4, where the stepwise process described above corresponds to the loop indexed by p.

Solving for the pth EMP $\{\mathbf{u}_p^{(n)}\}$ requires the determination of N projection vectors to maximize $S_{T_{y_p}}$ subject to the zero-correlation constraint. Similar to the TROD solution, we follow the APP method to determine the mode-n projection vector $\mathbf{u}_p^{(n)}$ conditioned on the projection vectors in the other modes $\{\mathbf{u}_p^{(j)}, j \neq n\}$.

To solve for $\mathbf{u}_p^{(n)}$ given $\{\mathbf{u}_p^{(j)}, j \neq n\}$, we do the mode-$n$ partial multilinear projection to obtain the vectors

$$\hat{\mathbf{y}}_{m_p}^{(n)} = \tilde{\mathcal{X}}_m \times_1 \mathbf{u}_p^{(1)^T} ... \times_{n-1} \mathbf{u}_p^{(n-1)^T} \times_{n+1} \mathbf{u}_p^{(n+1)^T} ... \times_N \mathbf{u}_p^{(N)^T}, \quad (6.62)$$

where $\hat{\mathbf{y}}_{m_p}^{(n)} \in \mathbb{R}^{I_n}$. The conditional subproblem then becomes to determine $\mathbf{u}_p^{(n)}$ that projects $\{\hat{\mathbf{y}}_{m_p}^{(n)}\}$ onto a line so that the variance (scatter) is maximized, subject to the zero-correlation constraint, which is a PCA problem with input samples $\{\hat{\mathbf{y}}_{m_p}^{(n)}\}$. The corresponding total scatter matrix $\hat{\mathbf{S}}_{T_p}^{(n)}$ is then defined as

$$\hat{\mathbf{S}}_{T_p}^{(n)} = \sum_{m=1}^{M} (\hat{\mathbf{y}}_{m_p}^{(n)} - \bar{\hat{\mathbf{y}}}_p^{(n)})(\hat{\mathbf{y}}_{m_p}^{(n)} - \bar{\hat{\mathbf{y}}}_p^{(n)})^T, \quad (6.63)$$

where

$$\bar{\hat{\mathbf{y}}}_p^{(n)} = \frac{1}{M} \sum_{m=1}^{M} \hat{\mathbf{y}}_{m_p}^{(n)} = \mathbf{0} \quad (6.64)$$

because $\{\hat{\mathbf{y}}_{m_p}^{(n)}\}$ are projections of centered (zero-mean) input samples. Thus, we have

$$\hat{\mathbf{S}}_{T_p}^{(n)} = \sum_{m=1}^{M} \hat{\mathbf{y}}_{m_p}^{(n)} \hat{\mathbf{y}}_{m_p}^{(n)^T}. \quad (6.65)$$

Algorithm 6.4 Uncorrelated multilinear PCA (UMPCA)

Input: A set of tensor samples $\{\mathcal{X}_m \in \mathbb{R}^{I_1 \times I_2 \times \dots \times I_N}, m = 1, \dots, M\}$, the desired feature vector length P, the maximum number of iterations K (or a small number η for testing convergence).

Process:

1: **Centering:** Center the input samples to get $\{\tilde{\mathcal{X}}_m\}$ according to Equation (6.21).

2: **for** $p = 1$ **to** P **do**

3:     [Step p: determine the pth EMP]

4:     Initialize $\mathbf{u}_{p_{(0)}}^{(n)} \in \mathbb{R}^{I_n}$ for $n = 1, \dots, N$.

5:     [**Local optimization:**]

6:     **for** $k = 1$ **to** K **do**

7:         **for** $n = 1$ **to** N **do**

8:             Calculate the mode-n partial multilinear projections $\{\hat{\mathbf{y}}_{m_p}^{(n)}, m = 1, \dots, M\}$ according to Equation (6.62).

9:             Calculate $\mathbf{\Upsilon}_p^{(n)}$ and $\hat{\mathbf{S}}_{T_p}^{(n)}$ according to Equations (6.71) and (6.65), respectively.

10:            Set $\mathbf{u}_{p_{(k)}}^{(n)}$ to be the eigenvector of $\mathbf{\Upsilon}_p^{(n)} \hat{\mathbf{S}}_{T_p}^{(n)}$ associated with the largest eigenvalue.

11:        **end for**

12:    **end for**

13:    Set $\tilde{\mathbf{u}}_p^{(n)} = \mathbf{u}_{p_{(K)}}^{(n)}$.

14:    Calculate the coordinate vector $\tilde{\mathbf{g}}_p$, with $\tilde{\mathbf{g}}_p(m) = \tilde{\mathcal{X}}_m \times_1 \tilde{\mathbf{u}}_p^{(1)^T} \dots \times_N \tilde{\mathbf{u}}_p^{(N)^T}$.

15: **end for**

Output: The TVP $\{\tilde{\mathbf{u}}_p^{(n)}\}_N^P$ that captures the most variance while producing uncorrelated features in the projected space.

With Equation (6.65), the P EMPs can be solved sequentially in the following way. For $p = 1$, the optimal $\tilde{\mathbf{u}}_1^{(n)}$ that maximizes the total scatter $\mathbf{u}_1^{(n)^T} \hat{\mathbf{S}}_{T_1}^{(n)} \mathbf{u}_1^{(n)}$ in the projected space is obtained as the eigenvector of $\hat{\mathbf{S}}_{T_1}^{(n)}$ associated with the largest eigenvalue. Next, the pth $(p > 1)$ EMP is determined. Given the first $(p - 1)$ EMPs, the pth EMP aims to maximize the total scatter $S_{T_{yp}}$, subject to the constraint that features projected by the pth EMP are uncorrelated with those projected by the first $(p - 1)$ EMPs. Let $\hat{\mathbf{Y}}_p^{(n)} \in \mathbb{R}^{I_n \times M}$ be a matrix with $\hat{\mathbf{y}}_{m_p}^{(n)}$ as its mth column, that is,

$$\hat{\mathbf{Y}}_p^{(n)} = \left[\hat{\mathbf{y}}_{1_p}^{(n)}, \hat{\mathbf{y}}_{2_p}^{(n)}, \dots, \hat{\mathbf{y}}_{M_p}^{(n)} \right], \tag{6.66}$$

then the pth coordinate vector is

$$\tilde{\mathbf{g}}_p = \hat{\mathbf{Y}}_p^{(n)^T} \mathbf{u}_p^{(n)}. \tag{6.67}$$

The constraint that $\tilde{\mathbf{g}}_p$ is uncorrelated (orthogonal after centering) with $\{\tilde{\mathbf{g}}_q, q = 1, ..., p-1\}$ can be written as

$$\tilde{\mathbf{g}}_p^T \tilde{\mathbf{g}}_q = \mathbf{u}_p^{(n)^T} \hat{\mathbf{Y}}_p^{(n)} \tilde{\mathbf{g}}_q = 0, q = 1, ..., p-1. \tag{6.68}$$

Thus, $\tilde{\mathbf{u}}_p^{(n)}$ $(p > 1)$ can be determined by solving the following constrained optimization problem:

$$\tilde{\mathbf{u}}_p^{(n)} \quad = \quad \arg\max_{\mathbf{u}_p^{(n)}} \mathbf{u}_p^{(n)^T} \hat{\mathbf{S}}_{T_p}^{(n)} \mathbf{u}_p^{(n)}, \tag{6.69}$$

$$\text{subject to } \mathbf{u}_p^{(n)^T} \mathbf{u}_p^{(n)} \quad = \quad 1 \text{ and } \mathbf{u}_p^{(n)^T} \hat{\mathbf{Y}}_p^{(n)} \tilde{\mathbf{g}}_q = 0, q = 1, ..., p-1.$$

The solution is given by the following theorem:

Theorem 6.3. *The solution to the problem in Equation (6.69) is the eigenvector corresponding to the largest eigenvalue of the following eigenvalue problem:*

$$\mathbf{\Upsilon}_p^{(n)} \hat{\mathbf{S}}_{T_p}^{(n)} \mathbf{u} = \lambda \mathbf{u}, \tag{6.70}$$

where

$$\mathbf{\Upsilon}_p^{(n)} = \mathbf{I}_{I_n} - \hat{\mathbf{Y}}_p^{(n)} \mathbf{G}_{p-1} \mathbf{\Gamma}_p^{-1} \mathbf{G}_{p-1}^T \hat{\mathbf{Y}}_p^{(n)^T}, \tag{6.71}$$

$$\mathbf{\Gamma}_p = \mathbf{G}_{p-1}^T \hat{\mathbf{Y}}_p^{(n)^T} \hat{\mathbf{Y}}_p^{(n)} \mathbf{G}_{p-1}, \tag{6.72}$$

$$\mathbf{G}_{p-1} = [\tilde{\mathbf{g}}_1 \quad \tilde{\mathbf{g}}_2 \quad \cdots \quad \tilde{\mathbf{g}}_{p-1}] \in \mathbb{R}^{M \times (p-1)}, \tag{6.73}$$

and $\mathbf{I}_{I_n}$ is an identity matrix of size $I_n \times I_n$.

Proof. First, Lagrange multipliers can be used to transform the problem Equation (6.69) to the following to include all the constraints:

$$\varphi_p = \mathbf{u}_p^{(n)^T} \hat{\mathbf{S}}_{T_p}^{(n)} \mathbf{u}_p^{(n)} - \nu\left(\mathbf{u}_p^{(n)^T} \mathbf{u}_p^{(n)} - 1\right) - \sum_{q=1}^{p-1} \mu_q \mathbf{u}_p^{(n)^T} \hat{\mathbf{Y}}_p^{(n)} \tilde{\mathbf{g}}_q, \tag{6.74}$$

where ν and $\{\mu_q, q = 1, ..., p-1\}$ are Lagrange multipliers.

The optimization is performed by setting the partial derivative of φ_p with respect to $\mathbf{u}_p^{(n)}$ to zero:

$$\frac{\partial \varphi_p}{\partial \mathbf{u}_p^{(n)}} = 2\hat{\mathbf{S}}_{T_p}^{(n)} \mathbf{u}_p^{(n)} - 2\nu \mathbf{u}_p^{(n)} - \sum_{q=1}^{p-1} \mu_q \hat{\mathbf{Y}}_p^{(n)} \tilde{\mathbf{g}}_q = 0. \tag{6.75}$$

Multiplying Equation (6.75) by $\mathbf{u}_p^{(n)^T}$ results in

$$2\mathbf{u}_p^{(n)^T} \hat{\mathbf{S}}_{T_p}^{(n)} \mathbf{u}_p^{(n)} - 2\nu \mathbf{u}_p^{(n)^T} \mathbf{u}_p^{(n)} = 0 \Rightarrow \nu = \frac{\mathbf{u}_p^{(n)^T} \hat{\mathbf{S}}_{T_p}^{(n)} \mathbf{u}_p^{(n)}}{\mathbf{u}_p^{(n)^T} \mathbf{u}_p^{(n)}}, \tag{6.76}$$

which indicates that ν is exactly the criterion to be maximized, with the constraint on the norm of the projection vector incorporated.

Next, a set of $(p-1)$ equations are obtained by multiplying Equation (6.75) by $\tilde{\mathbf{g}}_q^T \hat{\mathbf{Y}}_p^{(n)^T}$, $q = 1, ..., p - 1$, respectively:

$$2\tilde{\mathbf{g}}_q^T \hat{\mathbf{Y}}_p^{(n)^T} \hat{\mathbf{S}}_{T_p}^{(n)} \mathbf{u}_p^{(n)} - \sum_{s=1}^{p-1} \mu_s \tilde{\mathbf{g}}_q^T \hat{\mathbf{Y}}_p^{(n)^T} \hat{\mathbf{Y}}_p^{(n)} \tilde{\mathbf{g}}_s = 0. \tag{6.77}$$

Let

$$\boldsymbol{\mu}_{p-1} = [\mu_1 \ \mu_2 \ ... \ \mu_{p-1}]^T \tag{6.78}$$

and use Equations (6.72) and (6.73), then the $(p - 1)$ equations of Equation (6.77) can be represented in a single matrix equation as follows:

$$2\mathbf{G}_{p-1}^T \hat{\mathbf{Y}}_p^{(n)^T} \hat{\mathbf{S}}_{T_p}^{(n)} \mathbf{u}_p^{(n)} - \boldsymbol{\Gamma}_p \boldsymbol{\mu}_{p-1} = 0. \tag{6.79}$$

Thus,

$$\boldsymbol{\mu}_{p-1} = 2\boldsymbol{\Gamma}_p^{-1} \mathbf{G}_{p-1}^T \hat{\mathbf{Y}}_p^{(n)^T} \hat{\mathbf{S}}_{T_p}^{(n)} \mathbf{u}_p^{(n)}. \tag{6.80}$$

Because from Equations (6.73) and (6.78),

$$\sum_{q=1}^{p-1} \mu_q \hat{\mathbf{Y}}_p^{(n)} \tilde{\mathbf{g}}_q = \hat{\mathbf{Y}}_p^{(n)} \mathbf{G}_{p-1} \boldsymbol{\mu}_{p-1}, \tag{6.81}$$

then Equation (6.75) can be written as

$$\begin{aligned}
2\hat{\mathbf{S}}_{T_p}^{(n)} \mathbf{u}_p^{(n)} &- 2\nu \mathbf{u}_p^{(n)} - \hat{\mathbf{Y}}_p^{(n)} \mathbf{G}_{p-1} \boldsymbol{\mu}_{p-1} = 0 \\
\Rightarrow \quad \nu \mathbf{u}_p^{(n)} &= \hat{\mathbf{S}}_{T_p}^{(n)} \mathbf{u}_p^{(n)} - \hat{\mathbf{Y}}_p^{(n)} \mathbf{G}_{p-1} \frac{\boldsymbol{\mu}_{p-1}}{2} \\
&= \hat{\mathbf{S}}_{T_p}^{(n)} \mathbf{u}_p^{(n)} - \hat{\mathbf{Y}}_p^{(n)} \mathbf{G}_{p-1} \boldsymbol{\Gamma}_p^{-1} \mathbf{G}_{p-1}^T \hat{\mathbf{Y}}_p^{(n)^T} \hat{\mathbf{S}}_{T_p}^{(n)} \mathbf{u}_p^{(n)} \\
&= \left[\mathbf{I}_{I_n} - \hat{\mathbf{Y}}_p^{(n)} \mathbf{G}_{p-1} \boldsymbol{\Gamma}_p^{-1} \mathbf{G}_{p-1}^T \hat{\mathbf{Y}}_p^{(n)^T} \right] \hat{\mathbf{S}}_{T_p}^{(n)} \mathbf{u}_p^{(n)}.
\end{aligned}$$

Using the definition in Equation (6.71), an eigenvalue problem is obtained as

$$\boldsymbol{\Upsilon}_p^{(n)} \hat{\mathbf{S}}_{T_p}^{(n)} \mathbf{u} = \nu \mathbf{u}. \tag{6.82}$$

Because ν is the criterion to be maximized, the maximization is achieved by setting $\tilde{\mathbf{u}}_p^{(n)}$ to be the eigenvector corresponding to the largest eigenvalue of Equation (6.70). $\quad\square$

By setting $\boldsymbol{\Upsilon}_1^{(n)} = \mathbf{I}_{I_n}$ and from Theorem 6.3, a unified solution for UMPCA is obtained: for $p = 1, ..., P$, $\tilde{\mathbf{u}}_p^{(n)}$ is obtained as the eigenvector of $\boldsymbol{\Upsilon}_p^{(n)} \hat{\mathbf{S}}_{T_p}^{(n)}$ associated with the largest eigenvalue.

6.4.3 Discussions on UMPCA

Convergence: The derivation in Theorem 6.3 implies that, per iteration, the scatter $S_{T_{yp}}$ is a non-decreasing function because each update of $\mathbf{u}_p^{(n)}$ maximizes $S_{T_{yp}}$. On the other hand, $S_{T_{yp}}$ is upper-bounded by the scatter in the original samples. Therefore, UMPCA is expected to convergence over iterations. Empirical studies in [Lu et al., 2009d] indicate that the UMPCA algorithm converges within 10 iterations for typical tensor objects in practice. Furthermore, when the largest eigenvalues in each mode are with multiplicity 1, the projection vectors $\{\mathbf{u}_p^{(n)}\}$, which maximize the objective function $S_{T_{yp}}$, are expected to converge as well, where the convergence is up to sign. Experimental studies in [Lu et al., 2009d] show that $\{\mathbf{u}_p^{(n)}\}$ do converge over a number of iterations.

Nonetheless, UMPCA has a limitation as to the number of uncorrelated features that can be extracted, given by the following corollary.

Corollary 6.1. *The number of uncorrelated features that can be extracted by UMPCA, P, is upper-bounded by $\min\{\min_n I_n, M\}$, that is,*

$$P \leq \min\{\min_n I_n, M\}, \tag{6.83}$$

provided that the elements of $\hat{\mathbf{Y}}_p^{(n)}$ are not all zero.

Proof. To prove the corollary, it is only needed to show that for any mode n, the number of bases that can satisfy the zero-correlation constraint is upper-bounded by $\min\{I_n, M\}$.

Considering only one mode n, the zero-correlation constraint for mode n in Equation (6.69) is

$$\mathbf{u}_p^{(n)^T} \hat{\mathbf{Y}}_p^{(n)} \tilde{\mathbf{g}}_q = 0, q = 1, ..., p-1. \tag{6.84}$$

First, let $\hat{\tilde{\mathbf{g}}}_p^{(n)^T} = \mathbf{u}_p^{(n)^T} \hat{\mathbf{Y}}_p^{(n)} \in \mathbb{R}^{1 \times M}$ and the constraint becomes

$$\hat{\tilde{\mathbf{g}}}_p^{(n)^T} \tilde{\mathbf{g}}_q = 0, q = 1, ..., p-1. \tag{6.85}$$

Because $\tilde{\mathbf{g}}_q \in \mathbb{R}^{M \times 1}$, when $p = M+1$, the set $\{\tilde{\mathbf{g}}_q, q = 1, ..., M\}$ forms a basis for the M-dimensional space and there is no solution for Equation (6.85). Thus, $P \leq M$.

Second, let $\hat{\mathbf{u}}_q^{(n)} = \hat{\mathbf{Y}}_p^{(n)} \tilde{\mathbf{g}}_q \in \mathbb{R}^{I_n \times 1}$ and the constraint becomes

$$\mathbf{u}_p^{(n)^T} \hat{\mathbf{u}}_q^{(n)} = 0, q = 1, ..., p-1. \tag{6.86}$$

Because $\tilde{\mathbf{g}}_q, q = 1, ..., p-1$ are orthogonal, $\hat{\mathbf{u}}_q^{(n)}, q = 1, ..., p-1$ are linearly independent if the elements of $\hat{\mathbf{Y}}_p^{(n)}$ are not all zero. Because $\hat{\mathbf{u}}_q^{(n)} \in \mathbb{R}^{I_n \times 1}$, when $p = I_n + 1$, the set $\{\hat{\mathbf{u}}_q^{(n)}, q = 1, ..., p-1\}$ forms a basis for the I_n-dimensional space and there is no solution for Equation (6.86). Thus, $P \leq I_n$.

From the above, $P \leq \min\{\min_n I_n, M\}$ if the elements of $\hat{\mathbf{Y}}_p^{(n)}$ are not all zero, which is often the case as long as the projection basis is not initialized to zero and the elements of the training tensor samples are not all zero. $\square$

Remark 6.3. *The conclusion in Corollary 6.1 is expected because the EMPs to be solved in UMPCA correspond to highly constrained situations in the linear case where the features extracted are constrained by both their correlation property and the compactness of the projection. This implies that UMPCA may be more suitable for high-resolution/dimension tensor data where the dimensionality in each mode is high enough to enable the extraction of a sufficient number of (uncorrelated) features. UMPCA is also more useful for applications that need only a small number of features, such as clustering of a small number of classes.*

6.5 Boosting with MPCA

As reviewed in Section 2.8.2, boosting is an ensemble-based learning method that offers good generalization capability through combining weak learners repeatedly trained on weighted samples [Schapire, 2003]. This section examines the combination of MPCA (Section 6.2) with a LDA-based boosting scheme for better generalization [Lu et al., 2007a, 2009a], as shown in Figure 6.2. Input tensor samples are projected into the learned MPCA subspace, and a number of discriminative EMPs are selected (Section 5.5.1) to obtain feature vectors, which are fed into an LDA-based booster for learning and classification.

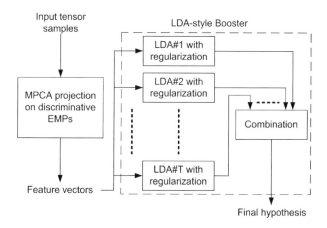

FIGURE 6.2: Illustration of recognition through LDA-style boosting with regularization on MPCA features.

6.5.1 Benefits of MPCA-Based Booster

The LDA-style base learners in this MPCA-based booster take $\{\mathbf{y}_m \in \mathbb{R}^{H_\mathbf{y}}, m = 1, ..., M\}$, the feature vectors extracted through MPCA, rather than the vectorized original data $\{\mathbf{x}_m \in \mathbb{R}^I, m = 1, ..., M\}$ ($I = \prod_{n=1}^{N} I_n$) as in [Lu et al., 2006e]. There are two benefits from this scheme:

1. The feature vector dimension $H_\mathbf{y}$, which is the number of discriminative EMPs selected, offers one more degree (in addition to the number of samples per class used for the LDA learners in [Lu et al., 2006e]) to control the weakness of the LDA learners. Similar to PCA+LDA [Belhumeur et al., 1997], where the performance is often affected by the number of principal components selected for LDA, $H_\mathbf{y}$ affects the performance of LDA on MPCA features as well. Therefore, by choosing an $H_\mathbf{y}$ that is not optimal for a single LDA learner, the obtained LDA learner is weakened. Of course, the LDA learner cannot be made "too weak" either. Otherwise, the boosting scheme will not work.

2. Using feature vectors of dimension $H_\mathbf{y}$ instead of the original data as the booster input is computationally advantageous. Because boosting is an iterative algorithm with T rounds, the computational cost is about T times that of a single learner with the same input, both in training and testing. By making the booster work on lower-dimensional features extracted by MPCA, the booster becomes much more efficient because it only needs to deal with low-dimensional vectors. Consequently, the computational cost can be reduced significantly.

6.5.2 LDA-Style Boosting on MPCA Features

The AdaBoost algorithm was originally developed for binary classification problems, and several methods have been proposed to extend it to the multiclass case [Freund and Schapire, 1997; Schapire and Singer, 1999; Allwein et al., 2000; Schapire, 1997]. The multi-class AdaBoost approach followed in [Lu et al., 2007a, 2009a] is the AdaBoost.M2 algorithm [Freund and Schapire, 1996]. Algorithm 6.5 is the pseudocode for the MPCA+boosting scheme.

AdaBoost.M2 aims to extend communication between the boosting algorithm and the weak learner by allowing the weak learner to generate more expressive hypotheses (a set of "plausible" labels rather than a single label) indicating a "degree of plausibility", that is, a hypothesis h takes a sample $\mathbf{y}$ and a class label c as the inputs and produces a "plausibility" score $h(\mathbf{y}, c) \in [0, 1]$ as the output. To achieve its objective, AdaBoost.M2 introduces a sophisticated error measure *pseudo-loss* $\hat{\epsilon}_t$ with respect to the *mislabel distribution* $\mathbf{D}_t(m, c)$ in [Freund and Schapire, 1996]. A *mislabel* is a pair (m, c), where m is the index of a training sample and c is an incorrect label associated with the sample $\mathbf{y}_m$. Let B be the set of all mislabels:

$$B = \{(m, c) : m = 1, ..., M, c \neq c_m\}. \tag{6.87}$$

Algorithm 6.5 LDA-style booster based on MPCA features

Input: The MPCA feature vectors $\{\mathbf{y}_m \in \mathbb{R}^{H_y}, m = 1, ..., M\}$ with class labels $\mathbf{c} \in \mathbb{R}^M$, the number of samples per class for LDA training ξ, the maximum number of iterations T.

Process:

1: **Initialize** $\mathbf{D}_1(m, c)$ as in Equation (6.88), $\hat{\mathbf{A}}_1(c_a, c_b) = \frac{1}{C^2}$, $\mathbf{D}_1(m, c_m) = 0$, $\hat{\mathbf{A}}_1(c_a, c_a) = 0$, and the first ξ samples from each class are selected to form the initial training set $\{\mathbf{y}_s, s = 1, ..., S\}_1$.

2: **for** $t = 1$ **to** T **do**

3:    Get $\hat{\mathbf{U}}_{LDA_t}$ from $\mathbf{S}_{B_t}$ and $\mathbf{S}_{W_t}$ constructed from $\{\mathbf{y}_s, s = 1, ..., S\}_t$ to project $\{\mathbf{y}_m\}$ to $\{\hat{\mathbf{z}}_m\}_t$.

4:    Get hypothesis $\{h_t(\mathbf{y}_m, c) \in [0, 1]\}$ by applying the nearest mean classifier on $\{\hat{\mathbf{z}}_m\}_t$.

5:    Calculate $\hat{\epsilon}_t$, the pseudo-loss of h_t, from Equation (6.90).

6:    Set $\beta_t = \hat{\epsilon}_t / (1 - \hat{\epsilon}_t)$.

7:    Update $\mathbf{D}_t$:

$$\mathbf{D}_{t+1}(m, c) = \mathbf{D}_t(m, c)\beta_t^{\frac{1}{2}(1 + h_t(\mathbf{y}_m, c_m) - h_t(\mathbf{y}_m, c))},$$

and normalize it:

$$\mathbf{D}_{t+1}(m, c) = \frac{\mathbf{D}_{t+1}(m, c)}{\sum_m \sum_c \mathbf{D}_{t+1}(m, c)}.$$

8:    Update $\mathbf{d}_{t+1}(m)$ and $\hat{\mathbf{A}}_{t+1}$ according to Equations (6.91) and (6.92), respectively, and update $\{\mathbf{y}_s\}_{t+1}$.

9: **end for**

Output: The final hypothesis:

$$h_{fin}(\mathbf{y}) = \arg \max_c \sum_{t=1}^{T} \left(\log \frac{1}{\beta_t} \right) h_t(\mathbf{y}, c).$$

The mislabel distribution is initialized as

$$\mathbf{D}_1(m, c) = \frac{1}{M \cdot (C - 1)} \quad \text{for} \quad (m, c) \in B. \tag{6.88}$$

Accordingly, the weak learner produces a hypothesis

$$h_t : \mathbb{R}^{H_y} \times C \to [0, 1], \tag{6.89}$$

where $h_t(\mathbf{y}, c)$ measures the degree to which it is believed that c is the correct label for $\mathbf{y}$. The pseudo-loss $\hat{\epsilon}_t$ of the hypothesis h_t with respect to $\mathbf{D}_t(m, c)$ is defined to measure the goodness of h_t and it is given by [Freund and Schapire,

1996]:

$$\hat{\epsilon}_t = \frac{1}{2} \sum_{(m,c) \in B} \mathbf{D}_t(m,c) \left(1 - h_t(\mathbf{y}_m, c_m) + h_t(\mathbf{y}_m, c)\right). \tag{6.90}$$

The introduction of the mislabel distribution enhances communication between the learner and the booster, so that AdaBoost.M2 can focus the weak learner not only on hard-to-classify samples, but also on the incorrect labels that are the hardest to discriminate [Freund and Schapire, 1996].

Another distribution $\mathbf{d}_t(m)$, named the *pseudo sample distribution* in [Lu et al., 2006e], is derived from $\mathbf{D}_t(m, c)$ as

$$\mathbf{d}_t(m) = \sum_{c \neq c_m} \mathbf{D}_t(m, c). \tag{6.91}$$

For the communication between the booster and the learner, the modified "pairwise class discriminant distribution" (PCDD) $\hat{\mathbf{A}}_t \in \mathbb{R}^{C \times C}$ introduced in [Lu et al., 2006e] is employed as

$$\hat{\mathbf{A}}_t(c_a, c_b) = \frac{1}{2} \left(\sum_{c_m = c_a, c_{m_t} = c_b} \mathbf{d}_t(m) + \sum_{c_m = c_b, c_{m_t} = c_a} \mathbf{d}_t(m) \right), \tag{6.92}$$

where

$$c_{m_t} = \arg\max_c h_t(\mathbf{y}_m, c) \tag{6.93}$$

and the diagonal elements of $\hat{\mathbf{A}}_t$ are set to zeros.

6.5.3 Modified LDA Learner

In building the LDA learner, the approach in [Lu et al., 2006e] is adopted with three modifications:

1. Only ξ samples per class are used as the input to the LDA learner in order to get weaker but more diverse LDA learners. The first ξ samples are taken for the first boosting step and the hardest ξ (with the largest $\mathbf{d}(m)$) samples are selected for subsequent steps. Let $\{\mathbf{y}_s, s = 1, ..., S\}_t$ denote the selected samples in round t, where $S = \xi \times C$.

2. For the between-class scatter matrix $\hat{\mathbf{S}}_{B_t}$, the pairwise between-class scatter in [Loog et al., 2001] is used instead of that used in [Lu et al., 2006e] for its simplicity and easy computation:

$$\hat{\mathbf{S}}_{B_t} = \sum_{c_a=1}^{C-1} \sum_{c_b=c_a+1}^{C} \hat{\mathbf{A}}_t(c_a, c_b)(\bar{\mathbf{y}}_{c_a} - \bar{\mathbf{y}}_{c_b})(\bar{\mathbf{y}}_{c_a} - \bar{\mathbf{y}}_{c_b})^T, \tag{6.94}$$

where

$$\bar{\mathbf{y}}_c = \frac{1}{\xi} \sum_{s, c_s = c} \mathbf{y}_s. \tag{6.95}$$

3. For the within-class scatter matrix, a regularized version of that in [Lu et al., 2006e] is used:

$$\hat{\mathbf{S}}_{W_t} = \sum_{s=1}^{S} \mathbf{d}_t(s)(\mathbf{y}_s - \bar{\mathbf{y}}_{c_s})(\mathbf{y}_s - \bar{\mathbf{y}}_{c_s})^T + \kappa \cdot \mathbf{I}_{H_\mathbf{y}}, \qquad (6.96)$$

where κ is a regularization parameter to increase the estimated within-class scatter and $\mathbf{I}_{H_\mathbf{y}}$ is an identity matrix of size $H_\mathbf{y} \times H_\mathbf{y}$. The regularization term is added because the actual within-class scatter of testing samples is expected to be greater than the within-class scatter estimated from the training samples.

With these definitions, the corresponding LDA projection $\hat{\mathbf{U}}_{LDA_t}$ consists of the generalized eigenvectors corresponding to the largest $H_{\hat{\mathbf{z}}}$ ($\leq C - 1$) generalized eigenvalues of the following generalized eigenvalue problem:

$$\hat{\mathbf{S}}_{B_t}\mathbf{u} = \lambda\hat{\mathbf{S}}_{W_t}\mathbf{u}. \qquad (6.97)$$

Thus, the LDA feature vector $\hat{\mathbf{z}}_{m_t}$ is obtained as

$$\hat{\mathbf{z}}_{m_t} = \hat{\mathbf{U}}_{LDA_t}^T \mathbf{y}_m \qquad (6.98)$$

for input to a classifier. To produce the hypothesis, the nearest mean classifier is used to assign the class label of the class mean nearest to the test sample, and the calculated distances between a test sample and the C class means are matched to the interval $[0, 1]$ as required by the AdaBoost.M2 algorithm.

6.6 Other Multilinear PCA Extensions

This last section of the chapter examines several other multilinear extensions of PCA.

6.6.1 Two-Dimensional PCA

A two-dimensional PCA (2DPCA) algorithm was proposed in [Yang et al., 2004] for image data $\{\mathbf{X}_m \in \mathbb{R}^{I_1 \times I_2}, m = 1, ..., M\}$. This algorithm solves for a linear transformation $\mathbf{U} \in \mathbb{R}^{I_2 \times P_2}$ ($P_2 < I_2$) that projects an image $\mathbf{X}_m$ to

$$\mathbf{Y}_m = \mathbf{X}_m\mathbf{U} = \mathbf{X}_m \times_2 \mathbf{U}^T \in \mathbb{R}^{I_1 \times P_2}, \qquad (6.99)$$

which maximizes the following scatter measure:

$$\sum_{m=1}^{M} \| \mathbf{Y}_m - \bar{\mathbf{Y}} \|_F^2 = \sum_{m=1}^{M} \text{tr}\left(\mathbf{U}^T(\mathbf{X}_m - \bar{\mathbf{X}})^T(\mathbf{X}_m - \bar{\mathbf{X}})\mathbf{U}\right), \qquad (6.100)$$

where $\bar{\mathbf{X}}$ is the mean image. This algorithm works directly on image matrices (second-order tensors) but there is only one linear transformation in mode 2 (the row mode). Thus, image data are projected in mode 2 only, while the projection in mode 1 (the column mode) is not considered. Therefore, 2DPCA is equivalent to GPCA with an identity matrix as the mode-1 projection matrix. The pseudocode for 2DPCA can be obtained by simple respective modifications of Algorithm 6.1 (initializing $\mathbf{U}^{(1)}$ to be an identity matrix and removing the iteration and mode-1 calculation). Note that 2DPCA is not iterative because it solves for only one projection matrix so the obtained solution is optimal with respect to maximization of the scatter measure defined in Equation (6.100).

6.6.2 Generalized Low Rank Approximation of Matrices

The generalized low rank approximation of matrices (GLRAM) introduced in [Ye, 2005a] applies two linear transformations to both the left and right sides of input image matrices $\{\mathbf{X}_m \in \mathbb{R}^{I_1 \times I_2}\}$. This algorithm solves for two linear transformations $\mathbf{U}^{(1)} \in \mathbb{R}^{I_1 \times P_1}$ $(P_1 \leq I_1)$ and $\mathbf{U}^{(2)} \in \mathbb{R}^{I_2 \times P_2}$ $(P_2 \leq I_2)$ that project an image $\mathbf{X}_m$ to $\mathbf{Y}_m$ as in Equation (6.1) while minimizing the least-square (reconstruction) error measure below:

$$\sum_{m=1}^{M} \| \mathbf{X}_m - \mathbf{U}^{(1)} \mathbf{Y}_m \mathbf{U}^{(2)^T} \|_F^2 = \sum_{m=1}^{M} \| \mathbf{X}_m - \mathbf{Y}_m \times_1 \mathbf{U}^{(1)} \times_2 \mathbf{U}^{(2)} \|_F^2 . \quad (6.101)$$

Thus, projections in both modes are involved and better dimensionality reduction results than 2DPCA are obtained in [Ye, 2005a]. Some further studies on this algorithm have been reported by [Liu et al., 2010]. As shown in Equation (6.17), GLRAM is equivalent to GPCA without centering the input data. In other words, GLRAM on centered data is equivalent to GPCA. The pseudocode for GLRAM can be obtained by removing the centering (step 1) in Algorithm 6.1 and replacing $\tilde{\mathbf{X}}_m$ with $\mathbf{X}_m$.

6.6.3 Concurrent Subspace Analysis

The concurrent subspaces analysis (CSA) algorithm was formulated in [Xu et al., 2008] for general tensor data. It can be considered a further generalization of GLRAM for higher-order tensors, aiming at minimizing reconstruction errors.

CSA is a TTP-based unsupervised MSL algorithm. As in MPCA, CSA solves for a TTP $\{\mathbf{U}^{(n)} \in \mathbb{R}^{I_n \times P_n}, P_n \leq I_n, n = 1, ..., N\}$ that projects a tensor $\mathcal{X}_m \in \mathbb{R}^{I_1 \times ... \times I_N}$ to

$$\mathcal{Y}_m = \mathcal{X}_m \times_1 \mathbf{U}^{(1)^T} \times_2 ... \times_N \mathbf{U}^{(N)^T} \in \mathbb{R}^{P_1 \times ... \times P_N} \quad (6.102)$$

while minimizing the following reconstruction error metric

$$\sum_{m=1}^{M} \| \mathcal{X}_m - \mathcal{Y}_m \times_1 \mathbf{U}^{(1)} \times_2 \dots \times_N \mathbf{U}^{(N)} \|_F^2, \tag{6.103}$$

where the projection matrices $\{\mathbf{U}^{(n)}\}$ have all orthonormal columns.

CSA is equivalent to MPCA without centering the input data. In other words, when the input data have zero-mean, MPCA and CSA are equivalent. This equivalence can be derived similarly by following the steps in Equation (6.17) or following Theorem 4.2 of [De Lathauwer et al., 2000a].

6.6.4   MPCA plus LDA

Following the method described in Section 5.5.1, discriminative features can be selected from $\{\mathcal{Y}_m\}$, the lower-dimensional tensors projected by MPCA, to form feature vectors $\{\mathbf{y}_m\}$. LDA can then be applied to $\{\mathbf{y}_m\}$, resulting in an MPCA+LDA approach, similar to the popular approach of PCA+LDA [Belhumeur et al., 1997]. Let $\mathbf{S}_{B_\mathbf{y}}$ and $\mathbf{S}_{W_\mathbf{y}}$ be the between-class scatter matrix and within-class scatter matrix based on $\{\mathbf{y}_m\}$, respectively. Then, the corresponding LDA projection $\mathbf{U}_{LDA}$ consists of the generalized eigenvectors associated with the $H_\mathbf{z}$ $(\leq C - 1)$ largest generalized eigenvalues of the following generalized eigenvalue problem:

$$\mathbf{S}_{B_\mathbf{y}}\mathbf{u} = \lambda\mathbf{S}_{W_\mathbf{y}}\mathbf{u}. \tag{6.104}$$

The MPCA+LDA feature vector $\mathbf{z}_m$ is then obtained as

$$\mathbf{z}_m = \mathbf{U}_{LDA}^T\mathbf{y}_m. \tag{6.105}$$

6.6.5   Nonnegative MPCA

Panagakis et al. [2010] extended MPCA to nonnegative MPCA (NMPCA) by imposing the constraint of nonnegative projection matrices. NMPCA aims to capture the total variations of nonnegative tensor input while preserving the non-negativity of auditory representations. This is considered important when the underlying data factors have physical or psychological interpretation.

To solve the NMPCA problem, Panagakis et al. [2010] first derived a multiplicative updating algorithm for maximizing homogenous functions over the Grassmann manifold subject to the non-negativity constraints. This is done through incorporating the natural gradient [Amari, 1998] of the Grassmann manifold [Edelman et al., 1998] into the multiplicative updates. The derived multiplicative updates are then used to maximize the captured variation in a manner similar to MPCA.

6.6.6   Robust Versions of MPCA

Inoue et al. [2009] proposed two robust MPCA (RMPCA) algorithms to handle two kinds of outliers, that is, sample outliers and intra-sample outliers. Iterative algorithms are derived on the basis of Lagrange multipliers using the error-minimization reformulation of MPCA. In RMPCA for sample outliers, the Frobenius norm minimization is modified using a robust M-estimator. In RMPCA for intra-sample outliers, the Frobenius norm minimization is modified to nested summation of robust M-estimators. The Welsch estimator [Hampel et al., 2011] is chosen in this work.

Another approach enhances the robustness of MPCA using the ℓ_1-norm, which can be considered a multilinear extension of the ℓ_1-norm-based PCA in [Kwak, 2008]. 2DPCA [Yang et al., 2004] was first extended in this way by Li et al. [2009b]. Later, ℓ_1-norm-based robust MPCA for higher-order tensors was developed by Pang et al. [2010].

6.6.7   Incremental Extensions of MPCA

Sun et al. [2008a] introduced an incremental tensor analysis (ITA) framework to summarize higher-order data streams represented as tensors and to reveal hidden correlations. While new tensors are arriving continuously over time, data summary is obtained through TTP and updated incrementally. Three variants of ITA were proposed in [Sun et al., 2008a]: dynamic tensor analysis (DTA), streaming tensor analysis (STA), and window-based tensor analysis (WTA). DTA incrementally maintains covariance matrices for all modes and uses the leading eigenvectors of covariance matrices as projection matrices [Sun et al., 2006]. STA directly updates the leading eigenvectors of covariance matrices using the SPIRIT algorithm [Papadimitriou et al., 2005]. WTA uses similar updates as DTA while performing alternating iteration to further improve the results. The ITA framework focuses on the approximation problem; hence, the objective is to minimize the least square error and it can be considered an incremental version of CSA for streaming data. Several design trade-offs were explored in [Sun, 2007], including space efficiency, computational cost, approximation accuracy, time dependency, and model complexity.

Wang et al. [2011a] proposed an incremental third-order tensor representation for robust tracking in video sequences. This method represents the target in each video frame as a third-order tensor to preserve the spatial correlation inside the target region (the column and row modes) and integrate multiple appearance cues for target description as the third mode, like in Figure 1.4. The MPCA model is learned online to model the target appearance variations during tracking. This work employs singular value decomposition (SVD) (instead of eigendecomposition) to get the projection matrices, which are updated using the recursive SVD [Brand, 2002; Levey and Lindenbaum, 2000; Ross et al., 2008].

6.6.8   Probabilistic Extensions of MPCA

Tao et al. [2008b] proposed a TTP-based MSL algorithm called Bayesian tensor analysis (BTA) to generalize the Bayesian PCA algorithm [Bishop, 1999] to the multilinear case for tensors. BTA can also be viewed as a Bayesian extension of MPCA. A probabilistic graphical model can be constructed based on BTA. Parameters in BTA are estimated using the expectation maximization (EM) algorithm with the maximum likelihood estimators [Moon and Stirling, 2000]. In addition, BTA determines the subspace dimensions automatically by setting a series of hyper parameters.

Probabilistic PCA (PPCA) [Tipping and Bishop, 1999b,a] is a linear latent variable model for probabilistic dimensionality reduction. Zhao et al. [2012] developed a bilinear PPCA (BPPCA) that performs PPCA in the row and column modes alternately. BPPCA can be seen as a bilinear (2D) extension of PPCA or a probabilistic extension of GLRAM (Section 6.6.2). The maximum likelihood estimation of its model parameters span the principal subspaces of the column and row covariance matrices, and it can be conducted in two ways, with or without the use of latent variables. One method is based on the conditional maximization [McLachlan and Krishnan, 2007], a special case of the coordinate ascent algorithm [Zangwill, 1969], which does not require the inclusion of latent variables. The other method is called alternating expectation conditional maximization (AECM) [Meng and Van Dyk, 2002], which is an EM-type algorithm with lower computational complexity but slower convergence.

6.6.9 Weighted MPCA and MPCA for Binary Tensors

In [Washizawa et al., 2010], sample weighting is incorporated into MPCA for electroencephalography (EEG) signal classification in Motor-Imagery Brain Computer Interface (MI-BCI). The motivation is to use a weight to reduce the effect of bad samples or outliers by assigning each mth sample a weight w_m. When there is no *a priori* information about the quality of a sample, the optimal weights are estimated through the iterative alternating estimation procedures of MPCA.

Mažgut et al. [2010] studied MPCA for binary tensors. Their method assumes a set of binary tensors independently generated from a Bernoulli distribution and aims to find a lower-dimensional representation to capture the data distribution well. The Bernoulli distribution is further parameterized through log-odds (natural parameters) to get the log-likelihood. The natural parameters are then collected into a tensor from which a subspace is to be learned. In addition, a bias tensor is added as a constraint. Next, the method employs the trick in [Schein et al., 2003] and follows the APP method to solve the problem.[3]

---

[3] Further readings for chapters in Part II are the respective references of the specific algorithms and applications.

Chapter 7

Multilinear Discriminant Analysis

After introducing various multilinear extensions of principal component analysis (PCA) in the previous chapter, this chapter presents several multilinear extensions of linear discriminant analysis (LDA) as shown in Figure 7.1 under the multilinear subspace learning (MSL) framework.

As in the previous chapter, we describe five algorithms in greater details and cover other algorithms briefly in the final section. We start with the two-dimensional LDA (2DLDA) algorithm [Ye et al., 2004b]. 2DLDA involves only linear algebra so it can be understood with ease. Then we discuss its higher-order generalization, the discriminant analysis with tensor representation (DATER) algorithm [Yan et al., 2007a], and a variant called the general tensor discriminant analysis (GTDA) algorithm [Tao et al., 2007b]. 2DLDA, DATER, and GTDA are based on tensor-to-tensor projection (TTP). Next, we study two algorithms based on tensor-to-vector projection (TVP): the tensor rank-one discriminant analysis (TR1DA) algorithm [Wang and Gong, 2006; Tao et al., 2008a] and the uncorrelated multilinear discriminant analysis (UMLDA) algorithm [Lu et al., 2009c]. In addition, we examine how to employ regularization and ensemble-based learning to enhance the performance

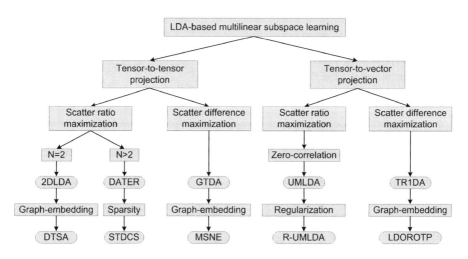

FIGURE 7.1: Multilinear discriminant analysis algorithms under the MSL framework.

of UMLDA. In the final section, we summarize several other multilinear discriminant analysis extensions under the MSL framework.

7.1 Two-Dimensional LDA[1]

We start off with a simple second-order case of $N = 2$, that is, a 2D extension of LDA for matrix data, which is easier for readers to follow with mainly linear algebra and matrix notations.

7.1.1 2DLDA Problem Formulation

The 2DLDA algorithm proposed in [Ye et al., 2004b] is a supervised algorithm targeting at maximizing the between-class scatter while minimizing the within-class scatter in the projected subspace as in classical LDA. From Definition 3.14 and the special cases in Equations (3.52) and (3.53), the 2DLDA problem is defined below, similar to the generalized PCA (GPCA) discussed in Section 6.1.

The 2DLDA Problem: A set of M labeled matrix samples $\{\mathbf{X}_1, \mathbf{X}_2, ..., \mathbf{X}_M\}$ are available for training with class labels $\mathbf{c} \in \mathbb{R}^M$. Each matrix sample $\mathbf{X}_m \in \mathbb{R}^{I_1 \times I_2}$ assumes values in a tensor space $\mathbb{R}^{I_1} \bigotimes \mathbb{R}^{I_2}$. The class label for the mth sample $\mathbf{X}_m$ is c_m ($= \mathbf{c}(m)$, the mth entry of $\mathbf{c}$) and there are C classes in total. The 2DLDA objective is to find two linear transformations (projection matrices) $\mathbf{U}^{(1)} \in \mathbb{R}^{I_1 \times P_1}$ ($P_1 \leq I_1$) and $\mathbf{U}^{(2)} \in \mathbb{R}^{I_2 \times P_2}$ ($P_2 \leq I_2$) that project a sample $\mathbf{X}_m$ to

$$\mathbf{Y}_m = \mathbf{U}^{(1)^T} \mathbf{X}_m \mathbf{U}^{(2)} = \mathbf{X}_m \times_1 \mathbf{U}^{(1)^T} \times_2 \mathbf{U}^{(2)^T} \in \mathbb{R}^{P_1 \times P_2} \quad (7.1)$$

such that the ratio of between-class scatter $\Psi_{B_\mathbf{Y}}$ to within-class scatter $\Psi_{W_\mathbf{Y}}$ is maximized, where $\Psi_{B_\mathbf{Y}}$ and $\Psi_{W_\mathbf{Y}}$ are defined as in Equations (3.52) and (3.53), respectively.

Thus, we can write the objective function for 2DLDA as

$$\{\tilde{\mathbf{U}}^{(1)}, \tilde{\mathbf{U}}^{(2)}\} = \arg \max_{\mathbf{U}^{(1)}, \mathbf{U}^{(2)}} \frac{\Psi_{B_\mathbf{Y}}}{\Psi_{W_\mathbf{Y}}} = \arg \max_{\mathbf{U}^{(1)}, \mathbf{U}^{(2)}} \frac{\sum_{c=1}^C M_c \parallel \bar{\mathbf{Y}}_c - \bar{\mathbf{Y}} \parallel_F^2}{\sum_{m=1}^M \parallel \mathbf{Y}_m - \bar{\mathbf{Y}}_{c_m} \parallel_F^2},$$
$$(7.2)$$

where $\bar{\mathbf{Y}}$ is the overall mean defined as

$$\bar{\mathbf{Y}} = \frac{1}{M} \sum_{m=1}^M \mathbf{Y}_m, \quad (7.3)$$

---

[1]In addition to the 2DLDA in [Ye et al., 2004b], there are several other 2D extensions of LDA, such as [Li and Yuan, 2005], [Yang et al., 2005], and [Kong et al., 2005].

$\bar{\mathbf{Y}}_c$ is the class mean defined as

$$\bar{\mathbf{Y}}_c = \frac{1}{M_c} \sum_{m=1,c_m=c}^{M} \mathbf{Y}_m, \tag{7.4}$$

and M_c is the number of samples with class label c.

Because $\mathrm{tr}(\mathbf{AA}^T) = \|\mathbf{A}\|_F^2$, $\Psi_{B\mathbf{Y}}$ and $\Psi_{W\mathbf{Y}}$ can be further written as

$$\Psi_{B\mathbf{Y}} = \sum_{c=1}^{C} M_c \|\bar{\mathbf{Y}}_c - \bar{\mathbf{Y}}\|_F^2 = \mathrm{tr}\left(\sum_{c=1}^{C} M_c \cdot (\bar{\mathbf{Y}}_c - \bar{\mathbf{Y}})(\bar{\mathbf{Y}}_c - \bar{\mathbf{Y}})^T\right) \tag{7.5}$$

and

$$\Psi_{W\mathbf{Y}} = \sum_{m=1}^{M} \|\mathbf{Y}_m - \bar{\mathbf{Y}}_{c_m}\|_F^2 = \mathrm{tr}\left(\sum_{m=1}^{M} (\mathbf{Y}_m - \bar{\mathbf{Y}}_{c_m})(\mathbf{Y}_m - \bar{\mathbf{Y}}_{c_m})^T\right). \tag{7.6}$$

From Equations (7.1), (7.5), and (7.6), the objective function Equation (7.2) becomes

$$
\begin{aligned}
\{\tilde{\mathbf{U}}^{(1)}, \tilde{\mathbf{U}}^{(2)}\} &= \arg\max_{\mathbf{U}^{(1)},\mathbf{U}^{(2)}} \frac{\mathrm{tr}\left(\sum_{c=1}^{C} M_c \cdot (\bar{\mathbf{Y}}_c - \bar{\mathbf{Y}})(\bar{\mathbf{Y}}_c - \bar{\mathbf{Y}})^T\right)}{\mathrm{tr}\left(\sum_{m=1}^{M} (\mathbf{Y}_m - \bar{\mathbf{Y}}_{c_m})(\mathbf{Y}_m - \bar{\mathbf{Y}}_{c_m})^T\right)} \\
&= \arg\max_{\mathbf{U}^{(1)},\mathbf{U}^{(2)}} \frac{\mathrm{tr}\left(\sum_{c=1}^{C} M_c \cdot \mathbf{U}^{(1)^T}(\bar{\mathbf{X}}_c - \bar{\mathbf{X}})\mathbf{U}^{(2)}\mathbf{U}^{(2)^T}(\bar{\mathbf{X}}_c - \bar{\mathbf{X}})^T\mathbf{U}^{(1)}\right)}{\mathrm{tr}\left(\sum_{m=1}^{M} \mathbf{U}^{(1)^T}(\mathbf{X}_m - \bar{\mathbf{X}}_{c_m})\mathbf{U}^{(2)}\mathbf{U}^{(2)^T}(\mathbf{X}_m - \bar{\mathbf{X}}_{c_m})^T\mathbf{U}^{(1)}\right)},
\end{aligned} \tag{7.7}
$$

where

$$\bar{\mathbf{X}} = \frac{1}{M} \sum_{m=1}^{M} \mathbf{X}_m, \tag{7.8}$$

and

$$\bar{\mathbf{X}}_c = \frac{1}{M_c} \sum_{m=1,c_m=c}^{M} \mathbf{X}_m. \tag{7.9}$$

7.1.2 2DLDA Algorithm Derivation

As in GPCA (Section 6.1), there is no closed-form solution for the two inter-dependent projection matrices. Thus, the solution follows the alternating partial projections (APP) method (Section 5.1) to solve for these two projection matrices one at a time through partial multilinear projections iteratively. In each iteration step, we assume one project matrix is given to find the other one that maximizes the scatter ratio, and then vice versa.

Let $\tilde{\mathbf{U}}^{(1)}$ and $\tilde{\mathbf{U}}^{(2)}$ be the matrices maximizing the scatter ratio $\frac{\Psi_{B\mathbf{Y}}}{\Psi_{W\mathbf{Y}}}$ in Equation (7.7).

1. For a given $\mathbf{U}^{(2)}$, define the conditioned mode-1 between-class and within-class scatter matrices as

$$\boldsymbol{\Psi}_{B_{\mathbf{Y}1}} = \sum_{c=1}^{C} M_c \left(\bar{\mathbf{X}}_c - \bar{\mathbf{X}}\right) \mathbf{U}^{(2)} \mathbf{U}^{(2)^T} \left(\bar{\mathbf{X}}_c - \bar{\mathbf{X}}\right)^T \qquad (7.10)$$

and

$$\boldsymbol{\Psi}_{W_{\mathbf{Y}1}} = \sum_{m=1}^{M} \left(\mathbf{X}_m - \bar{\mathbf{X}}_{c_m}\right) \mathbf{U}^{(2)} \mathbf{U}^{(2)^T} \left(\mathbf{X}_m - \bar{\mathbf{X}}_{c_m}\right)^T, \qquad (7.11)$$

respectively. We can write Equation (7.7) further as

$$\tilde{\mathbf{U}}^{(1)} = \arg\max_{\mathbf{U}^{(1)}} \frac{\operatorname{tr}\left(\mathbf{U}^{(1)^T} \boldsymbol{\Psi}_{B_{\mathbf{Y}1}} \mathbf{U}^{(1)}\right)}{\operatorname{tr}\left(\mathbf{U}^{(1)^T} \boldsymbol{\Psi}_{W_{\mathbf{Y}1}} \mathbf{U}^{(1)}\right)}. \qquad (7.12)$$

This is a classical LDA problem and the ratio above is maximized only if $\tilde{\mathbf{U}}^{(1)}$ consists of the P_1 generalized eigenvectors of $\boldsymbol{\Psi}_{B_{\mathbf{Y}1}}$ and $\boldsymbol{\Psi}_{W_{\mathbf{Y}1}}$ associated with the P_1 largest generalized eigenvalues.

2. Similarly, for a given $\mathbf{U}^{(1)}$, define the conditioned mode-2 between-class and within-class scatter matrices as

$$\boldsymbol{\Psi}_{B_{\mathbf{Y}2}} = \sum_{c=1}^{C} M_c \left(\bar{\mathbf{X}}_c - \bar{\mathbf{X}}\right)^T \mathbf{U}^{(1)} \mathbf{U}^{(1)^T} \left(\bar{\mathbf{X}}_c - \bar{\mathbf{X}}\right) \qquad (7.13)$$

and

$$\boldsymbol{\Psi}_{W_{\mathbf{Y}2}} = \sum_{m=1}^{M} \left(\mathbf{X}_m - \bar{\mathbf{X}}_{c_m}\right)^T \mathbf{U}^{(1)} \mathbf{U}^{(1)^T} \left(\mathbf{X}_m - \bar{\mathbf{X}}_{c_m}\right), \qquad (7.14)$$

respectively. We can write Equation (7.7) further as

$$\tilde{\mathbf{U}}^{(2)} = \arg\max_{\mathbf{U}^{(2)}} \frac{\operatorname{tr}\left(\mathbf{U}^{(2)^T} \boldsymbol{\Psi}_{B_{\mathbf{Y}2}} \mathbf{U}^{(2)}\right)}{\operatorname{tr}\left(\mathbf{U}^{(2)^T} \boldsymbol{\Psi}_{W_{\mathbf{Y}2}} \mathbf{U}^{(2)}\right)}. \qquad (7.15)$$

This is a classical LDA problem and the ratio above is maximized only if $\tilde{\mathbf{U}}^{(2)}$ consists of the P_2 generalized eigenvectors of $\boldsymbol{\Psi}_{B_{\mathbf{Y}2}}$ and $\boldsymbol{\Psi}_{W_{\mathbf{Y}2}}$ associated with the P_2 largest generalized eigenvalues.

Thus, the 2DLDA problem can be solved using an iterative procedure as summarized in Algorithm 7.1. The 2DLDA algorithm uses the pseudo-identity matrices (see Section 5.2) for initialization [Ye et al., 2004b]. However, unlike GPCA, the 2DLDA algorithm does not necessarily increase its objective function monotonically so it may not converge over iterations [Inoue and Urahama, 2006; Luo et al., 2009]. Therefore, the iteration is terminated by specifying a maximum number of iterations K in Algorithm 7.1.

Algorithm 7.1 Two-dimensional LDA (2DLDA)

Input: A set of matrix samples $\{\mathbf{X}_m \in \mathbb{R}^{I_1 \times I_2}, m = 1, ..., M\}$ with class labels $\mathbf{c} \in \mathbb{R}^M$, the desired subspace dimensions P_1 and P_2, and the maximum number of iterations K.

Process:

1: Initialize $\mathbf{U}^{(1)}$ and $\mathbf{U}^{(2)}$ (initialization of $\mathbf{U}^{(1)}$ can be skipped).

2: [**Local optimization:**]

3: **for** $k = 1$ **to** K **do**

4: [**Mode 1**]

5: Calculate $\mathbf{\Psi}_{B_{\mathbf{Y1}}}$ and $\mathbf{\Psi}_{W_{\mathbf{Y1}}}$ according to Equations (7.10) and (7.11).

6: Set the matrix $\mathbf{U}^{(1)}$ to consist of the P_1 generalized eigenvectors of $\mathbf{\Psi}_{B_{\mathbf{Y1}}}$ and $\mathbf{\Psi}_{W_{\mathbf{Y1}}}$, corresponding to the largest P_1 generalized eigenvalues.

7: [**Mode 2**]

8: Calculate $\mathbf{\Psi}_{B_{\mathbf{Y2}}}$ and $\mathbf{\Psi}_{W_{\mathbf{Y2}}}$ according to Equations (7.13) and (7.14).

9: Set the matrix $\mathbf{U}^{(2)}$ to consist of the P_2 generalized eigenvectors of $\mathbf{\Psi}_{B_{\mathbf{Y2}}}$ and $\mathbf{\Psi}_{W_{\mathbf{Y2}}}$, corresponding to the largest P_2 generalized eigenvalues.

10: **end for**

Output: The projection matrices $\tilde{\mathbf{U}}^{(1)} \in \mathbb{R}^{I_1 \times P_1}$ and $\tilde{\mathbf{U}}^{(2)} \in \mathbb{R}^{I_2 \times P_2}$ that maximize the ratio of between-class scatter over within-class scatter in the projected subspace.

Remark 7.1. $\mathbf{\Psi}_{B_{\mathbf{Y1}}}$ *and* $\mathbf{\Psi}_{W_{\mathbf{Y1}}}$ *can be seen as the between-class and within-class scatter matrices of the mode-1 partial projection of the input samples* $\{\mathbf{X}_m \mathbf{U}^{(2)}\}$. $\mathbf{\Psi}_{B_{\mathbf{Y2}}}$ *and* $\mathbf{\Psi}_{W_{\mathbf{Y2}}}$ *can be seen as the between-class and within-class scatter matrices of the mode-2 partial projection of the input samples* $\{\mathbf{X}_m^T \mathbf{U}^{(1)}\}$.

7.2 Discriminant Analysis with Tensor Representation

The DATER [Yan et al., 2005, 2007a] algorithm performs discriminant analysis on general tensor inputs . As in MPCA (Section 6.2), DATER aims to solve for a TTP $\{\mathbf{U}^{(n)} \in \mathbb{R}^{I_n \times P_n}, P_n \leq I_n, n = 1, ..., N\}$ that projects a tensor $\mathcal{X}_m \in \mathbb{R}^{I_1 \times ... \times I_N}$ to a low-dimensional tensor $\mathcal{Y}_m \in \mathbb{R}^{P_1 \times ... \times P_N}$. The difference is that it maximizes a discriminative objective criterion, the tensor-based scatter ratio instead of the total scatter (variation). It is a higher-order extension of 2DLDA presented in the previous section.

7.2.1 DATER Problem Formulation

From Definition 3.14, the DATER problem is defined as follows.

The DATER Problem: A set of M labeled tensor data samples $\{\mathcal{X}_1,$ $\mathcal{X}_2, ..., \mathcal{X}_M\}$ are available for training with class labels $\mathbf{c} \in \mathbb{R}^M$. Each sample $\mathcal{X}_m \in \mathbb{R}^{I_1 \times I_2 \times ... \times I_N}$ assumes values in a tensor space $\mathbb{R}^{I_1} \otimes \mathbb{R}^{I_2} ... \otimes \mathbb{R}^{I_N}$, where I_n is the mode-n dimension. The class label for the mth sample $\mathcal{X}_m$ is $c_m = \mathbf{c}(m)$ and there are C classes in total. The DATER objective is to find a TTP $\{\mathbf{U}^{(n)} \in \mathbb{R}^{I_n \times P_n}, n = 1, ..., N\}$ that maps the original tensor space $\mathbb{R}^{I_1} \otimes \mathbb{R}^{I_2} ... \otimes \mathbb{R}^{I_N}$ into a tensor subspace $\mathbb{R}^{P_1} \otimes \mathbb{R}^{P_2} ... \otimes \mathbb{R}^{P_N}$ (with $P_n \leq I_n$, for $n = 1, ..., N$) as

$$\mathcal{Y}_m = \mathcal{X}_m \times_1 \mathbf{U}^{(1)^T} \times_2 \mathbf{U}^{(2)^T} ... \times_N \mathbf{U}^{(N)^T}, m = 1, ..., M, \qquad (7.16)$$

such that the ratio of the between-class and within-class scatters for $\{\mathcal{Y}_m\}$, as defined in Definition 3.14, is maximized.

We write the objective function of DATER as

$$\{\tilde{\mathbf{U}}^{(n)}\} = \arg \max_{\{\mathbf{U}^{(n)}\}} \frac{\Psi_{B\mathcal{Y}}}{\Psi_{W\mathcal{Y}}} = \arg \max_{\{\mathbf{U}^{(n)}\}} \frac{\sum_{c=1}^{C} M_c \parallel \bar{\mathcal{Y}}_c - \bar{\mathcal{Y}} \parallel_F^2}{\sum_{m=1}^{M} \parallel \mathcal{Y}_m - \bar{\mathcal{Y}}_{c_m} \parallel_F^2}, \qquad (7.17)$$

where the overall mean $\bar{\mathcal{Y}}$ is defined as

$$\bar{\mathcal{Y}} = \frac{1}{M} \sum_{m=1}^{M} \mathcal{Y}_m \qquad (7.18)$$

and the class mean $\bar{\mathcal{Y}}_c$ is defined as

$$\bar{\mathcal{Y}}_c = \frac{1}{M_c} \sum_{m=1, c_m=c}^{M} \mathcal{Y}_m. \qquad (7.19)$$

7.2.2 DATER Algorithm Derivation

As in MPCA, the N projection matrices to be determined are inter-dependent (except for $N = 1$), so we use the APP method to solve this problem. Because the projection to an Nth-order tensor subspace consists of N projections, we solve N optimization subproblems by finding $\mathbf{U}^{(n)}$ that maximizes the mode-n scatter ratio, conditioned on the projection matrices in the other modes. The mode-n partial multilinear projection of $\tilde{\mathcal{X}}_m$ is

$$\acute{\mathcal{Y}}_m^{(n)} = \mathcal{X}_m \times_1 \mathbf{U}^{(1)^T} ... \times_{(n-1)} \mathbf{U}^{(n-1)^T} \times_{(n+1)} \mathbf{U}^{(n+1)^T} ... \times_N \mathbf{U}^{(N)^T}. \quad (7.20)$$

From Definition 3.15, the mode-n between-class and within-class scatter matrices can be obtained from $\acute{\mathbf{Y}}_{m(n)}$, the mode-$n$ unfolding of $\acute{\mathcal{Y}}_m^{(n)}$ as

$$\mathbf{S}_{B\acute{y}}^{(n)} = \sum_{c=1}^{C} M_c \cdot \left(\bar{\acute{\mathbf{Y}}}_{c(n)} - \bar{\acute{\mathbf{Y}}}_{(n)} \right) \left(\bar{\acute{\mathbf{Y}}}_{c(n)} - \bar{\acute{\mathbf{Y}}}_{(n)} \right)^T, \qquad (7.21)$$

and

$$\mathbf{S}_{W_{\acute{y}}}^{(n)} = \sum_{m=1}^{M} \left(\acute{\mathbf{Y}}_{m(n)} - \bar{\bar{\mathbf{Y}}}_{c_m(n)} \right) \left(\acute{\mathbf{Y}}_{m(n)} - \bar{\bar{\mathbf{Y}}}_{c_m(n)} \right)^{T}, \qquad (7.22)$$

respectively, where

$$\bar{\bar{\mathbf{Y}}}_{(n)} = \frac{1}{M} \sum_{m=1}^{M} \acute{\mathbf{Y}}_{m(n)} \qquad (7.23)$$

and

$$\bar{\bar{\mathbf{Y}}}_{c(n)} = \frac{1}{M_c} \sum_{m=1,c_m=c}^{M} \acute{\mathbf{Y}}_{m(n)}. \qquad (7.24)$$

Following Equations (7.12) and (7.15) in the derivation of 2DLDA, we solve for the mode-n projection matrix $\mathbf{U}^{(n)}$ in this conditional optimization problem as

$$\tilde{\mathbf{U}}^{(n)} = \arg\max_{\mathbf{U}^{(n)}} \frac{\text{tr}\left(\mathbf{U}^{(n)^{T}} \mathbf{S}_{B_{\acute{y}}}^{(n)} \mathbf{U}^{(n)}\right)}{\text{tr}\left(\mathbf{U}^{(n)^{T}} \mathbf{S}_{W_{\acute{y}}}^{(n)} \mathbf{U}^{(n)}\right)}. \qquad (7.25)$$

This is a classical LDA problem and the ratio above is maximized only if $\tilde{\mathbf{U}}^{(n)}$ consists of the P_n generalized eigenvectors of $\mathbf{S}_{B_{\acute{y}}}^{(n)}$ and $\mathbf{S}_{W_{\acute{y}}}^{(n)}$ associated with the P_n largest generalized eigenvalues.

Therefore, similar to MPCA and 2DLDA, an iterative procedure can be utilized to solve the DATER problem. The pseudocode is summarized in Algorithm 7.2. The projection matrices can be initialized using one of the methods discussed in Section 5.2. However, as 2DLDA, this algorithm does not converge either [Xu et al., 2006; Lu et al., 2008b].

Feature extraction with DATER is similar to that with MPCA. As a TTP-based algorithm, DATER produces tensor output. Each entry of the output tensor can be viewed as a feature produced by an elementary multilinear projection (EMP) and feature selection strategies (Section 5.5) can then be applied to get vector-valued features for input to conventional classifiers.

7.3 General Tensor Discriminant Analysis

Similar to DATER, GTDA [Tao et al., 2007b] solves for a TTP $\{\mathbf{U}^{(n)} \in \mathbb{R}^{I_n \times P_n}, P_n \leq I_n, n = 1, ..., N\}$ that projects a tensor $\mathcal{X}_m \in \mathbb{R}^{I_1 \times ... \times I_N}$ to a low-dimensional tensor $\mathcal{Y}_m \in \mathbb{R}^{P_1 \times ... \times P_N}$. The difference with DATER is that it maximizes a tensor-based scatter difference (rather than scatter ratio) criterion and the projection matrices in GTDA have orthonormal columns.

Algorithm 7.2 Discriminant analysis with tensor representation (DATER)

Input: A set of tensor samples $\{\mathcal{X}_m \in \mathbb{R}^{I_1 \times I_2 \times ... \times I_N}, m = 1, ..., M\}$ with class labels $\mathbf{c} \in \mathbb{R}^M$, the desired tensor subspace dimensions $\{P_n, n = 1, 2, ..., N\}$, and the maximum number of iterations K.

Process:

1: Initialize $\mathbf{U}^{(n)}$ for $n = 1, ..., N$.

2: [**Local optimization:**]

3: **for** $k = 1$ **to** K **do**

4:    **for** $n = 1$ **to** N **do**

5:       Calculate the mode-n partial multilinear projections $\{\acute{\mathcal{Y}}_m^{(n)}, m = 1, ..., M\}$ according to Equation (7.20).

6:       Calculate $\mathbf{S}_{B_{\acute{y}}}^{(n)}$ and $\mathbf{S}_{W_{\acute{y}}}^{(n)}$ according to Equations (7.21) and (7.22), respectively.

7:       Set the matrix $\mathbf{U}^{(n)}$ to be consisting of the P_n generalized eigenvectors of $\mathbf{S}_{B_{\acute{y}}}^{(n)}$ and $\mathbf{S}_{W_{\acute{y}}}^{(n)}$, corresponding to the largest P_n generalized eigenvalues.

8:    **end for**

9: **end for**

Output: The N projection matrices $\{\tilde{\mathbf{U}}^{(n)} \in \mathbb{R}^{I_n \times P_n}, n = 1, 2, ..., N\}$ that maximize the ratio of between-class scatter over within-class scatter in the projected subspace.

GTDA aims to maximize a multilinear extension of the scatter-difference-based discriminant criterion in [Liu et al., 2006], which corresponds to the J_3 criterion in Section 10.2 of [Fukunaga, 1990]. The objective function of GTDA can be written following Equation (7.17) as

$$\{\tilde{\mathbf{U}}^{(n)}\} = \arg \max_{\{\mathbf{U}^{(n)}\}} \Psi_{B_{\mathcal{Y}}} - \zeta \cdot \Psi_{W_{\mathcal{Y}}}$$

$$= \arg \max_{\{\mathbf{U}^{(n)}\}} \sum_{c=1}^{C} M_c \parallel \bar{\mathcal{Y}}_c - \bar{\mathcal{Y}} \parallel_F^2 - \zeta \sum_{m=1}^{M} \parallel \mathcal{Y}_m - \bar{\mathcal{Y}}_{c_m} \parallel_F^2, (7.26)$$

where ζ is a tuning parameter.

The GTDA problem also employs the APP method to solve for the N projection matrices iteratively by solving for the mode-n projection matrix conditioned on the projection matrices in all the other modes. First, we get the mode-n partial multilinear projection $\acute{\mathcal{Y}}_m^{(n)}$ for $\mathcal{X}_m$ using Equation (7.20). Then, we form the mode-n between-class and within-class scatter matrices as defined in Equation (7.21) and (7.22) using $\acute{\mathbf{Y}}_{m(n)}$, the mode-$n$ unfolding of $\acute{\mathcal{Y}}_m^{(n)}$. Next, we solve for the mode-n projection matrix $\mathbf{U}^{(n)}$ in this conditional

---

Algorithm 7.3 General tensor discriminant analysis (GTDA)

Input: A set of tensor samples $\{\mathcal{X}_m \in \mathbb{R}^{I_1 \times I_2 \times ... \times I_N}, m = 1, ..., M\}$ with class labels $\mathbf{c} \in \mathbb{R}^M$, the desired tensor subspace dimensions $\{P_n, n = 1, 2, ..., N\}$, and the maximum number of iterations K.

Process:

1: Initialize $\mathbf{U}^{(n)}$ for $n = 1, ..., N$.

2: [**Local optimization:**]

3: **for** $k = 1$ **to** K **do**

4:     **for** $n = 1$ **to** N **do**

5:         Calculate the mode-n partial multilinear projections $\{\acute{\mathcal{Y}}_m^{(n)}, m = 1, ..., M\}$ according to Equation (7.20).

6:         Calculate $\mathbf{S}_{B_{\acute{y}}}^{(n)}$ and $\mathbf{S}_{W_{\acute{y}}}^{(n)}$ according to Equations (7.21) and (7.22), respectively.

7:         Set the tuning parameter ζ to the largest eigenvalue of $(\mathbf{S}_{W_{\acute{y}}}^{(n)})^{-1}\mathbf{S}_{B_{\acute{y}}}^{(n)}$.

8:         Set the matrix $\mathbf{U}^{(n)}$ to consist of the P_n eigenvectors of $(\mathbf{S}_{B_{\acute{y}}}^{(n)} - \zeta \cdot \mathbf{S}_{W_{\acute{y}}}^{(n)})$, corresponding to the largest P_n eigenvalues.

9:     **end for**

10: **end for**

Output: The N projection matrices $\{\tilde{\mathbf{U}}^{(n)} \in \mathbb{R}^{I_n \times P_n}, n = 1, 2, ..., N\}$ that maximize the difference between between-class scatter and within-class scatter in the projected subspace.

---

optimization problem as

$$
\begin{aligned}
\tilde{\mathbf{U}}^{(n)} &= \arg\max_{\mathbf{U}^{(n)}} \mathrm{tr}\left(\mathbf{U}^{(n)^T}\mathbf{S}_{B_{\acute{y}}}^{(n)}\mathbf{U}^{(n)}\right) - \zeta \cdot \mathrm{tr}\left(\mathbf{U}^{(n)^T}\mathbf{S}_{W_{\acute{y}}}^{(n)}\mathbf{U}^{(n)}\right) \\
&= \arg\max_{\mathbf{U}^{(n)}} \mathrm{tr}\left(\mathbf{U}^{(n)^T}(\mathbf{S}_{B_{\acute{y}}}^{(n)} - \zeta \cdot \mathbf{S}_{W_{\acute{y}}}^{(n)})\mathbf{U}^{(n)}\right).
\end{aligned}
\tag{7.27}
$$

This can be treated as an eigenvalue problem and the objective function above is maximized only if $\tilde{\mathbf{U}}^{(n)}$ consists of the P_n eigenvectors of $(\mathbf{S}_{B_{\acute{y}}}^{(n)} - \zeta \cdot \mathbf{S}_{W_{\acute{y}}}^{(n)})$ associated with the P_n largest eigenvalues.

The subspace dimensions in GTDA can be determined in a similar manner as the Q-based method discussed in Section 6.2.4. The tuning parameter ζ is set to the largest eigenvalue of the $(\mathbf{S}_{W_{\acute{y}}}^{(n)})^{-1}\mathbf{S}_{B_{\acute{y}}}^{(n)}$ in each iteration. The pseudocode is summarized in Algorithm 7.3. This algorithm is shown to have good convergence property in [Tao et al., 2007b]. It should be noted that the criterion used (J_3) is dependent on the coordinate system [Fukunaga, 1990].

7.4 Tensor Rank-One Discriminant Analysis

The TR1DA algorithm is a TVP-based supervised MSL algorithm introduced in [Wang and Gong, 2006; Tao et al., 2006, 2008a]. It can be considered a combination of TROD (Section 6.3) and GTDA, aiming to solve for a TVP to maximize the difference between the between-class scatter and within-class scatter.

7.4.1 TR1DA Problem Formulation

As a TVP-based MSL approach, TR1DA finds P EMPs one at a time as in TROD and UMPCA. Given M training tensor samples $\{\mathcal{X}_m\}$, the pth EMP $\{\mathbf{u}_p^{(n)}\}$ projects the mth sample $\mathcal{X}_m$ to a scalar feature y_{m_p} as

$$y_{m_p} = \mathcal{X}_m \times_1 \mathbf{u}_p^{(1)^T} \times_2 \mathbf{u}_p^{(2)^T} \ldots \times_N \mathbf{u}_p^{(N)^T} = \mathcal{X}_m \times_{n=1}^N \{\mathbf{u}_p^{(n)}\}. \tag{7.28}$$

The TVP formed with these P EMPs $\{\mathbf{u}_p^{(n)}\}_N^P$ projects $\mathcal{X}_m$ to a vector $\mathbf{y}_m \in \mathbb{R}^{P \times 1}$ as

$$\mathbf{y}_m = \mathcal{X}_m \times_{n=1}^N \{\mathbf{u}_p^{(n)}\}_N^P. \tag{7.29}$$

From Definition 3.17, the between-class scatter of $\{y_{m_p}\}$ for the pth EMP is

$$S_{B_{yp}} = \sum_{c=1}^C M_c (\bar{y}_{c_p} - \bar{y}_p)^2, \tag{7.30}$$

and the within-class scatter of $\{y_{m_p}\}$ for the pth EMP is

$$S_{W_{yp}} = \sum_{m=1}^M (y_{m_p} - \bar{y}_{c_{m_p}})^2, \tag{7.31}$$

where $\bar{y}_p$ is the overall mean defined as

$$\bar{y}_p = \frac{1}{M} \sum_{m=1}^M y_{m_p}, \tag{7.32}$$

and $\bar{y}_{c_p}$ is the class mean defined as

$$\bar{y}_{c_p} = \frac{1}{M_c} \sum_{m=1, c_m=c}^M y_{m_p}. \tag{7.33}$$

The TR1DA problem: A set of M labeled tensor data samples $\{\mathcal{X}_1, \mathcal{X}_2, \ldots, \mathcal{X}_M\}$ are available for training with class labels $\mathbf{c} \in \mathbb{R}^M$. Each sample $\mathcal{X}_m \in \mathbb{R}^{I_1 \times I_2 \times \ldots \times I_N}$ assumes values in a tensor space $\mathbb{R}^{I_1} \bigotimes \mathbb{R}^{I_2} \ldots \bigotimes \mathbb{R}^{I_N}$, where I_n is the mode-n dimension. The class label for the mth sample $\mathcal{X}_m$ is $c_m = \mathbf{c}(m)$ and there are C classes in total. The objective of TR1DA is to find a TVP, which consists of P EMPs, $\{\mathbf{u}_p^{(n)} \in \mathbb{R}^{I_n \times 1}, n = 1, ..., N\}_{p=1}^{P}$, that maps the original tensor space $\mathbb{R}^{I_1} \bigotimes \mathbb{R}^{I_2} ... \bigotimes \mathbb{R}^{I_N}$ into a vector subspace $\mathbb{R}^P$ (with $P < \prod_{n=1}^{N} I_n$) as

$$\mathbf{y}_m = \mathcal{X}_m \times_{n=1}^{N} \{\mathbf{u}_p^{(n)}\}_N^P, m = 1, ..., M, \qquad (7.34)$$

to maximize the scatter-difference-based discriminant criterion $(S_{B_{yp}} - \zeta \cdot S_{W_{yp}})$ in the projected space, where ζ is a tuning parameter.

Thus, we solve TR1DA in P steps. At each step p, we solve for the pth EMP $\{\mathbf{u}_p^{(n)}\}$ and the objective function of TR1DA can be written as

$$\{\tilde{\mathbf{u}}_p^{(n)}\} = \arg\max_{\{\mathbf{u}_p^{(n)}\}} S_{B_{yp}} - \zeta \cdot S_{W_{yp}}$$

$$= \arg\max_{\{\mathbf{u}_p^{(n)}\}} \sum_{c=1}^{C} M_c (\bar{y}_{c_p} - \bar{y}_p)^2 - \zeta \sum_{m=1}^{M} (y_{m_p} - \bar{y}_{c_{m_p}})^2. \quad (7.35)$$

Its solution follows the greedy approach of successive residue calculation in TROD (Section 6.3.2), where after obtaining the pth EMP $\{\tilde{\mathbf{u}}_p^{(n)}\}$, an input tensor is replaced by its residue as

$$\mathcal{X}_m \leftarrow \mathcal{X}_m - y_{m_p} \cdot (\tilde{\mathbf{u}}_p^{(1)} \circ \tilde{\mathbf{u}}_p^{(2)} \circ ... \circ \tilde{\mathbf{u}}_p^{(N)}). \qquad (7.36)$$

Algorithm 7.4 is the pseudocode for this algorithm.

7.4.2 Solving for the pth EMP

As in TROD and UMPCA, we employ the APP method to solve for $\mathbf{u}_p^{(n)}$, the mode-n projection vector of the pth EMP, conditioned on the projection vectors in all the other modes $\{\mathbf{u}_p^{(j)}, j \neq n\}$.

To solve for $\mathbf{u}_p^{(n)}$ given $\{\mathbf{u}_p^{(j)}, j \neq n\}$, the M tensor samples are projected in the other $(N-1)$ modes $\{j \neq n\}$ first to obtain the following vectors through partial multilinear projection as

$$\acute{\mathbf{y}}_{m_p}^{(n)} = \mathcal{X}_m \times_1 \mathbf{u}_p^{(1)^T} ... \times_{n-1} \mathbf{u}_p^{(n-1)^T} \times_{n+1} \mathbf{u}_p^{(n+1)^T} ... \times_N \mathbf{u}_p^{(N)^T}, \quad (7.37)$$

where $\acute{\mathbf{y}}_{m_p}^{(n)} \in \mathbb{R}^{I_n}$. Thus, we obtain a conditional subproblem of determining $\mathbf{u}_p^{(n)}$ to maximize the scatter difference of the projections of vector samples $\{\acute{\mathbf{y}}_{m_p}^{(n)}\}$. This is a linear and simpler discriminant analysis problem with input samples $\{\acute{\mathbf{y}}_{m_p}^{(n)}\}$. The corresponding between-class scatter matrix $\acute{\mathbf{S}}_{B_p}^{(n)}$ and within-class scatter matrix $\acute{\mathbf{S}}_{W_p}^{(n)}$ are then defined as

$$\acute{\mathbf{S}}_{B_p}^{(n)} = \sum_{c=1}^{C} M_c (\bar{\acute{\mathbf{y}}}_{c_p}^{(n)} - \bar{\acute{\mathbf{y}}}_p^{(n)})(\bar{\acute{\mathbf{y}}}_{c_p}^{(n)} - \bar{\acute{\mathbf{y}}}_p^{(n)})^T, \qquad (7.38)$$

Algorithm 7.4 Tensor rank-one discriminant analysis (TR1DA)

Input: A set of tensor samples $\{\mathcal{X}_m \in \mathbb{R}^{I_1 \times I_2 \times \dots \times I_N}, m = 1, \dots, M\}$ with class labels $\mathbf{c} \in \mathbb{R}^M$, the desired feature vector length P, a tuning parameter ζ, the maximum number of iterations K (or a small number η for testing convergence).

Process:

1: **for** $p = 1$ **to** P **do**
2:    [Step p: determine the pth EMP]
3:    Initialize $\mathbf{u}_{p_{(0)}}^{(n)} \in \mathbb{R}^{I_n}$ for $n = 1, \dots, N$.
4:    **[Local optimization:]**
5:    **for** $k = 1$ **to** K **do**
6:       **for** $n = 1$ **to** N **do**
7:          Calculate the mode-n partial multilinear projection $\acute{\mathbf{y}}_{m_p}^{(n)}$ according to Equation (7.37), for $m = 1, \dots, M$.
8:          Calculate $\acute{\mathbf{S}}_{B_p}^{(n)}$ and $\acute{\mathbf{S}}_{W_p}^{(n)}$ according to Equations (7.38) and (7.39).
9:          Set $\mathbf{u}_{p_{(k)}}^{(n)}$ to be the eigenvector of $(\acute{\mathbf{S}}_{B_p}^{(n)} - \zeta \cdot \acute{\mathbf{S}}_{W_p}^{(n)})$ associated with the largest eigenvalue.
10:      **end for**
11:    **end for**
12:    Set $\tilde{\mathbf{u}}_p^{(n)} = \mathbf{u}_{p_{(K)}}^{(n)}$.
13:    Calculate the projection y_{m_p} according to Equation (7.28) for $m = 1, \dots, M$.
14:    Replace $\mathcal{X}_m$ with its residue according to Equation (7.36).
15: **end for**

Output: The TVP $\{\tilde{\mathbf{u}}_p^{(n)}\}_N^P$ that maximizes the difference between between-class scatter and within-class scatter in the projected subspace.

---

and

$$\acute{\mathbf{S}}_{W_p}^{(n)} = \sum_{m=1}^{M} (\acute{\mathbf{y}}_{m_p}^{(n)} - \bar{\acute{\mathbf{y}}}_{c_{m_p}}^{(n)})(\acute{\mathbf{y}}_{m_p}^{(n)} - \bar{\acute{\mathbf{y}}}_{c_{m_p}}^{(n)})^T, \tag{7.39}$$

respectively, where the class mean is

$$\bar{\acute{\mathbf{y}}}_{c_p}^{(n)} = \frac{1}{M_c} \sum_{m=1, c_m=c}^{M} \acute{\mathbf{y}}_{m_p}^{(n)}, \tag{7.40}$$

and the overall mean is

$$\bar{\acute{\mathbf{y}}}_p^{(n)} = \frac{1}{M} \sum_{m=1}^{M} \acute{\mathbf{y}}_{m_p}^{(n)}. \tag{7.41}$$

Thus, $\tilde{\mathbf{u}}_p^{(n)}$ is obtained as the eigenvector of the scatter difference $(\acute{\mathbf{S}}_{B_p}^{(n)} - \zeta \acute{\mathbf{S}}_{W_p}^{(n)})$ corresponding to the largest eigenvalue.

In testing, TR1DA extracts feature vector $\mathbf{y}$ from a test sample $\mathcal{X}$ through the TVP obtained in the same way as TROD/UMPCA as

$$\mathbf{y} = \mathcal{X} \times_{n=1}^{N} \{\tilde{\mathbf{u}}_p^{(n)}\}_N^P. \tag{7.42}$$

The criterion used in TR1DA can be considered a degenerate version of that in GTDA. Thus, the criterion is also dependent on the coordinate system as GTDA. The parameter ζ can be set heuristically or determined through cross-validation. This algorithm is shown to have good convergence property in [Tao et al., 2008a].

7.5   Uncorrelated Multilinear Discriminant Analysis

UMLDA [Lu et al., 2009c] shares similarity with UMPCA (Section 6.4) in the motivation of deriving uncorrelated features. It extracts uncorrelated discriminative features directly from tensor data by solving a TVP that maximizes a scatter-ratio-based criterion, in contrast with the scatter difference criterion in TR1DA. In addition, to enhance performance in the small sample size (SSS) scenario, an adaptive regularization factor is incorporated to reduce the variance of the within-class scatter estimation through a data-independent regularization parameter, resulting in regularized UMLDA (R-UMLDA). Furthermore, as different initialization or regularization of UMLDA results in different features, an aggregation scheme is adopted to combine several differently initialized and differently regularized UMLDA learners, leading to regularized UMLDA with aggregation (R-UMLDA-A). The aggregation also alleviates the regularization parameter selection problem.

7.5.1   UMLDA Problem Formulation

As in TR1DA, TROD and UMPCA, y_{m_p} is the scalar feature of sample $\mathcal{X}_m$ mapped through the pth EMP $\{\mathbf{u}_p^{(n)}\}$. The corresponding between-class scatter $S_{B_{yp}}$ and the within-class scatter $S_{W_{yp}}$ are as defined in Equations (7.30) and (7.31), respectively. Thus, the Fisher's discrimination criterion for the pth scalar samples is

$$F_{yp} = \frac{S_{B_{yp}}}{S_{W_{yp}}}. \tag{7.43}$$

Also, as in UMPCA, let $\mathbf{g}_p$ denote the pth coordinate vector. Its mth component $\mathbf{g}_p(m) = g_{p_m} = y_{m_p}$.

The following gives a formal definition of the UMLDA problem.

The UMLDA Problem: A set of M labeled tensor data samples $\{\mathcal{X}_1, \mathcal{X}_2, ..., \mathcal{X}_M\}$ are available for training with class labels $\mathbf{c} \in \mathbb{R}^M$. Each sample $\mathcal{X}_m \in \mathbb{R}^{I_1 \times I_2 \times ... \times I_N}$ assumes values in a tensor space $\mathbb{R}^{I_1} \bigotimes \mathbb{R}^{I_2} ... \bigotimes \mathbb{R}^{I_N}$, where I_n is the mode-n dimension. The class label for the mth sample $\mathcal{X}_m$ is $c_m = \mathbf{c}(m)$ and there are C classes in total. The objective of UMLDA is to find a TVP, which consists of P EMPs $\{\mathbf{u}_p^{(n)} \in \mathbb{R}^{I_n \times 1}, n = 1, ..., N\}_{p=1}^P$, mapping from the original tensor space $\mathbb{R}^{I_1} \bigotimes \mathbb{R}^{I_2} ... \bigotimes \mathbb{R}^{I_N}$ into a vector subspace $\mathbb{R}^P$ (with $P < \prod_{n=1}^N I_n$):

$$\mathbf{y}_m = \mathcal{X}_m \times_{n=1}^N \{\mathbf{u}_p^{(n)}\}_N^P, m = 1, ..., M, \tag{7.44}$$

such that the Fisher's discrimination criterion F_{yp} in Equation (7.43) is maximized in each EMP direction, subject to the constraint that the P coordinate vectors $\{\mathbf{g}_p \in \mathbb{R}^M, p = 1, ..., P\}$ are uncorrelated.

Centering: As in UMPCA, we center the input data first to obtain $\{\tilde{\mathcal{X}}_m\}$ as in Equation (6.21). The corresponding projections become $\{\tilde{\mathbf{y}}_m\}$ and the pth element of $\tilde{\mathbf{y}}_m$ is $\tilde{y}_{m_p}$. Thus, the between-class scatter in Equation (7.30) becomes

$$S_{B_{yp}} = \sum_{c=1}^C M_c \cdot \bar{\tilde{y}}_{c_p}^2, \tag{7.45}$$

and the within-class scatter in Equation (7.31) becomes

$$S_{W_{yp}} = \sum_{m=1}^M (\tilde{y}_{m_p} - \bar{\tilde{y}}_{c_{m_p}})^2. \tag{7.46}$$

The corresponding coordinate vectors become $\{\tilde{\mathbf{g}}_p\}$ so that the constraint of uncorrelated features is equivalent to that of orthogonal features (see Equation (6.59)).

Thus, the UMLDA objective function for the pth EMP is

$$\{\tilde{\mathbf{u}}_p^{(n)}\} = \arg \max_{\{\mathbf{u}_p^{(n)}\}} F_{yp} = \arg \max_{\{\mathbf{u}_p^{(n)}\}} \frac{\sum_{c=1}^C M_c \cdot \bar{\tilde{y}}_{c_p}^2}{\sum_{m=1}^M (\tilde{y}_{m_p} - \bar{\tilde{y}}_{c_{m_p}})^2}, \tag{7.47}$$

$$\text{subject to} \quad \frac{\tilde{\mathbf{g}}_p^T \tilde{\mathbf{g}}_q}{\| \tilde{\mathbf{g}}_p \| \| \tilde{\mathbf{g}}_q \|} = \delta_{pq}, p, q = 1, ..., P,$$

where δ_{pq} is the Kronecker delta defined in Equation (6.61).

7.5.2 R-UMLDA Algorithm Derivation

The solution follows the successive determination approach in the derivation of the uncorrelated LDA (ULDA) in [Jin et al., 2001a], and it is similar to the successive approach in UMPCA. The P EMPs $\{\mathbf{u}_p^{(n)}\}_N^P$ are determined sequentially in P steps, with the pth step obtaining the pth EMP:

Step 1: Determine the first EMP $\{\mathbf{u}_1^{(n)}\}$ by maximizing F_{y1} without any constraint.

Algorithm 7.5 Regularized uncorrelated multilinear discriminant analysis (R-UMLDA)

Input: A set of tensor samples $\{\mathcal{X}_m \in \mathbb{R}^{I_1 \times I_2 \times \dots \times I_N}, m = 1, \dots, M\}$ with class labels $\mathbf{c} \in \mathbb{R}^M$, the desired feature vector length P, the regularization parameter γ, the maximum number of iterations K (or a small number η for testing convergence).

Process:

1: **Centering:** Center the input samples to get $\{\tilde{\mathcal{X}}_m\}$ according to Equation (6.21).

2: **for** $p = 1$ **to** P **do**

3:     [Step p: determine the pth EMP]

4:     Initialize $\mathbf{u}_{p_{(0)}}^{(n)} \in \mathbb{R}^{I_n}$ for $n = 1, \dots, N$.

5:     [**Local optimization:**]

6:     **for** $k = 1$ **to** K **do**

7:         **for** $n = 1$ **to** N **do**

8:             Calculate the mode-n partial multilinear projection $\hat{\mathbf{y}}_{m_p}^{(n)}$ according to Equation (7.48), for $m = 1, \dots, M$.

9:             Calculate $\mathbf{R}_p^{(n)}$, $\hat{\mathbf{S}}_{B_p}^{(n)}$, and $\hat{\mathbf{S}}_{W_p}^{(n)}$ according to Equations (7.62), (7.49), and (7.52), respectively.

10:             Set $\mathbf{u}_{p_{(k)}}^{(n)}$ to be the eigenvector of $\left(\hat{\mathbf{S}}_{W_p}^{(n)}\right)^{-1} \mathbf{R}_p^{(n)} \hat{\mathbf{S}}_{B_p}^{(n)}$ associated with the largest eigenvalue.

11:         **end for**

12:     **end for**

13:     Set $\tilde{\mathbf{u}}_p^{(n)} = \mathbf{u}_{p_{(K)}}^{(n)}$.

14:     Calculate the coordinate vector $\tilde{\mathbf{g}}_p$: $\tilde{\mathbf{g}}_p(m) = \tilde{\mathcal{X}}_m \times_1 \tilde{\mathbf{u}}_p^{(1)^T} \dots \times_N \tilde{\mathbf{u}}_p^{(N)^T}$.

15: **end for**

Output: The TVP $\{\tilde{\mathbf{u}}_p^{(n)}\}_N^P$ that maximizes the ratio of between-class scatter over within-class scatter in the projected space.

Step 2: Determine the second EMP $\{\mathbf{u}_2^{(n)}\}$ by maximizing F_{y2} subject to the constraint that $\tilde{\mathbf{g}}_2^T \tilde{\mathbf{g}}_1 = 0$.

Step 3: Determine the third EMP $\{\mathbf{u}_3^{(n)}\}$ by maximizing F_{y3} subject to the constraint that $\tilde{\mathbf{g}}_3^T \tilde{\mathbf{g}}_1 = 0$ and $\tilde{\mathbf{g}}_3^T \tilde{\mathbf{g}}_2 = 0$.

Step p $(p = 4, \dots, P)$**:** Determine the pth EMP $\{\mathbf{u}_p^{(n)}\}$ by maximizing F_{yp} subject to the constraint that $\tilde{\mathbf{g}}_p^T \tilde{\mathbf{g}}_q = 0$ for $q = 1, \dots, p - 1$.

The algorithm to compute these EMPs is summarized in the pseudocode in Algorithm 7.5. The detailed derivation is presented below.

Following the APP method, to determine the pth EMP $\{\mathbf{u}_p^{(n)}\}$, the mode-n projection vector $\mathbf{u}_p^{(n)}$ is estimated one by one separately, conditioned on $\{\mathbf{u}_p^{(j)}, j \neq n\}$, the projection vectors in all the other modes.

To solve for $\mathbf{u}_p^{(n)}$ given $\{\mathbf{u}_p^{(j)}, j \neq n\}$, the tensor samples are projected in these $(N-1)$ modes $\{j \neq n\}$ first to obtain the mode-n partial multilinear projection

$$\hat{\mathbf{y}}_{m_p}^{(n)} = \tilde{\mathcal{X}}_m \times_1 \mathbf{u}_p^{(1)^T} ... \times_{n-1} \mathbf{u}_p^{(n-1)^T} \times_{n+1} \mathbf{u}_p^{(n+1)^T} ... \times_N \mathbf{u}_p^{(N)^T}. \quad (7.48)$$

The conditional subproblem then becomes to determine $\mathbf{u}_p^{(n)}$ that maximizes the scatter ratio of the vector samples $\{\hat{\mathbf{y}}_{m_p}^{(n)}\}$, subject to the zero-correlation constraint. This is the ULDA problem in [Jin et al., 2001a] with input samples $\{\hat{\mathbf{y}}_{m_p}^{(n)}\}$. The corresponding between-class scatter matrix $\hat{\mathbf{S}}_{B_p}^{(n)}$ is defined similar to Equation (7.38) as

$$\hat{\mathbf{S}}_{B_p}^{(n)} = \sum_{c=1}^{C} M_c \cdot \bar{\hat{\mathbf{y}}}_{c_p}^{(n)} \bar{\hat{\mathbf{y}}}_{c_p}^{(n)^T}, \quad (7.49)$$

where

$$\bar{\hat{\mathbf{y}}}_{c_p}^{(n)} = \frac{1}{M_c} \sum_{m=1, c_m=c}^{M} \hat{\mathbf{y}}_{m_p}^{(n)}, \quad (7.50)$$

and the overall mean is

$$\bar{\hat{\mathbf{y}}}_p^{(n)} = \frac{1}{M} \sum_{m=1}^{M} \hat{\mathbf{y}}_{m_p}^{(n)} = \mathbf{0}. \quad (7.51)$$

Motivation of regularization: Empirical study of the iterative UMLDA algorithm (i.e., $\gamma = 0$) under the SSS scenario in [Lu et al., 2009c] indicates that the iterations tend to minimize the within-class scatter toward zero in order to maximize the scatter ratio because the scatter ratio reaches a maximum of infinity when the within-class scatter is zero and the between-class scatter is non-zero. However, the estimated within-class scatter on limited training data is usually much smaller than the real within-class scatter, due to limited number of samples for each class. Therefore, regularization is adopted in [Lu et al., 2009c] to improve the generalization capability of UMLDA under the SSS scenario, leading to the R-UMLDA algorithm. A regularization term is introduced in Equation (7.52) so that during the iteration, less focus is put on shrinking the within-class scatter. Moreover, the regularization introduced is adaptive because γ is the only regularization parameter, and the regularization term in mode-n is scaled by $\lambda_{max}(\check{\mathbf{S}}_W^{(n)})$, an approximate estimate of the mode-n within-class scatter in the training data. The basic UMLDA is obtained by setting $\gamma = 0$.

Next, we define a regularized version of the corresponding within-class scatter matrix as

$$\hat{\mathbf{S}}_{W_p}^{(n)} = \sum_{m=1}^{M} (\hat{\mathbf{y}}_{m_p}^{(n)} - \bar{\hat{\mathbf{y}}}_{c_{m_p}}^{(n)})(\hat{\mathbf{y}}_{m_p}^{(n)} - \bar{\hat{\mathbf{y}}}_{c_{m_p}}^{(n)})^T + \gamma \cdot \lambda_{max}(\check{\mathbf{S}}_W^{(n)}) \cdot \mathbf{I}_{I_n}, (7.52)$$

where $\gamma \geq 0$ is a regularization parameter, $\mathbf{I}_{I_n}$ is an identity matrix of size $I_n \times I_n$, and $\lambda_{max}(\breve{\mathbf{S}}_W^{(n)})$ is the maximum eigenvalue of $\breve{\mathbf{S}}_W^{(n)}$. $\breve{\mathbf{S}}_W^{(n)}$ is the within-class scatter matrix for the mode-n vectors of training samples defined as

$$\breve{\mathbf{S}}_W^{(n)} = \sum_{m=1}^{M} \left(\tilde{\mathbf{X}}_{m(n)} - \bar{\tilde{\mathbf{X}}}_{c_m(n)} \right) \left(\tilde{\mathbf{X}}_{m(n)} - \bar{\tilde{\mathbf{X}}}_{c_m(n)} \right)^T, \qquad (7.53)$$

where $\bar{\tilde{\mathbf{X}}}_{c(n)}$ is the mode-n unfolding of the class mean

$$\bar{\tilde{\mathcal{X}}}_c = \frac{1}{M_c} \sum_{m=1, c_m=c}^{M} \tilde{\mathcal{X}}_m. \qquad (7.54)$$

With Equations (7.49) and (7.52), we are ready to solve for the P EMPs. For $p = 1$, the optimal projection vector $\tilde{\mathbf{u}}_1^{(n)}$ that maximizes the Fisher's discrimination criterion

$$\frac{\mathbf{u}_1^{(n)^T} \hat{\mathbf{S}}_{B_1}^{(n)} \mathbf{u}_1^{(n)}}{\mathbf{u}_1^{(n)^T} \hat{\mathbf{S}}_{W_1}^{(n)} \mathbf{u}_1^{(n)}} \qquad (7.55)$$

in the projected subspace is obtained as the eigenvector of

$$\left(\hat{\mathbf{S}}_{W_1}^{(n)} \right)^{-1} \hat{\mathbf{S}}_{B_1}^{(n)} \qquad (7.56)$$

associated with the largest eigenvalue for a nonsingular $\hat{\mathbf{S}}_{W_1}^{(n)}$. Next, given the first $(p-1)$ EMPs, where $p > 1$, the pth EMP aims to maximize the scatter ratio F_{yp}, subject to the constraint that features projected by the pth EMP are uncorrelated with those projected by the first $(p-1)$ EMPs. Again, let $\hat{\mathbf{Y}}_p^{(n)} \in \mathbb{R}^{I_n \times M}$ be a matrix with its mth column to be $\hat{\mathbf{y}}_{m_p}^{(n)}$, that is,

$$\hat{\mathbf{Y}}_p^{(n)} = \left[\hat{\mathbf{y}}_{1_p}^{(n)}, \hat{\mathbf{y}}_{2_p}^{(n)}, ..., \hat{\mathbf{y}}_{M_p}^{(n)} \right]. \qquad (7.57)$$

The pth coordinate vector is then obtained as

$$\tilde{\mathbf{g}}_p = \hat{\mathbf{Y}}_p^{(n)^T} \mathbf{u}_p^{(n)}. \qquad (7.58)$$

The constraint that $\tilde{\mathbf{g}}_p$ is uncorrelated with $\{\tilde{\mathbf{g}}_q, q = 1, ..., p-1\}$ can be written as

$$\tilde{\mathbf{g}}_p^T \tilde{\mathbf{g}}_q = \mathbf{u}_p^{(n)^T} \hat{\mathbf{Y}}_p^{(n)} \tilde{\mathbf{g}}_q = 0, q = 1, ..., p-1. \qquad (7.59)$$

Thus, $\mathbf{u}_p^{(n)}$ $(p > 1)$ can be determined by solving the following constrained optimization problem:

$$\tilde{\mathbf{u}}_p^{(n)} = \underset{\mathbf{u}_p^{(n)}}{\arg\max} \frac{\mathbf{u}_p^{(n)^T} \hat{\mathbf{S}}_{B_p}^{(n)} \mathbf{u}_p^{(n)}}{\mathbf{u}_p^{(n)^T} \hat{\mathbf{S}}_{W_p}^{(n)} \mathbf{u}_p^{(n)}},$$

$$\texttt{subject to} \quad \mathbf{u}_p^{(n)^T} \hat{\mathbf{Y}}_p^{(n)} \tilde{\mathbf{g}}_q = 0, q = 1, ..., p-1. \qquad (7.60)$$

The solution is given by Theorem 7.1 for nonsingular $\hat{\mathbf{S}}_{W_p}^{(n)}$.

Theorem 7.1. *When* $\hat{\mathbf{S}}_{W_p}^{(n)}$ *is nonsingular, the solution to the problem Equation (7.60) is the generalized eigenvector corresponding to the largest generalized eigenvalue of the following generalized eigenvalue problem:*

$$\mathbf{R}_p^{(n)}\hat{\mathbf{S}}_{B_p}^{(n)}\mathbf{u} = \lambda\hat{\mathbf{S}}_{W_p}^{(n)}\mathbf{u}, \tag{7.61}$$

where

$$\mathbf{R}_p^{(n)} = \mathbf{I}_{I_n} -$$

$$\hat{\mathbf{Y}}_p^{(n)}\mathbf{G}_{p-1}\left(\mathbf{G}_{p-1}^T\hat{\mathbf{Y}}_p^{(n)^T}\hat{\mathbf{S}}_{W_p}^{(n)^{-1}}\hat{\mathbf{Y}}_p^{(n)}\mathbf{G}_{p-1}\right)^{-1}\mathbf{G}_{p-1}^T\hat{\mathbf{Y}}_p^{(n)^T}\hat{\mathbf{S}}_{W_p}^{(n)^{-1}}, \tag{7.62}$$

$$\mathbf{G}_{p-1} = [\tilde{\mathbf{g}}_1 \quad \tilde{\mathbf{g}}_2 \quad \cdots \quad \tilde{\mathbf{g}}_{p-1}] \in \mathbb{R}^{M\times(p-1)}. \tag{7.63}$$

Proof. For a nonsingular $\hat{\mathbf{S}}_{W_p}^{(n)}$, any $\mathbf{u}_p^{(n)}$ can be normalized such that

$$\mathbf{u}_p^{(n)^T}\hat{\mathbf{S}}_{W_p}^{(n)}\mathbf{u}_p^{(n)} = 1 \tag{7.64}$$

and the ratio $\dfrac{\mathbf{u}_p^{(n)^T}\hat{\mathbf{S}}_{B_p}^{(n)}\mathbf{u}_p^{(n)}}{\mathbf{u}_p^{(n)^T}\hat{\mathbf{S}}_{W_p}^{(n)}\mathbf{u}_p^{(n)}}$ remains unchanged. Therefore, the maximization of this ratio is equivalent to the maximization of $\mathbf{u}_p^{(n)^T}\hat{\mathbf{S}}_{B_p}^{(n)}\mathbf{u}_p^{(n)}$ with the constraint Equation (7.64). Lagrange multipliers can be used to transform the problem Equation (7.60) to the following to include all the constraints:

$$\varphi_p = \mathbf{u}_p^{(n)^T}\hat{\mathbf{S}}_{B_p}^{(n)}\mathbf{u}_p^{(n)} - \nu\left(\mathbf{u}_p^{(n)^T}\hat{\mathbf{S}}_{W_p}^{(n)}\mathbf{u}_p^{(n)} - 1\right) - \sum_{q=1}^{p-1}\mu_q\mathbf{u}_p^{(n)^T}\hat{\mathbf{Y}}_p^{(n)}\tilde{\mathbf{g}}_q, \tag{7.65}$$

where ν and $\{\mu_q, q = 1, ..., p-1\}$ are Lagrange multipliers.

The optimization is performed by setting the partial derivative of φ_p with respect to $\mathbf{u}_p^{(n)}$ to zero:

$$\frac{\partial\varphi_p}{\partial\mathbf{u}_p^{(n)}} = 2\hat{\mathbf{S}}_{B_p}^{(n)}\mathbf{u}_p^{(n)} - 2\nu\hat{\mathbf{S}}_{W_p}^{(n)}\mathbf{u}_p^{(n)} - \sum_{q=1}^{p-1}\mu_q\hat{\mathbf{Y}}_p^{(n)}\tilde{\mathbf{g}}_q = 0. \tag{7.66}$$

Multiplying Equation (7.66) by $\mathbf{u}_p^{(n)^T}$ results in

$$2\mathbf{u}_p^{(n)^T}\hat{\mathbf{S}}_{B_p}^{(n)}\mathbf{u}_p^{(n)} - 2\nu\mathbf{u}_p^{(n)^T}\hat{\mathbf{S}}_{W_p}^{(n)}\mathbf{u}_p^{(n)} = 0 \Rightarrow \nu = \frac{\mathbf{u}_p^{(n)^T}\hat{\mathbf{S}}_{B_p}^{(n)}\mathbf{u}_p^{(n)}}{\mathbf{u}_p^{(n)^T}\hat{\mathbf{S}}_{W_p}^{(n)}\mathbf{u}_p^{(n)}}, \tag{7.67}$$

which indicates that ν is exactly the criterion to be maximized.

Next, a set of $(p-1)$ equations are obtained by multiplying Equation (7.66) by $\tilde{\mathbf{g}}_q^T\hat{\mathbf{Y}}_p^{(n)^T}\hat{\mathbf{S}}_{W_p}^{(n)^{-1}}$, $q = 1, ..., p-1$, respectively:

$$2\tilde{\mathbf{g}}_q^T\hat{\mathbf{Y}}_p^{(n)^T}\hat{\mathbf{S}}_{W_p}^{(n)^{-1}}\hat{\mathbf{S}}_{B_p}^{(n)}\mathbf{u}_p^{(n)} - \sum_{s=1}^{p-1}\mu_s\tilde{\mathbf{g}}_q^T\hat{\mathbf{Y}}_p^{(n)^T}\hat{\mathbf{S}}_{W_p}^{(n)^{-1}}\hat{\mathbf{Y}}_p^{(n)}\tilde{\mathbf{g}}_s = 0. \tag{7.68}$$

Let

$$\boldsymbol{\mu}_{p-1} = [\mu_1 \ \mu_2 \ \cdots \ \mu_{p-1}]^T \tag{7.69}$$

and use Equation (7.63), then the $(p-1)$ equations of Equation (7.68) can be represented in a single matrix equation as follows:

$$2\mathbf{G}_{p-1}^T \hat{\mathbf{Y}}_p^{(n)^T} \hat{\mathbf{S}}_{W_p}^{(n)^{-1}} \hat{\mathbf{S}}_{B_p}^{(n)} \mathbf{u}_p^{(n)} - \mathbf{G}_{p-1}^T \hat{\mathbf{Y}}_p^{(n)^T} \hat{\mathbf{S}}_{W_p}^{(n)^{-1}} \hat{\mathbf{Y}}_p^{(n)} \mathbf{G}_{p-1} \boldsymbol{\mu}_{p-1} = 0. \tag{7.70}$$

Thus,

$$\boldsymbol{\mu}_{p-1} = 2 \left(\mathbf{G}_{p-1}^T \hat{\mathbf{Y}}_p^{(n)^T} \hat{\mathbf{S}}_{W_p}^{(n)^{-1}} \hat{\mathbf{Y}}_p^{(n)} \mathbf{G}_{p-1} \right)^{-1} \mathbf{G}_{p-1}^T \hat{\mathbf{Y}}_p^{(n)^T} \hat{\mathbf{S}}_{W_p}^{(n)^{-1}} \hat{\mathbf{S}}_{B_p}^{(n)} \mathbf{u}_p^{(n)}. \tag{7.71}$$

Because from Equations (7.63) and (7.69),

$$\sum_{q=1}^{p-1} \mu_q \hat{\mathbf{Y}}_p^{(n)} \tilde{\mathbf{g}}_q = \hat{\mathbf{Y}}_p^{(n)} \mathbf{G}_{p-1} \boldsymbol{\mu}_{p-1}, \tag{7.72}$$

Equation (7.66) can be written as

$$2\hat{\mathbf{S}}_{B_p}^{(n)} \mathbf{u}_p^{(n)} - 2\nu \hat{\mathbf{S}}_{W_p}^{(n)} \mathbf{u}_p^{(n)} - \hat{\mathbf{Y}}_p^{(n)} \mathbf{G}_{p-1} \boldsymbol{\mu}_{p-1} = 0$$

$$\Rightarrow \quad \nu \hat{\mathbf{S}}_{W_p}^{(n)} \mathbf{u}_p^{(n)} = \hat{\mathbf{S}}_{B_p}^{(n)} \mathbf{u}_p^{(n)} - \hat{\mathbf{Y}}_p^{(n)} \mathbf{G}_{p-1} \frac{\boldsymbol{\mu}_{p-1}}{2}$$

$$= \quad \hat{\mathbf{S}}_{B_p}^{(n)} \mathbf{u}_p^{(n)} - \hat{\mathbf{Y}}_p^{(n)} \mathbf{G}_{p-1} \left(\mathbf{G}_{p-1}^T \hat{\mathbf{Y}}_p^{(n)^T} \hat{\mathbf{S}}_{W_p}^{(n)^{-1}} \hat{\mathbf{Y}}_p^{(n)} \mathbf{G}_{p-1} \right)^{-1}$$

$$\mathbf{G}_{p-1}^T \hat{\mathbf{Y}}_p^{(n)^T} \hat{\mathbf{S}}_{W_p}^{(n)^{-1}} \hat{\mathbf{S}}_{B_p}^{(n)} \mathbf{u}_p^{(n)}$$

$$= \quad \left[\mathbf{I}_{I_n} - \hat{\mathbf{Y}}_p^{(n)} \mathbf{G}_{p-1} \left(\mathbf{G}_{p-1}^T \hat{\mathbf{Y}}_p^{(n)^T} \hat{\mathbf{S}}_{W_p}^{(n)^{-1}} \hat{\mathbf{Y}}_p^{(n)} \mathbf{G}_{p-1} \right)^{-1} \right.$$

$$\left. \mathbf{G}_{p-1}^T \hat{\mathbf{Y}}_p^{(n)^T} \hat{\mathbf{S}}_{W_p}^{(n)^{-1}} \right] \hat{\mathbf{S}}_{B_p}^{(n)} \mathbf{u}_p^{(n)}. \tag{7.73}$$

Using the definition in Equation (7.62), a generalized eigenvalue problem is obtained as

$$\mathbf{R}_p^{(n)} \hat{\mathbf{S}}_{B_p}^{(n)} \mathbf{u} = \nu \hat{\mathbf{S}}_{W_p}^{(n)} \mathbf{u}. \tag{7.74}$$

Because ν is the criterion to be maximized, the maximization is achieved by setting $\tilde{\mathbf{u}}_p^{(n)}$ to be the generalized eigenvector corresponding to the largest generalized eigenvalue of Equation (7.61). $\qquad\qquad\square$

By setting $\mathbf{R}_1^{(n)} = \mathbf{I}_{I_n}$ and from Theorem 7.1, a unified solution for R-UMLDA is obtained when $\hat{\mathbf{S}}_{W_p}^{(n)}$ is nonsingular: for $p = 1, \ldots, P$, $\tilde{\mathbf{u}}_p^{(n)}$ is obtained as the eigenvector of

$$\left(\hat{\mathbf{S}}_{W_p}^{(n)} \right)^{-1} \mathbf{R}_p^{(n)} \hat{\mathbf{S}}_{B_p}^{(n)} \tag{7.75}$$

associated with the largest eigenvalue. In addition, as in UMPCA, the maximum number of features that can be extracted by R-UMLDA does not exceed

$\min\{\min_n I_n, M\}$, similarly from Corollary 6.1 (page 130). In addition, in the implementation of R-UMLDA, in order to get a better conditioned matrix for the inverse computation, a small term $(\varrho \cdot \mathbf{I}_{p-1})$ is added in computing the matrix inverse of $\left(\mathbf{G}_{p-1}^T \hat{\mathbf{Y}}_p^{(n)^T} \hat{\mathbf{S}}_{W_p}^{(n)^{-1}} \hat{\mathbf{Y}}_p^{(n)} \mathbf{G}_{p-1} \right)$ in Equation (7.62), with $\varrho = 10^{-3}$ [Lu et al., 2009c].

The derivation of Theorem 7.1 implies that per iteration, the scatter ratio F_{yp} is a non-decreasing function because each update of mode-n projection vector $\mathbf{u}^{(n)}$ maximizes F_{yp}, while the projection vectors in all the other modes $\{\mathbf{u}^{(j)}, j \neq n\}$ are considered fixed. However, the ratio F_{yp} may not have an upper-bound as in MPCA and UMPCA because it may reach infinity when a projection leads to zero within-class scatter, especially when there are only a small number of samples and the regularization is not strong enough. Therefore, R-UMLDA may not converge in terms of F_{yp}. Experimental studies in [Lu et al., 2009c] indicate that in practice, the R-UMLDA algorithm has better convergence with strong regularization.

7.5.3 Aggregation of R-UMLDA Learners

For better generalization, Lu et al. [2009c] further proposed aggregation of several differently initialized and regularized UMLDA learners, motivated from two properties of the basic R-UMLDA learner. On one hand, the number of useful discriminative features that can be extracted by a single R-UMLDA is limited. On the other hand, because R-UMLDA is affected by initialization and regularization [Lu et al., 2009c], which cannot be optimally determined, different initialization or regularization could result in different discriminative features. From the generalization theory explaining the success of the random subspace method [Ho, 1998], bagging, and boosting [Schapire, 2003; Skurichina and Duin, 2002; Breiman, 1998], the sensitivity of R-UMLDA to initialization and regularization suggests that R-UMLDA is not a stable learner and it is good for ensemble-based learning (Section 2.8). Therefore, several differently initialized and regularized UMLDA learners can be aggregated to get the regularized UMLDA with aggregation (R-UMLDA-A) so that multiple R-UMLDA learners can work together for better performance.

Remark 7.2. *Different projection orders can also result in different features so R-UMLDA with different projection orders can be aggregated as well. However, the effects of projection orders are similar to those of initializations and the number of possible projection orders (N!) is much less than the number of possible initializations (infinite) [Lu et al., 2009c].*

Several learners can be combined (or fused) at the feature level [Ross and Govindarajan, 2005], or at the matching score level [Ross et al., 2003; Kittler et al., 1998]. In R-UMLDA-A [Lu et al., 2009c], the simple sum rule is used with matching score level fusion.

Because ensemble-based learning prefers high diversity of the learners to be

---

Algorithm 7.6 Regularized UMLDA with aggregation (R-UMLDA-A)

Input: A set of tensor samples $\{\mathcal{X}_m \in \mathbb{R}^{I_1 \times I_2 \times \ldots \times I_N}, m = 1, ..., M\}$ with class labels $\mathbf{c} \in \mathbb{R}^M$, a test tensor sample $\mathcal{X}$, the desired feature vector length P, the R-UMLDA learner (Algorithm 7.5), the maximum number of iterations K, the number of R-UMLDA learners to be aggregated A.

Process:

1: **Step 1: Feature extraction**
2: **for** $a = 1$ **to** A **do**
3:     Obtain the ath TVP $\{\tilde{\mathbf{u}}_p^{(n)}\}_{N_{[a]}}^P$ from the ath R-UMLDA (Algorithm 7.5) with the input: $\{\mathcal{X}_m\}$, $\mathbf{c}$, P, γ_a, K, using random or uniform initialization.
4:     Project $\{\mathcal{X}_m\}$ and $\mathcal{X}$ to $\{\mathbf{y}_{m_{[a]}}, m = 1, ..., M\}$ and $\mathbf{y}_{[a]}$, respectively, using $\{\tilde{\mathbf{u}}_p^{(n)}\}_{N_{[a]}}^P$.
5: **end for**
6: **Step 2: Aggregation** at the matching score level for classification
7: **for** $a = 1$ **to** A **do**
8:     **for** $c = 1$ **to** C **do**
9:         Obtain the nearest-neighbor distance $d(\mathcal{X}, c, a)$ according to Equation (7.77).
10:     **end for**
11:     Normalize $d(\mathcal{X}, c, a)$ to $[0, 1]$ to get $\tilde{d}(\mathcal{X}, c, a)$ according to Equation (7.78).
12: **end for**
13: Obtain the aggregated distance $d(\mathcal{X}, c)$ according to Equation (7.79).
14: Obtain the class label for the test sample c^* according to Equation (7.80).
Output: The class label c^* for $\mathcal{X}$.

---

combined [Lu et al., 2006e], both uniform and random initializations are used in R-UMLDA-A for more diversity [Lu et al., 2009c]. In this way, although the best initialization cannot be determined, several R-UMLDA learners with different initializations are aggregated to make complementary discriminative features working together to separate classes better. Furthermore, to introduce more diversity and alleviate the problem of regularization parameter selection, the regularization parameter γ_a is sampled from an interval $[10^{-7}, 10^{-2}]$ (to cover a wide range of γ) uniformly in log scale so that each learner is differently regularized, where $a = 1, ..., A$ is the index of the individual R-UMLDA learner and A is the number of R-UMLDA learners to be aggregated.

Algorithm 7.6 provides the pseudocode implementation for R-UMLDA-A. The input training samples $\{\mathcal{X}_m\}$ are fed into A differently initialized and regularized UMLDA learners described in Algorithm 7.5 with parameters $\mathbf{c}$, P, γ_a, and K to obtain a set of A TVPs

$$\{\tilde{\mathbf{u}}_p^{(n)}\}_{N_{[a]}}^P, a = 1, ..., A. \tag{7.76}$$

The training samples $\{\mathcal{X}_m\}$ are then projected to R-UMLDA feature vectors $\{\mathbf{y}_{m_{[a]}}, m = 1, ..., M\}$ using the obtained TVPs for $a = 1, ..., A$. To classify a test sample $\mathcal{X}$, it is projected to A feature vectors $\{\mathbf{y}_{[a]}, a = 1, ..., A\}$ using the A TVPs first. Next, for the ath R-UMLDA learner, the nearest-neighbor distance of the test sample $\mathcal{X}$ to each candidate class c is calculated as

$$d(\mathcal{X}, c, a) = \min_{m, c_m = c} \| \mathbf{y}_{[a]} - \mathbf{y}_{m_{[a]}} \|. \tag{7.77}$$

The range of $d(\mathcal{X}, c, a)$ is then matched to the interval $[0, 1]$ as

$$\tilde{d}(\mathcal{X}, c, a) = \frac{d(\mathcal{X}, c, a) - \min_c d(\mathcal{X}, c, a)}{\max_c d(\mathcal{X}, c, a) - \min_c d(\mathcal{X}, c, a)}. \tag{7.78}$$

Finally, the aggregated nearest-neighbor distance is obtained employing the simple sum rule as

$$d(\mathcal{X}, c) = \sum_{a=1}^{A} \tilde{d}(\mathcal{X}, c, a), \tag{7.79}$$

and the test sample $\mathcal{X}$ is assigned the label

$$c^* = \arg \min_c d(\mathcal{X}, c). \tag{7.80}$$

7.6   Other Multilinear Extensions of LDA

This section examines several other multilinear extensions of LDA. As many of them are extensions based on graph-embedding, we give it a brief review first.

7.6.1   Graph-Embedding for Dimensionality Reduction

Dimensionality reduction methods based on *graph-embedding* [Yan et al., 2007b] trace their origins to manifold learning algorithms including isometric feature mapping (ISOMAP) [Tenenbaum et al., 2000], locally linear embedding (LLE) [Roweis and Saul, 2000], Laplacian eigenmaps (LE) [Belkin and Niyogi, 2001], and locality preserving projection (LPP) [He et al., 2005b].

Graph-embedding [Yan et al., 2007b] represents each vertex of a graph as a low-dimensional vector preserving geometric relationships between the vertex pairs. Similarities are commonly measured by a graph similarity or weight matrix $\mathbf{W}$ to characterize statistical or geometric properties of the data. For M training data samples, $\mathbf{W}$ is of size $M \times M$. Each element of $\mathbf{W}$ describes the similarity or relationship between a pair of data samples (vertices). Commonly used similarity functions include Gaussian similarity based on Euclidean distance [Belkin and Niyogi, 2001], local neighborhood

Chapter 8

Multilinear ICA, CCA, and PLS

The previous two chapters have discussed various multilinear extensions of principal component analysis (PCA) and linear discriminant analysis (LDA). This chapter deals with multilinear extensions of independent component analysis (ICA), canonical correlation analysis (CCA), and partial least squares (PLS) analysis, as shown in Figure 8.1 under the multilinear subspace learning (MSL) framework.

As in the previous two chapters, we describe five algorithms in greater detail and cover other algorithms briefly. We start with multilinear ICA algorithms and discuss the multilinear modewise ICA (MMICA) [Lu, 2013b] in detail, which is based on tensor-to-tensor projection (TTP). Then we study multilinear extensions of CCA and examine two of them in detail: the two-dimensional CCA (2D-CCA) [Lee and Choi, 2007], which is TTP-based, and the multilinear CCA (MCCA) [Lu, 2013a], which is based on tensor-to-vector projection (TVP). Finally, we cover two multilinear PLS algorithms in detail: the TVP-based N-way PLS (N-PLS) [Bro, 1996] and the TTP-based higher-order PLS (HOPLS) [Zhao et al., 2011].

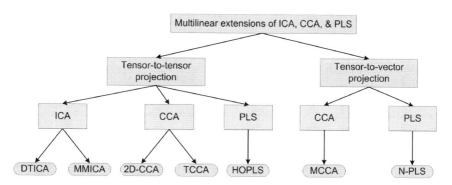

FIGURE 8.1: Multilinear ICA, CCA, and PLS algorithms under the MSL framework.

8.1   Overview of Multilinear ICA Algorithms

ICA aims to find representational components of data with maximum statistical independence [Hyvärinen et al., 2001]. The following is a summary of works dealing with the ICA problem using multilinear approaches.

8.1.1   Multilinear Approaches for ICA on Vector-Valued Data

In (linear) ICA for vector-valued observation data, the mixing matrix can be identified through higher-order *cumulant tensors* [Cardoso, 1991; Hyvärinen et al., 2001]. Cumulant tensors are higher-order generalizations of the covariance matrices and they capture higher-order statistics in data. The ICA problem can be solved either through the best Rank-$(R_1, R_2, ..., R_N)$ approximation of higher-order tensors [De Lathauwer and Vandewalle, 2004; De Lathauwer et al., 1997] or simultaneous third-order tensor diagonalization [De Lathauwer et al., 2001].

Beckmann and Smith [2005] developed a tensor probabilistic ICA algorithm as a higher-order extension of the probabilistic ICA [Beckmann and Smith, 2004] for multisubject or multisession fMRI analysis. It is based on the parallel factors (PARAFAC) decomposition but selected voxels are represented as very-high-dimensional vectors rather than tensors [Harshman, 1970]. Thus, it belongs to multiple factor analysis (Section 4.4.3).

The *multilinear ICA* (MICA) model in [Vasilescu and Terzopoulos, 2005] is another multiple factor analysis method for analyzing multiple factors for image ensembles organized into a tensor according to different image formation factors such as people, views, and illuminations. It requires a large number of samples for training, for example, 36 well-selected samples per class in [Vasilescu and Terzopoulos, 2005]. Moreover, it represents images as vectors rather than tensors and needs to know image forming factors. Thus, as discussed in Section 4.4.3, MICA is a *supervised learning* method requiring data to be labeled with such information. In *unsupervised learning* without labels, it *degenerates* to classical ICA.

Another work with the same name MICA in [Raj and Bovik, 2009] uses a multilinear expansion of the probability density function of source statistics but represents data as vectors too. This method does not involve tensors at all.

8.1.2   Multilinear Approaches for ICA on Tensor-Valued Data

The *directional tensor ICA* (DTICA) in [Zhang et al., 2008; Gao et al., 2010a] is a multilinear ICA formulation for tensor data. For images, it estimates *two* mixing matrices. It forms row and column directional images by shifting the rows/columns and applies FastICA [Hyvärinen, 1999] to row/column vectors. As in [Bartlett et al., 2002], DTICA uses PCA for dimensionality reduction. DTICA models tensor mixtures with N *mixing matrices and one single source tensor*, so it is built from a *factor-analysis* point of view and cannot be used for *blind source separation*.

The multilinear modewise ICA (MMICA) [Lu, 2013b] is another multilinear ICA formulation for tensors. MMICA models tensor data as mixtures generated from modewise *source matrices* that encode statistically independent information. It embeds ICA into multilinear PCA (MPCA) (Section 6.2) to solve for each source matrix. Mixing tensors are then obtained through regularized inverses of source matrices. MMICA can estimate hidden sources accurately from structured tensor data. In the next section, we study this method in detail.

8.2   Multilinear Modewise ICA

MMICA [Lu, 2013b] is an unsupervised MSL method that extracts modewise independent sources, that is, hidden sources, directly from tensor representations using a *multilinear mixing model*: the MMICA model. It assumes that tensor observation data have rich structures and they are mixtures generated from simple *modewise sources*. Figure 8.2 is an illustration showing ten mixtures generated from two simple binary patterns with the MMICA model. The objective of MMICA is to estimate the sources in Figure 8.2(b) from the observed mixtures in Figure 8.2(a). MMICA can be viewed as an extension of MPCA to non-Gaussian case by embedding ICA into MPCA to deal with non-Gaussianity of data in each mode. It has been shown to have *blind source separation* capability and it can recover hidden sources from their mixtures on synthetic data [Lu, 2013b].

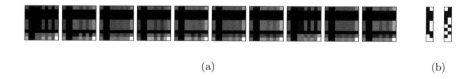

(a)   (b)

FIGURE 8.2: The structured data in (a) are all mixtures generated from the source data in (b) with a *multilinear mixing model*. MMICA can recover the sources in (b) from observed mixtures in (a).

8.2.1 Multilinear Mixing Model for Tensors

As discussed in Section 2.2, the simplified noise-free ICA model [Hyvärinen and Oja, 2000] assumes that we observe P linear mixtures $\{x_q, q = 1, ..., P\}$ of P sources $\{s_p\}$ (the latent variables):

$$x_q = a_{q_1} s_1 + a_{q_2} s_2 + ... + a_{q_P} s_P, \tag{8.1}$$

where each mixture x_q and each source s_p are random scalar variables. The P sources $\{s_p\}$ are assumed to be independent, that is, they are assumed to be independent components (ICs). In ICA for random vector variables $\{\mathbf{x}_q\}$, each $\mathbf{x}_q$ is a mixture of P independent vector sources $\{\mathbf{s}_p\}$:

$$\mathbf{x}_q = a_{q_1} \mathbf{s}_1 + a_{q_2} \mathbf{s}_2 + ... + a_{q_P} \mathbf{s}_P. \tag{8.2}$$

For random Nth-order tensor variables $\{\mathcal{X}_q\}$ of dimension $I_1 \times ... \times I_N$, we have the following mixing model similar to Equations (8.1) and (8.2) assuming P tensor variables $\{\mathcal{S}_p\}$ as the sources:

$$\mathcal{X}_q = a_{q_1} \mathcal{S}_1 + a_{q_2} \mathcal{S}_2 + ... + a_{q_P} \mathcal{S}_P. \tag{8.3}$$

Real-world tensor data often have rich structures. Therefore, we assume that the source tensors have compact representation as rank-one tensors (Definition 3.4). For a simpler model, we further assume that these simple rank-one tensors are formed by $P_1 \times P_2 \times ... \times P_N = P$ vectors with one set in each mode, where the mode-n set has P_n *independent column vectors*: $\{\mathbf{s}_{p_n}^{(n)}, p_n = 1, ..., P_n\}$, and each source tensor is the outer product of N vectors, one from each mode, that is,

$$\mathcal{S}_p = \mathbf{s}_{p_1}^{(1)} \circ \mathbf{s}_{p_2}^{(2)} \circ ... \circ \mathbf{s}_{p_N}^{(N)}. \tag{8.4}$$

Next, we form an Nth-order *mixing tensor* $\mathcal{A}_q \in \mathbb{R}^{P_1 \times P_2 \times ... \times P_N}$ by stacking all the P mixing parameters $\{a_{q_1}, a_{q_2}, ..., a_{q_P}\}$ in Equation (8.3) into an Nth-order tensor so its size $P_1 \times P_2 \times ... \times P_N = P$. Respectively, we form the *mode-n source matrix* $\mathbf{S}^{(n)} \in \mathbb{R}^{I_n \times P_n}$ with *independent columns* $\{\mathbf{s}_{p_n}^{(n)}, p_n = 1, ..., P_n\}$. We can then write the multilinear mixing model Equation (8.3) in a form of TTP as

$$\mathcal{X}_q = \mathcal{A}_q \times_1 \mathbf{S}^{(1)} \times_2 \mathbf{S}^{(2)} \times ... \times_N \mathbf{S}^{(N)}. \tag{8.5}$$

It models tensor mixtures using one single mixing tensor $\mathcal{A}_q$ and N mode-wise source matrices $\{\mathbf{S}^{(n)}\}$ with independent columns, and assumes that the sources are structured tensors formed from $\{\mathbf{S}^{(n)}\}$ according to Equation (8.4). This is the MMICA model [Lu, 2013b].

8.2.2 Regularized Estimation of Mixing Tensor

When applying MMICA to learning and recognition, we estimate the source matrices $\{\mathbf{S}^{(n)}\}$ from P observed mixtures $\{\mathcal{X}_q\}, q = 1, .., P$ (to be described in following sections). To get mixing tensor $\mathcal{A}_q$ from an observation tensor $\mathcal{X}_q$ based on $\{\mathbf{S}^{(n)}\}$, we use Equation (8.5) to get

$$\mathcal{A}_q = \mathcal{X}_q \times_1 \mathbf{S}^{(1)^+} \times_2 \mathbf{S}^{(2)^+} \times \ldots \times_N \mathbf{S}^{(N)^+}, \tag{8.6}$$

where $\mathbf{S}^{(n)^+} = (\mathbf{S}^{(n)^T} \mathbf{S}^{(n)})^{-1} \mathbf{S}^{(n)^T}$ indicates the *left inverse* of $\mathbf{S}^{(n)}$. As $\mathbf{S}^{(n)^T} \mathbf{S}^{(n)}$ can be poorly conditioned in practice, a *regularized* left inverse of $\mathbf{S}^{(n)}$ was introduced in [Lu, 2013b] to reduce the estimation variance by adding small bias as

$$\mathbf{S}_r^{(n)^+} = (\mathbf{S}^{(n)^T} \mathbf{S}^{(n)} + \eta \mathbf{I}_{P_n})^{-1} \mathbf{S}^{(n)^T}, \tag{8.7}$$

where η is a small *regularization* parameter and $\mathbf{I}_{P_n}$ is an identity matrix of size $P_n \times P_n$. Thus, the mixing tensor is approximated as

$$\hat{\mathcal{A}}_q = \mathcal{X}_q \times_1 \mathbf{S}_r^{(1)^+} \times_2 \mathbf{S}_r^{(2)^+} \times \ldots \times_N \mathbf{S}_r^{(N)^+}. \tag{8.8}$$

8.2.3 MMICA Algorithm Derivation

The MMICA algorithm embeds ICA into MPCA, following [Bartlett et al., 2002] that performs PCA before ICA. The procedures involved are *centering, initialization of source matrices, partial multilinear projection, modewise PCA, and modewise ICA*. The modewise ICA can be carried out in two architectures as in [Bartlett et al., 2002], where Architecture I is commonly used for traditional blind source separation task of ICA and Architecture II is for estimation of ICs for images. The MMICA algorithm is summarized in Algorithm 8.1, with details described below [Lu, 2013b].

The input to MMICA is a set of M tensor data samples $\{\mathcal{X}_m \in \mathbb{R}^{I_1 \times \ldots \times I_N}, m = 1, \ldots, M\}$. Two parameters need to be specified: one is Q, the percentage of energy to be kept in PCA, and the other is K, the maximum number of iterations. Input data are centered first as

$$\tilde{\mathcal{X}}_m = \mathcal{X}_m - \bar{\mathcal{X}}, m = 1, \ldots, M, \tag{8.9}$$

where

$$\bar{\mathcal{X}} = \frac{1}{M} \sum_{m=1}^{M} \mathcal{X}_m. \tag{8.10}$$

In the MMICA model Equation (8.5), data are generated from all N source matrices $\{\mathbf{S}^{(n)}, n = 1, \ldots, N\}$ rather than any one of them individually. We can not determine these N matrices simultaneously, except when $N = 1$ where it is degenerated to classical ICA. Therefore, the solution follows the alternating partial projections (APP) method (Section 5.1) to estimate $\mathbf{S}^{(n)}$ conditioned on all the other source matrices $\{\mathbf{S}^{(j)}, j \neq n\}$. The mode-$n$ source matrix $\mathbf{S}^{(n)}$ is initialized to an identity matrix $\mathbf{I}_{I_n}$ of size $I_n \times I_n$. Thus, the mode-n source dimension P_n is initialized to I_n.

Algorithm 8.1 Multilinear modewise ICA (MMICA)

Input: M tensor samples $\{\mathcal{X}_m \in \mathbb{R}^{I_1 \times \cdots \times I_N}, m = 1, ..., M\}$, the percentage of energy to be kept in PCA Q, the maximum number of iterations K.

Process:

1: $\diamond$ Center the input samples by subtracting the mean $\bar{\mathcal{X}}$ as in Equation (8.9).

2: $\diamond$ Initialize source matrices $\mathbf{S}^{(n)} = \mathbf{I}_{I_n}$ and $P_n = I_n$ for $n = 1, ..., N$.

3: [**Local optimization:**]

4: **for** $k = 1$ **to** K **do**

5:    **for** $n = 1$ **to** N **do**

6:       • Calculate the mode-n partial multilinear projection $\tilde{\mathcal{A}}_m^{(n)}$ according to Equation (8.11) for $m = 1, ..., M$.

7:       • Form $\tilde{\mathbf{A}}^{(n)}$ with columns consisting of mode-n vectors from $\{\tilde{\mathcal{A}}_m^{(n)}, m = 1, ..., M\}$.

8:       • Perform PCA on $\tilde{\mathbf{A}}^{(n)}$ and keep $Q\%$ of the total energy. Obtain $\mathbf{U}$ with the first R eigenvectors as its columns. Update $P_n = R$.

9:       • **Architecture I:** Perform FastICA on $\mathbf{U}^T$ to get $\mathbf{A}$ and $\mathbf{W}$. Set $\mathbf{S}^{(n)} = \mathbf{U}\mathbf{A}$.

10:      • **Architecture II:** Get $\mathbf{V} = \mathbf{U}^T\tilde{\mathbf{A}}^{(n)}$ and perform FastICA on $\mathbf{V}^T$ to get $\mathbf{A}$ and $\mathbf{W}$. Set $\mathbf{S}^{(n)} = \mathbf{U}\mathbf{W}^T$.

11:    **end for**

12: **end for**

Output: $\{\mathbf{S}^{(n)}, n = 1, ..., N\}$

For a particular mode n, we have all the other modewise source matrices $\{\mathbf{S}^{(j)}, j \neq n\}$ fixed and estimate $\mathbf{S}^{(n)}$ by first calculating the mode-n *partial multilinear projection* based on Equation (8.8) as

$$\tilde{\mathcal{A}}_m^{(n)} = \tilde{\mathcal{X}}_m \times_1 \mathbf{S}_r^{(1)^+} ... \times_{n-1} \mathbf{S}_r^{(n-1)^+} \times_{n+1} \mathbf{S}_r^{(n+1)^+} ... \times_N \mathbf{S}_r^{(N)^+}. \quad (8.11)$$

Next, we form a matrix $\tilde{\mathbf{A}}^{(n)} \in \mathbb{R}^{I_n \times (M \times \prod_{j=1, j \neq n}^{N} P_j)}$ by concatenating $\{\tilde{\mathbf{A}}_{m(n)}, m = 1, ..., M\}$, the mode-$n$ unfolding of $\{\tilde{\mathcal{A}}_m^{(n)}\}$, so that the columns of $\tilde{\mathbf{A}}^{(n)}$ consist of mode-n vectors from $\{\tilde{\mathcal{A}}_m^{(n)}\}$. We then perform PCA on $\tilde{\mathbf{A}}^{(n)}$ and keep Q percent of the total energy/variations, resulting in a PCA basis matrix $\mathbf{U} \in \mathbb{R}^{I_n \times R}$ with R leading eigenvectors. Subsequently, we update the mode-n source dimension as $P_n = R$ [Lu, 2013b].

8.2.4 Architectures and Discussions on MMICA

Two architectures: ICA can be performed under two architectures in [Bartlett et al., 2002] using FastICA [Hyvärinen, 1999]. FastICA takes a data matrix in and returns a mixing matrix $\mathbf{A}$ and a separating matrix $\mathbf{W}$. They can be used under two architectures in MMICA in the following ways [Lu, 2013b]:

- **Architecture I:** FastICA on $\mathbf{U}^T$ gives mixing matrix $\mathbf{A}$ and separating matrix $\mathbf{W}$. Thus, we set the mode-n source matrix as

$$\mathbf{S}^{(n)} = \mathbf{U}\mathbf{A}. \tag{8.12}$$

- **Architecture II:** We first obtain the PCA projection as $\mathbf{V} = \mathbf{U}^T \tilde{\mathbf{A}}^{(n)}$, and FastICA on $\mathbf{V}^T$ gives $\mathbf{A}$ and $\mathbf{W}$. Hence, we set the mode-n source matrix as

$$\mathbf{S}^{(n)} = \mathbf{U}\mathbf{W}^T. \tag{8.13}$$

Identifiability and number of ICs: Following ICA [Hyvärinen et al., 2001], the independent column vectors of modewise source matrices in MMICA are identifiable up to permutation and scaling if they (except one at most) have non-Gaussian distributions and the number of mixtures is no smaller than the number of column vectors (ICs) to be estimated. However, as in the general case of ICA, MMICA cannot estimate the number of modewise ICs. When this number is unknown, we determine it by specifying Q in PCA, as described above.

Convergence and termination: Formal convergence analysis for FastICA is only available for the so-called one-unit case, which considers only one row of the separating matrix [Oja and Yuan, 2006]. Empirical results are shown in [Lu, 2013b], where MMICA converges in one iteration in studies on synthetic data while its classification accuracy stabilizes in just a few iterations in face recognition experiments. Thus, the iteration is terminated by setting K, the maximum number of iterations, to a small number for efficiency in [Lu, 2013b].

Feature selection for classification: After obtaining the separated sources matrices $\{\mathbf{S}^{(n)}\}$, we have the MMICA representation (coordinates in the mixing tensor) $\hat{\mathcal{A}}$ of a sample $\mathcal{X}$ from Equation (8.8). We can further select features from $\hat{\mathcal{A}}$ using the strategies in Section 5.5 to get vector-valued features for input to conventional classifiers and also for better classification accuracy.

8.2.5 Blind Source Separation on Synthetic Data

Data generation: To study whether MMICA can estimate hidden source matrices, synthetic mixture data were generated according to Equation (8.5) in [Lu, 2013b]. The source matrices used are as shown in Figure 8.2(b), which are reproduced in Figures. 8.3(a) and 8.3(d). Each source matrix is a randomly generated simple *binary pattern* of size 10×2 ($I_n = 10, P_n = 2$). A hundred such mixtures ($M = 100$) were generated by drawing the elements of mixing tensors randomly from a uniform distribution on the unit interval. Figure 8.2(a) shows ten such mixtures as 8-bit gray images.

Hidden source recovery: In [Lu, 2013b], MMICA was applied with

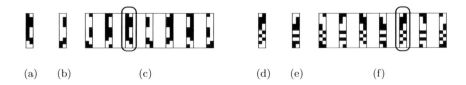

(a) (b)   (c)   (d) (e)   (f)

FIGURE 8.3: Blind source separation by MMICA on synthetic data: (a) true mode-1 source, (b) MMICA estimate of mode-1 source, (c) equivalent patterns of mode-1 MMICA estimate, (d) true mode-2 source, (e) MMICA estimate of mode-2 source, (f) equivalent patterns of mode-2 MMICA estimate. (The pattern matched with the true source is enclosed with an oval.)

$Q = 100$ using Architecture I for this blind source separation task, followed by binarization to obtain binary source patterns in Figures. 8.3(b) and 8.3(e). Because ICA estimation is only unique up to sign and permutation [Hyvärinen and Oja, 2000], the estimated MMICA sources in Figures. 8.3(b) and 8.3(e) are equivalent to the patterns in Figures. 8.3(c) and 8.3(f), respectively. One pattern in Figures. 8.3(c) and 8.3(f) matches Figures. 8.3(a) and 8.3(d) exactly, respectively, showing that independent modewise source patterns are estimated correctly.

8.3 Overview of Multilinear CCA Algorithms

CCA assumes that two datasets are two views of the same set of objects. It analyzes the relations between two sets of vector-valued variables by projecting them into low-dimensional spaces where each pair is maximally correlated subject to being uncorrelated with other pairs. There are two approaches in generalization of CCA to tensor data.

One approach analyzes the relationship between two tensor datasets rather than vector datasets, as in classical CCA. The two-dimensional CCA (2D-CCA) [Lee and Choi, 2007] is the first work along this line to analyze relations between two sets of image data without reshaping into vectors. It was further extended to local 2D-CCA in [Wang, 2010], sparse 2D-CCA in [Yan et al., 2012], and 3D CCA (3D-CCA) [Gang et al., 2011]. These CCA extensions are TTP based. However, they do not extract uncorrelated features for different pairs of projections like CCA. The multilinear CCA (MCCA) [Lu, 2013a] algorithm, which is based on TVP, was designed to learn canonical correlations of paired tensor datasets with uncorrelated features under two architectures.

Another approach studies the correlations between two data tensors with one or more shared modes because CCA can be viewed as taking two data matrices as input. This idea was first introduced in [Harshman, 2006]. *Tensor CCA* (TCCA) [Kim and Cipolla, 2009] was proposed later with two architectures for matching (3D) video volume data, represented as tensors of the same dimensions. The joint-shared-mode TCCA requires vectorization of two modes so it is equivalent to classical CCA for vector sets. The single-shared-mode TCCA solves for transformations in two modes so it can be viewed as several transposed versions of 2D-CCA, that is, 2D-CCA on paired sets consisting of slices of input tensors. They apply canonical transformations to the non-shared one/two axes of 3D data, which can be considered partial TTPs under MSL. Following TCCA, the *multiway CCA* in [Zhang et al., 2011] uses partial projections to find correlations between second-order and third-order tensors.

The next two sections will discuss two multilinear extensions of CCA in the first classical approach: 2D-CCA and MCCA. To reduce notation complexity, they will be presented only for second-order tensors, that is, matrices, as a special case. Further generalizations to higher-order tensors are left to the readers as exercises.

8.4 Two-Dimensional CCA

The 2D-CCA algorithm [Lee and Choi, 2007] extracts features directly from paired matrix (image) datasets while maximizing paired correlations. It is a TTP-based 2D extension of CCA for matrix data. It can also be considered an extension of GPCA [Ye et al., 2004a], 2D extension of PCA, to two matrix data pairs.

8.4.1 2D-CCA Problem Formulation

In 2D-CCA, we consider two matrix datasets with M samples each: $\left\{\mathbf{X}_m \in \mathbb{R}^{I_1 \times I_2}\right\}$ and $\left\{\mathbf{Y}_m \in \mathbb{R}^{J_1 \times J_2}\right\}$, $m = 1, ..., M$. I_1 and I_2 may not equal to J_1 and J_2, respectively. These samples are centered to $\{\tilde{\mathbf{X}}_m\}$ and $\{\tilde{\mathbf{Y}}_m\}$ by subtracting their means as

$$\tilde{\mathbf{X}}_m = \mathbf{X}_m - \bar{\mathbf{X}}, \ \tilde{\mathbf{Y}}_m = \mathbf{Y}_m - \bar{\mathbf{Y}}, \tag{8.14}$$

where the means

$$\bar{\mathbf{X}} = \sum_{m=1}^{M} \mathbf{X}_m, \ \bar{\mathbf{Y}} = \sum_{m=1}^{M} \mathbf{Y}_m. \tag{8.15}$$

2D-CCA aims to find two pairs of linear transformations (projection matrices) $\mathbf{U}_x \in \mathbb{R}^{I_1 \times P_1}$ $(P_1 \leq I_1)$ and $\mathbf{V}_x \in \mathbb{R}^{I_2 \times P_2}$ $(P_2 \leq I_2)$, and $\mathbf{U}_y \in \mathbb{R}^{J_1 \times P_1}$ $(P_1 \leq J_1)$ and $\mathbf{V}_y \in \mathbb{R}^{J_2 \times P_2}$ $(P_2 \leq J_2)$, that project a pair of samples $\mathbf{X}_m$

and $\mathbf{Y}_m$ to

$$\mathbf{R}_m = \mathbf{U}_x^T \tilde{\mathbf{X}}_m \mathbf{V}_x, \text{ and } \mathbf{S}_m = \mathbf{U}_y^T \tilde{\mathbf{Y}}_m \mathbf{V}_y, \tag{8.16}$$

so that the correlations between the projections in $\{\mathbf{R}_m\}$ and the corresponding projections in $\{\mathbf{S}_m\}$ are maximized.

To simplify the problem, it is reformulated by first considering only two pairs of projection vectors $\mathbf{u}_x \in \mathbb{R}^{I_1}$ and $\mathbf{v}_x \in \mathbb{R}^{I_2}$, and $\mathbf{u}_y \in \mathbb{R}^{J_1}$ and $\mathbf{v}_y \in \mathbb{R}^{J_2}$, that project a pair of samples $\tilde{\mathbf{X}}_m$ and $\tilde{\mathbf{Y}}_m$ to

$$r_m = \mathbf{u}_x^T \tilde{\mathbf{X}}_m \mathbf{v}_x, \text{ and } s_m = \mathbf{u}_y^T \tilde{\mathbf{Y}}_m \mathbf{v}_y, \tag{8.17}$$

respectively, such that the correlation between corresponding projections in $\{r_m\}$ and $\{s_m\}$ is maximized.

8.4.2 2D-CCA Algorithm Derivation

As in 2DLDA and GPCA, there is no closed-form solution for the four projection vectors in the reformulated problem. Therefore, the solution again follows the APP method to solve for them one pair at a time through partial multilinear projections iteratively. In each iteration step, we assume one pair of projection vectors is given to find the other pair that maximizes the correlation, and then vice versa.

Let $\tilde{\mathbf{u}}_x$ and $\tilde{\mathbf{v}}_x$, and $\tilde{\mathbf{u}}_y$ and $\tilde{\mathbf{v}}_y$, be the projection vectors maximizing the correlation between corresponding projections in $\{r_m\}$ and $\{s_m\}$.

1. For a given pair $\mathbf{v}_x$ and $\mathbf{v}_y$, define the conditioned sample covariance and cross-covariance matrices following Equations (2.42), (2.43), and (2.44) as:

$$\Sigma_{\mathbf{xx1}} = \frac{1}{M-1} \sum_{m=1}^{M} \tilde{\mathbf{X}}_m \mathbf{v}_x \mathbf{v}_x^T \tilde{\mathbf{X}}_m^T, \tag{8.18}$$

$$\Sigma_{\mathbf{yy1}} = \frac{1}{M-1} \sum_{m=1}^{M} \tilde{\mathbf{Y}}_m \mathbf{v}_y \mathbf{v}_y^T \tilde{\mathbf{Y}}_m^T, \tag{8.19}$$

$$\Sigma_{\mathbf{xy1}} = \frac{1}{M-1} \sum_{m=1}^{M} \tilde{\mathbf{X}}_m \mathbf{v}_x \mathbf{v}_y^T \tilde{\mathbf{Y}}_m^T, \tag{8.20}$$

and $\Sigma_{\mathbf{yx1}} = \Sigma_{\mathbf{xy1}}^T$. These matrices can be viewed as the sample covariance and cross-covariance matrices of the mode-1 partial projections of the input samples, $\{\tilde{\mathbf{X}}_m \mathbf{v}_x\}$ and $\{\tilde{\mathbf{Y}}_m \mathbf{v}_y\}$. Then, we have a classical CCA problem to solve for these partial projections. Following the derivations from Equations (2.49) to (2.56), the correlation is maximized only if $\tilde{\mathbf{u}}_x$ and $\tilde{\mathbf{u}}_y$ are from the generalized eigenvector associated with the largest

---

Algorithm 8.2 Two-dimensional CCA (2D-CCA)

Input: Two paired *zero-mean* matrix sets: $\left\{ \tilde{\mathbf{X}}_m \in \mathbb{R}^{I_1 \times I_2} \right\}$ and $\left\{ \tilde{\mathbf{Y}}_m \in \mathbb{R}^{J_1 \times J_2} \right\}$, $m = 1, ..., M$, the desired subspace dimensions P_1 and P_2, and the maximum number of iterations K.

Process:

1: Initialize $\tilde{\mathbf{U}}_x$, $\tilde{\mathbf{U}}_y$, $\tilde{\mathbf{V}}_x$ and $\tilde{\mathbf{V}}_y$ to identity matrices.

2: [**Local optimization:**]

3: **for** $k = 1$ **to** K **do**

4:     [**Mode 1**]

5:     Calculate $\boldsymbol{\Sigma}_{\mathbf{xx1}}$, $\boldsymbol{\Sigma}_{\mathbf{yy1}}$ and $\boldsymbol{\Sigma}_{\mathbf{xy1}}$ according to Equations (8.18), (8.19), and (8.20).

6:     Set the matrices $\tilde{\mathbf{U}}_x$ and $\tilde{\mathbf{U}}_y$ to consist of the P_1 generalized eigenvectors of Equation (8.21), associated with the largest P_1 generalized eigenvalues.

7:     [**Mode 2**]

8:     Calculate $\boldsymbol{\Sigma}_{\mathbf{xx2}}$, $\boldsymbol{\Sigma}_{\mathbf{yy2}}$ and $\boldsymbol{\Sigma}_{\mathbf{xy2}}$ according to Equations (8.22), (8.23), and (8.24).

9:     Set the matrices $\tilde{\mathbf{V}}_x$ and $\tilde{\mathbf{V}}_y$ to consist of the P_2 generalized eigenvectors of Equation (8.25), associated with the largest P_2 generalized eigenvalues.

10: **end for**

Output: The two pairs of projection matrices $\tilde{\mathbf{U}}_x$ and $\tilde{\mathbf{V}}_x$, and $\tilde{\mathbf{U}}_y$ and $\tilde{\mathbf{V}}_y$, that maximize the correlations of corresponding projections in the projected subspace.

---

generalized eigenvalue of the following generalized eigenvalue problem as in Equation (2.87):

$$\begin{bmatrix} \mathbf{0} & \boldsymbol{\Sigma}_{\mathbf{xy1}} \\ \boldsymbol{\Sigma}_{\mathbf{yx1}} & \mathbf{0} \end{bmatrix} \begin{bmatrix} \mathbf{u}_x \\ \mathbf{u}_y \end{bmatrix} = \lambda \begin{bmatrix} \boldsymbol{\Sigma}_{\mathbf{xx1}} & \mathbf{0} \\ \mathbf{0} & \boldsymbol{\Sigma}_{\mathbf{yy1}} \end{bmatrix} \begin{bmatrix} \mathbf{u}_x \\ \mathbf{u}_y \end{bmatrix}. \qquad (8.21)$$

2. Similarly, for a given pair $\mathbf{u}_x$ and $\mathbf{u}_y$, define the conditioned sample covariance and cross-covariance matrices as

$$\boldsymbol{\Sigma}_{\mathbf{xx2}} = \frac{1}{M-1} \sum_{m=1}^{M} \tilde{\mathbf{X}}_m^T \mathbf{u}_x \mathbf{u}_x^T \tilde{\mathbf{X}}_m, \qquad (8.22)$$

$$\boldsymbol{\Sigma}_{\mathbf{yy2}} = \frac{1}{M-1} \sum_{m=1}^{M} \tilde{\mathbf{Y}}_m^T \mathbf{u}_y \mathbf{u}_y^T \tilde{\mathbf{Y}}_m, \qquad (8.23)$$

$$\boldsymbol{\Sigma}_{\mathbf{xy2}} = \frac{1}{M-1} \sum_{m=1}^{M} \tilde{\mathbf{X}}_m^T \mathbf{u}_x \mathbf{u}_y^T \tilde{\mathbf{Y}}_m, \qquad (8.24)$$

and $\boldsymbol{\Sigma}_{\mathbf{yx2}} = \boldsymbol{\Sigma}_{\mathbf{xy2}}^{T}$. These matrices can be viewed as the sample covariance and cross-covariance matrices of the mode-2 partial projections of the input samples, $\{\tilde{\mathbf{X}}_m^T \mathbf{u}_x\}$ and $\{\tilde{\mathbf{Y}}_m^T \mathbf{u}_y\}$. Then, we have a classical CCA problem to solve for these partial projections. The correlation is maximized only if $\tilde{\mathbf{v}}_x$ and $\tilde{\mathbf{v}}_y$ are from the generalized eigenvector associated with the largest generalized eigenvalue of the following generalized eigenvalue problem:

$$\begin{bmatrix} \mathbf{0} & \boldsymbol{\Sigma}_{\mathbf{xy2}} \\ \boldsymbol{\Sigma}_{\mathbf{yx2}} & \mathbf{0} \end{bmatrix} \begin{bmatrix} \mathbf{v}_x \\ \mathbf{v}_y \end{bmatrix} = \lambda \begin{bmatrix} \boldsymbol{\Sigma}_{\mathbf{xx2}} & \mathbf{0} \\ \mathbf{0} & \boldsymbol{\Sigma}_{\mathbf{yy2}} \end{bmatrix} \begin{bmatrix} \mathbf{v}_x \\ \mathbf{v}_y \end{bmatrix}. \qquad (8.25)$$

Based on the above derivation, the two pairs of projection matrices $\tilde{\mathbf{U}}_x$ and $\tilde{\mathbf{U}}_y$, and $\tilde{\mathbf{V}}_x$ and $\tilde{\mathbf{V}}_y$, are (heuristically) set to the generalized eigenvectors of Equations (8.21) and (8.25) associated with the largest P_1 and P_2 generalized eigenvalues, respectively, in [Lee and Choi, 2007]. The solution is an iterative procedure as summarized in Algorithm 8.2. Random initialization was suggested in [Lee and Choi, 2007], while it was found in [Lu, 2013a] that better performance can be achieved by initializing the projection matrices to identity matrices in facial image matching experiments.

Remark 8.1. *As 2D-CCA uses TTP to project a matrix (tensor) to another matrix (tensor) without enforcing the zero-correlation constraint, the correlations are non-zero in general for all same-set and cross-set 2D-CCA features.*

8.5   Multilinear CCA

MCCA [Lu, 2013a] extracts uncorrelated features directly from paired tensor datasets under two architectures while maximizing correlations of paired features. Through a pair of TVPs, one architecture enforces zero-correlation within each set while the other enforces zero-correlation between different pairs of the two sets. It is a TVP-based MSL extension of CCA. It can also be considered an extension of uncorrelated MPCA (UMPCA) [Lu et al., 2009d] to paired tensor datasets. In the following, we consider only the second-order MCCA for analyzing the correlations between two matrix datasets for simpler notations.

8.5.1   MCCA Problem Formulation

As in 2D-CCA, the second-order MCCA [Lu, 2013a] considers two matrix datasets $\{\mathbf{X}_m \in \mathbb{R}^{I_1 \times I_2}\}$ and $\{\mathbf{Y}_m \in \mathbb{R}^{J_1 \times J_2}\}$. To simplify notations, we assume that they have *zero-mean* so that

$$\sum_{m=1}^{M} \mathbf{X}_m = \mathbf{0}, \quad \sum_{m=1}^{M} \mathbf{Y}_m = \mathbf{0}. \tag{8.26}$$

We are interested in paired projection vectors $[(\mathbf{u}_{x_p}, \mathbf{v}_{x_p}), (\mathbf{u}_{y_p}, \mathbf{v}_{y_p})]$, $\mathbf{u}_{x_p} \in \mathbb{R}^{I_1}$, $\mathbf{v}_{x_p} \in \mathbb{R}^{I_2}$, $\mathbf{u}_{y_p} \in \mathbb{R}^{J_1}$, $\mathbf{v}_{y_p} \in \mathbb{R}^{J_2}$, $p = 1, ..., P$, for dimensionality reduction to vector subspaces to reveal correlations between the two datasets. These projection vectors form matrices

$$\mathbf{U}_x = [\mathbf{u}_{x_1}, ..., \mathbf{u}_{x_P}], \quad \mathbf{V}_x = [\mathbf{v}_{x_1}, ..., \mathbf{v}_{x_P}], \tag{8.27}$$

and

$$\mathbf{U}_y = [\mathbf{u}_{y_1}, ..., \mathbf{u}_{y_P}], \quad \mathbf{V}_y = [\mathbf{v}_{y_1}, ..., \mathbf{v}_{y_P}]. \tag{8.28}$$

The matrix sets $\{\mathbf{X}_m\}$ and $\{\mathbf{Y}_m\}$ are projected to $\{\mathbf{r}_m \in \mathbb{R}^P\}$ and $\{\mathbf{s}_m \in \mathbb{R}^P\}$, respectively, through TVP as in Equation (3.39) using the matrices in Equations (8.27) and (8.28).

$$\mathbf{r}_m = diag\left(\mathbf{U}_x^T \mathbf{X}_m \mathbf{V}_x\right), \quad \mathbf{s}_m = diag\left(\mathbf{U}_y^T \mathbf{Y}_m \mathbf{V}_y\right). \tag{8.29}$$

We can write the above projections following elementary multilinear projection (EMP) in Equation (3.28) as

$$r_{m_p} = \mathbf{u}_{x_p}^T \mathbf{X}_m \mathbf{v}_{x_p}, \quad s_{m_p} = \mathbf{u}_{y_p}^T \mathbf{Y}_m \mathbf{v}_{y_p}, \tag{8.30}$$

where r_{m_p} and s_{m_p} are the pth elements of $\mathbf{r}_m$ and $\mathbf{s}_m$, respectively. We then define two coordinate vectors $\mathbf{w}_p \in \mathbb{R}^M$ and $\mathbf{z}_p \in \mathbb{R}^M$ for the pth EMP, where their pth elements $w_{p_m} = r_{m_p}$ and $z_{p_m} = s_{m_p}$. Denote

$$\mathbf{R} = [\mathbf{r}_1, ..., \mathbf{r}_M], \quad \mathbf{S} = [\mathbf{s}_1, ..., \mathbf{s}_M], \tag{8.31}$$

and

$$\mathbf{W} = [\mathbf{w}_1, ..., \mathbf{w}_P], \quad \mathbf{Z} = [\mathbf{z}_1, ..., \mathbf{z}_P]. \tag{8.32}$$

We have $\mathbf{W} = \mathbf{R}^T$ and $\mathbf{Z} = \mathbf{S}^T$. Here, $\mathbf{w}_p$ and $\mathbf{z}_p$ are analogous to the *canonical variates* in CCA [Anderson, 2003].

MCCA can be formulated with two architectures [Lu, 2013a].

- **Architecture I:** $\mathbf{w}_p$ and $\mathbf{z}_p$ are maximally correlated, while for $q \neq p$ $(p, q = 1, ..., P)$, $\mathbf{w}_q$ and $\mathbf{w}_p$, and $\mathbf{z}_q$ and $\mathbf{z}_p$, respectively, are uncorrelated.

- **Architecture II:** $\mathbf{w}_p$ and $\mathbf{z}_p$ are maximally correlated, while for $q \neq p$ $(p, q = 1, ..., P)$, $\mathbf{w}_p$ and $\mathbf{z}_q$ are uncorrelated.

Figure 8.4 is a schematic diagram showing MCCA and its two different architectures.

The sample-based estimation of canonical correlations and canonical variates in CCA are extended to the multilinear case for matrix data using Architectures I and II [Lu, 2013a].

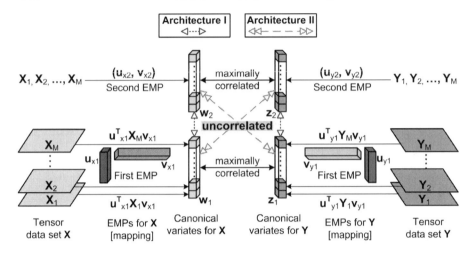

FIGURE 8.4: Schematic of multilinear CCA for paired (second-order) tensor datasets with two architectures.

Definition 8.1. *Canonical correlations and canonical variates for matrix data (Architecture I).* *For two matrix datasets with M samples each $\{\mathbf{X}_m \in \mathbb{R}^{I_1 \times I_2}\}$ and $\{\mathbf{Y}_m \in \mathbb{R}^{J_1 \times J_2}\}$, $m = 1, ..., M$, the pth pair of canonical variates is the pair $\{w_{p_m} = \mathbf{u}_{x_p}^T \mathbf{X}_m \mathbf{v}_{x_p}\}$ and $\{z_{p_m} = \mathbf{u}_{y_p}^T \mathbf{Y}_m \mathbf{v}_{y_p}\}$, where $\mathbf{w}_p$ and $\mathbf{z}_p$ have maximum correlation ρ_p while for $q \neq p$ $(p, q = 1, ..., P)$, $\mathbf{w}_q$ and $\mathbf{w}_p$, and $\mathbf{z}_q$ and $\mathbf{z}_p$, respectively, are uncorrelated. The correlation ρ_p is the pth canonical correlation.*

Definition 8.2. *Canonical correlations and canonical variates for matrix data (Architecture II).* *For two matrix datasets with M samples each $\{\mathbf{X}_m \in \mathbb{R}^{I_1 \times I_2}\}$ and $\{\mathbf{Y}_m \in \mathbb{R}^{J_1 \times J_2}\}$, $m = 1, ..., M$, the pth pair of canonical variates is the pair $\{w_{p_m} = \mathbf{u}_{x_p}^T \mathbf{X}_m \mathbf{v}_{x_p}\}$ and $\{z_{p_m} = \mathbf{u}_{y_p}^T \mathbf{Y}_m \mathbf{v}_{y_p}\}$, where $\mathbf{w}_p$ and $\mathbf{z}_p$ have maximum correlation ρ_p while for $q \neq p$ $(p, q = 1, ..., P)$, $\mathbf{w}_p$ and $\mathbf{z}_q$ are uncorrelated. The correlation ρ_p is the pth canonical correlation.*

8.5.2 MCCA Algorithm Derivation

Here, we derive MCCA with *Architecture I* (MCCA1) [Lu, 2013a]. MCCA with *Architecture II* (MCCA2) differs from MCCA1 only in the constraints enforced, so it can be obtained in an analogous way by swapping $\mathbf{w}_q$ and $\mathbf{z}_q$ $(q = 1, ..., p - 1)$ below, which is left to readers as an exercise. Following the successive derivation of CCA in [Anderson, 2003], we consider the canonical correlations one by one. The MCCA objective function for the pth EMP pair is

$$[(\tilde{\mathbf{u}}_{x_p}, \tilde{\mathbf{v}}_{x_p}), (\tilde{\mathbf{u}}_{y_p}, \tilde{\mathbf{v}}_{y_p})] = \arg \max_{[(\mathbf{u}_{x_p}, \mathbf{v}_{x_p}), (\mathbf{u}_{y_p}, \mathbf{v}_{y_p})]} \rho_p,$$

$$\text{subject to } \mathbf{w}_p^T \mathbf{w}_q = 0, \mathbf{z}_p^T \mathbf{z}_q = 0, \text{ for } p \neq q, \ p, q = 1, ..., P, \quad (8.33)$$

where ρ_p is the pth canonical correlation between $\mathbf{w}_p$ and $\mathbf{z}_p$ given by the *sample Pearson correlation*[1]:

$$\rho_p = \frac{\mathbf{w}_p^T \mathbf{z}_p}{\|\mathbf{w}_p\|\|\mathbf{z}_p\|}. \quad (8.34)$$

The P EMP pairs in MCCA are determined one pair at a time in P steps, with the pth step obtaining the pth EMP pair:

Step 1: Determine $[(\mathbf{u}_{x_1}, \mathbf{v}_{x_1}), (\mathbf{u}_{y_1}, \mathbf{v}_{y_1})]$ that maximizes ρ_1 without any constraint.

Step 2: Determine $[(\mathbf{u}_{x_2}, \mathbf{v}_{x_2}), (\mathbf{u}_{y_2}, \mathbf{v}_{y_2})]$ that maximizes ρ_2 subject to the constraint that $\mathbf{w}_2^T \mathbf{w}_1 = \mathbf{z}_2^T \mathbf{z}_1 = 0$.

Step p ($= 3, ..., P$): Determine $[(\mathbf{u}_{x_p}, \mathbf{v}_{x_p}), (\mathbf{u}_{y_p}, \mathbf{v}_{y_p})]$ that maximizes ρ_p subject to the constraint that $\mathbf{w}_p^T \mathbf{w}_q = \mathbf{z}_p^T \mathbf{z}_q = 0$ for $q = 1, ..., p - 1$.

To solve for the pth EMP pair, we need to determine four projection vectors, $\mathbf{u}_{x_p}, \mathbf{v}_{x_p}, \mathbf{u}_{y_p}$, and $\mathbf{v}_{y_p}$. As their simultaneous determination is difficult, we follow the APP method again. To determine each EMP pair, we estimate $\mathbf{u}_{x_p}$ and $\mathbf{u}_{y_p}$ conditioned on $\mathbf{v}_{x_p}$ and $\mathbf{v}_{y_p}$ first; then we estimate $\mathbf{v}_{x_p}$ and $\mathbf{v}_{y_p}$ conditioned on $\mathbf{u}_{x_p}$ and $\mathbf{u}_{y_p}$.

Conditional subproblem: To estimate $\mathbf{u}_{x_p}$ and $\mathbf{u}_{y_p}$ conditioned on $\mathbf{v}_{x_p}$ and $\mathbf{v}_{y_p}$, we assume that $\mathbf{v}_{x_p}$ and $\mathbf{v}_{y_p}$ are given and they can project input samples to obtain vectors

$$\tilde{\mathbf{r}}_{m_p} = \mathbf{X}_m \mathbf{v}_{x_p} \in \mathbb{R}^{I_1}, \quad \tilde{\mathbf{s}}_{m_p} = \mathbf{Y}_m \mathbf{v}_{y_p} \in \mathbb{R}^{J_1}, \quad (8.35)$$

which are the mode-1 partial multilinear projections. The conditional subproblem then becomes to determine $\mathbf{u}_{x_p}$ and $\mathbf{u}_{y_p}$ that project the samples $\{\tilde{\mathbf{r}}_{m_p}\}$ and $\{\tilde{\mathbf{s}}_{m_p}\}$ ($m = 1, ..., M$) to maximize the pairwise correlation subject to the zero-correlation constraint in Equation (8.33), which is a CCA problem with input samples $\{\tilde{\mathbf{r}}_{m_p}\}$ and $\{\tilde{\mathbf{s}}_{m_p}\}$ [Lu, 2013a].

The corresponding *maximum likelihood estimator* of the covariance and cross-covariance matrices are defined as in [Anderson, 2003]

$$\tilde{\boldsymbol{\Sigma}}_{\mathbf{rr}_p} = \frac{1}{M} \sum_{m=1}^{M} \tilde{\mathbf{r}}_{m_p} \tilde{\mathbf{r}}_{m_p}^T, \quad \tilde{\boldsymbol{\Sigma}}_{\mathbf{ss}_p} = \frac{1}{M} \sum_{m=1}^{M} \tilde{\mathbf{s}}_{m_p} \tilde{\mathbf{s}}_{m_p}^T, \quad (8.36)$$

$$\tilde{\boldsymbol{\Sigma}}_{\mathbf{rs}_p} = \frac{1}{M} \sum_{m=1}^{M} \tilde{\mathbf{r}}_{m_p} \tilde{\mathbf{s}}_{m_p}^T, \quad \tilde{\boldsymbol{\Sigma}}_{\mathbf{sr}_p} = \frac{1}{M} \sum_{m=1}^{M} \tilde{\mathbf{s}}_{m_p} \tilde{\mathbf{r}}_{m_p}^T, \quad (8.37)$$

---

[1]Note that the means of $\mathbf{w}_p$ and $\mathbf{z}_p$ are both zero because the input matrix sets are assumed to be *zero-mean*.

where $\tilde{\Sigma}_{\mathbf{rs}_p} = \tilde{\Sigma}_{\mathbf{sr}_p}^T$.

For $p = 1$, we solve for $\mathbf{u}_{x_1}$ and $\mathbf{u}_{y_1}$ that maximize the correlation between the projections of $\{\tilde{\mathbf{r}}_{m_p}\}$ and $\{\tilde{\mathbf{s}}_{m_p}\}$ without any constraint. We get $\tilde{\mathbf{u}}_{x_1}$ and $\tilde{\mathbf{u}}_{y_1}$ as the leading eigenvectors of the following eigenvalue problems as in Equations (2.57) and (2.58):

$$\tilde{\Sigma}_{\mathbf{rr}_p}^{-1} \tilde{\Sigma}_{\mathbf{rs}_p} \tilde{\Sigma}_{\mathbf{ss}_p}^{-1} \tilde{\Sigma}_{\mathbf{sr}_p} \mathbf{u}_{x_1} = \lambda \mathbf{u}_{x_1}, \tag{8.38}$$

$$\tilde{\Sigma}_{\mathbf{ss}_p}^{-1} \tilde{\Sigma}_{\mathbf{sr}_p} \tilde{\Sigma}_{\mathbf{rr}_p}^{-1} \tilde{\Sigma}_{\mathbf{rs}_p} \mathbf{u}_{y_1} = \lambda \mathbf{u}_{y_1}, \tag{8.39}$$

assuming that $\tilde{\Sigma}_{\mathbf{ss}_p}$ and $\tilde{\Sigma}_{\mathbf{rr}_p}$ are nonsingular.

Constrained optimization for $p > 1$: Next, we determine the pth $(p > 1)$ EMP pair given the first $(p - 1)$ EMP pairs by maximizing the correlation ρ_p, subject to the constraint that features projected by the pth EMP pair are uncorrelated with those projected by the first $(p - 1)$ EMP pairs as in Equation (8.33). Due to fundamental differences between linear and multilinear projections, the remaining eigenvectors of Equations (8.38) and (8.39), which are the solutions of CCA for $p > 1$, are not the solutions for MCCA. Instead, we have to solve a new constrained optimization problem in the multilinear setting.

Let $\tilde{\mathbf{R}}_p \in \mathbb{R}^{I_1 \times M}$ and $\tilde{\mathbf{S}}_p \in \mathbb{R}^{J_1 \times M}$ be matrices with $\tilde{\mathbf{r}}_{m_p}$ and $\tilde{\mathbf{s}}_{m_p}$ as their mth columns, respectively, that is,

$$\tilde{\mathbf{R}}_p = \left[\tilde{\mathbf{r}}_{1_p}, \tilde{\mathbf{r}}_{2_p}, ..., \tilde{\mathbf{r}}_{M_p}\right], \quad \tilde{\mathbf{S}}_p = \left[\tilde{\mathbf{s}}_{1_p}, \tilde{\mathbf{s}}_{2_p}, ..., \tilde{\mathbf{s}}_{M_p}\right]; \tag{8.40}$$

then the pth coordinate vectors are $\mathbf{w}_p = \tilde{\mathbf{R}}_p^T \mathbf{u}_{x_p}$ and $\mathbf{z}_p = \tilde{\mathbf{S}}_p^T \mathbf{u}_{y_p}$. The constraint that $\mathbf{w}_p$ and $\mathbf{z}_p$ are uncorrelated with $\{\mathbf{w}_q\}$ and $\{\mathbf{z}_q\}$, respectively, can be written as

$$\mathbf{u}_{x_p}^T \tilde{\mathbf{R}}_p \mathbf{w}_q = 0, \text{ and } \mathbf{u}_{y_p}^T \tilde{\mathbf{S}}_p \mathbf{z}_q = 0, q = 1, ..., p - 1. \tag{8.41}$$

The *coordinate matrices* are formed as

$$\mathbf{W}_{p-1} = [\mathbf{w}_1 \quad \mathbf{w}_2 \quad ...\mathbf{w}_{p-1}] \in \mathbb{R}^{M \times (p-1)}, \tag{8.42}$$

$$\mathbf{Z}_{p-1} = [\mathbf{z}_1 \quad \mathbf{z}_2 \quad ...\mathbf{z}_{p-1}] \in \mathbb{R}^{M \times (p-1)}. \tag{8.43}$$

Thus, we determine $\mathbf{u}_{x_p}$ and $\mathbf{u}_{y_p}$ $(p > 1)$ by solving the following constrained optimization problem:

$$(\tilde{\mathbf{u}}_{x_p}, \tilde{\mathbf{u}}_{y_p}) = \arg \max_{\mathbf{u}_{x_p}, \mathbf{u}_{y_p}} \tilde{\rho}_p, \text{ subject to} \tag{8.44}$$

$$\mathbf{u}_{x_p}^T \tilde{\mathbf{R}}_p \mathbf{w}_q = 0, \mathbf{u}_{y_p}^T \tilde{\mathbf{S}}_p \mathbf{z}_q = 0, q = 1, ..., p - 1,$$

where the pth canonical correlation for partial multilinear projections

$$\tilde{\rho}_p = \frac{\mathbf{u}_{x_p}^T \tilde{\Sigma}_{\mathbf{rs}_p} \mathbf{u}_{y_p}}{\sqrt{\mathbf{u}_{x_p}^T \tilde{\Sigma}_{\mathbf{rr}_p} \mathbf{u}_{x_p} \mathbf{u}_{y_p}^T \tilde{\Sigma}_{\mathbf{ss}_p} \mathbf{u}_{y_p}}}. \tag{8.45}$$

The solution is given by the following theorem [Lu, 2013a]:

Theorem 8.1. *For nonsingular $\tilde{\boldsymbol{\Sigma}}_{\mathbf{rr}_p}$ and $\tilde{\boldsymbol{\Sigma}}_{\mathbf{ss}_p}$, the solutions for $\mathbf{u}_{x_p}$ and $\mathbf{u}_{y_p}$ in the problem Equation (8.44) are the eigenvectors corresponding to the largest eigenvalue λ of the following eigenvalue problems:*

$$\tilde{\boldsymbol{\Sigma}}_{\mathbf{rr}_p}^{-1} \boldsymbol{\Phi}_p \tilde{\boldsymbol{\Sigma}}_{\mathbf{rs}_p} \tilde{\boldsymbol{\Sigma}}_{\mathbf{ss}_p}^{-1} \boldsymbol{\Psi}_p \tilde{\boldsymbol{\Sigma}}_{\mathbf{sr}_p} \mathbf{u}_{x_p} = \lambda \mathbf{u}_{x_p}, \tag{8.46}$$

$$\tilde{\boldsymbol{\Sigma}}_{\mathbf{ss}_p}^{-1} \boldsymbol{\Psi}_p \tilde{\boldsymbol{\Sigma}}_{\mathbf{sr}_p} \tilde{\boldsymbol{\Sigma}}_{\mathbf{rr}_p}^{-1} \boldsymbol{\Phi}_p \tilde{\boldsymbol{\Sigma}}_{\mathbf{rs}_p} \mathbf{u}_{y_p} = \lambda \mathbf{u}_{y_p}. \tag{8.47}$$

where

$$\boldsymbol{\Phi}_p = \mathbf{I} - \tilde{\mathbf{R}}_p \mathbf{W}_{p-1} \boldsymbol{\Theta}_p^{-1} \mathbf{W}_{p-1}^T \tilde{\mathbf{R}}_p^T \tilde{\boldsymbol{\Sigma}}_{\mathbf{rr}_p}^{-1}, \tag{8.48}$$

$$\boldsymbol{\Psi}_p = \mathbf{I} - \tilde{\mathbf{S}}_p \mathbf{Z}_{p-1} \boldsymbol{\Omega}_p^{-1} \mathbf{Z}_{p-1}^T \tilde{\mathbf{S}}_p^T \tilde{\boldsymbol{\Sigma}}_{\mathbf{ss}_p}^{-1}, \tag{8.49}$$

$$\boldsymbol{\Theta}_p = \mathbf{W}_{p-1}^T \tilde{\mathbf{R}}_p^T \tilde{\boldsymbol{\Sigma}}_{\mathbf{rr}_p}^{-1} \tilde{\mathbf{R}}_p \mathbf{W}_{p-1}, \tag{8.50}$$

$$\boldsymbol{\Omega}_p = \mathbf{Z}_{p-1}^T \tilde{\mathbf{S}}_p^T \tilde{\boldsymbol{\Sigma}}_{\mathbf{ss}_p}^{-1} \tilde{\mathbf{S}}_p \mathbf{Z}_{p-1}, \tag{8.51}$$

and $\mathbf{I}$ is an identity matrix.

Proof. For nonsingular $\tilde{\boldsymbol{\Sigma}}_{\mathbf{rr}_p}$ and $\tilde{\boldsymbol{\Sigma}}_{\mathbf{ss}_p}$, any $\mathbf{u}_{x_p}$ and $\mathbf{u}_{y_p}$ can be normalized such that $\mathbf{u}_{x_p}^T \tilde{\boldsymbol{\Sigma}}_{\mathbf{rr}_p} \mathbf{u}_{x_p} = \mathbf{u}_{y_p}^T \tilde{\boldsymbol{\Sigma}}_{\mathbf{ss}_p} \mathbf{u}_{y_p} = 1$ and the correlation $\tilde{\rho}_p$ remains unchanged. Therefore, maximizing $\tilde{\rho}_p$ is equivalent to maximizing $\mathbf{u}_{x_p}^T \tilde{\boldsymbol{\Sigma}}_{\mathbf{rs}_p} \mathbf{u}_{y_p}$ with the constraint

$$\mathbf{u}_{x_p}^T \tilde{\boldsymbol{\Sigma}}_{\mathbf{rr}_p} \mathbf{u}_{x_p} = 1 \text{ and } \mathbf{u}_{y_p}^T \tilde{\boldsymbol{\Sigma}}_{\mathbf{ss}_p} \mathbf{u}_{y_p} = 1. \tag{8.52}$$

Lagrange multipliers can be used to transform the problem Equation (8.44) to the following to include all the constraints:

$$\varphi_p = \mathbf{u}_{x_p}^T \tilde{\boldsymbol{\Sigma}}_{\mathbf{rs}_p} \mathbf{u}_{y_p} - \frac{\phi}{2} \left(\mathbf{u}_{x_p}^T \tilde{\boldsymbol{\Sigma}}_{\mathbf{rr}_p} \mathbf{u}_{x_p} - 1 \right) - \frac{\psi}{2} \left(\mathbf{u}_{y_p}^T \tilde{\boldsymbol{\Sigma}}_{\mathbf{ss}_p} \mathbf{u}_{y_p} - 1 \right)$$

$$- \sum_{q=1}^{p-1} \theta_q \mathbf{u}_{x_p}^T \tilde{\mathbf{R}}_p \mathbf{w}_q - \sum_{q=1}^{p-1} \omega_q \mathbf{u}_{y_p}^T \tilde{\mathbf{S}}_p \mathbf{z}_q, \tag{8.53}$$

where ϕ, ψ, $\{\theta_q\}$, and $\{\omega_q\}$ are Lagrange multipliers. Let

$$\boldsymbol{\theta}_{p-1} = [\theta_1 \ \theta_2 \ ... \ \theta_{p-1}]^T, \quad \boldsymbol{\omega}_{p-1} = [\omega_1 \ \omega_2 \ ... \ \omega_{p-1}]^T; \tag{8.54}$$

then we have the two summations in Equation (8.53) as

$$\sum_{q=1}^{p-1} \theta_q \mathbf{u}_{x_p}^T \tilde{\mathbf{R}}_p \mathbf{w}_q = \mathbf{u}_{x_p}^T \tilde{\mathbf{R}}_p \mathbf{W}_{p-1} \boldsymbol{\theta}_{p-1} \tag{8.55}$$

and

$$\sum_{q=1}^{p-1} \omega_q \mathbf{u}_{y_p}^T \tilde{\mathbf{S}}_p \mathbf{z}_q = \mathbf{u}_{y_p}^T \tilde{\mathbf{S}}_p \mathbf{Z}_{p-1} \boldsymbol{\omega}_{p-1}. \tag{8.56}$$

The vectors of partial derivatives of φ_p with respect to the elements of $\mathbf{u}_{x_p}$ and $\mathbf{u}_{y_p}$ are set equal to zero, giving

$$\frac{\partial \varphi_p}{\partial \mathbf{u}_{x_p}} = \tilde{\boldsymbol{\Sigma}}_{\mathbf{rs}_p} \mathbf{u}_{y_p} - \phi \tilde{\boldsymbol{\Sigma}}_{\mathbf{rr}_p} \mathbf{u}_{x_p} - \tilde{\mathbf{R}}_p \mathbf{W}_{p-1} \boldsymbol{\theta}_{p-1} = \mathbf{0}, \qquad (8.57)$$

$$\frac{\partial \varphi_p}{\partial \mathbf{u}_{y_p}} = \tilde{\boldsymbol{\Sigma}}_{\mathbf{sr}_p} \mathbf{u}_{x_p} - \psi \tilde{\boldsymbol{\Sigma}}_{\mathbf{ss}_p} \mathbf{u}_{y_p} - \tilde{\mathbf{S}}_p \mathbf{Z}_{p-1} \boldsymbol{\omega}_{p-1} = \mathbf{0}. \qquad (8.58)$$

Multiplication of Equation (8.57) on the left by $\mathbf{u}_{x_p}^T$ and Equation (8.58) on the left by $\mathbf{u}_{y_p}^T$ results in

$$\mathbf{u}_{x_p}^T \tilde{\boldsymbol{\Sigma}}_{\mathbf{rs}_p} \mathbf{u}_{y_p} - \phi \mathbf{u}_{x_p}^T \tilde{\boldsymbol{\Sigma}}_{\mathbf{rr}_p} \mathbf{u}_{x_p} = 0, \qquad (8.59)$$

$$\mathbf{u}_{y_p}^T \tilde{\boldsymbol{\Sigma}}_{\mathbf{rs}_p}^T \mathbf{u}_{x_p} - \psi \mathbf{u}_{y_p}^T \tilde{\boldsymbol{\Sigma}}_{\mathbf{ss}_p} \mathbf{u}_{y_p} = 0. \qquad (8.60)$$

Because $\mathbf{u}_{x_p}^T \tilde{\boldsymbol{\Sigma}}_{\mathbf{rr}_p} \mathbf{u}_{x_p} = 1$ and $\mathbf{u}_{y_p}^T \tilde{\boldsymbol{\Sigma}}_{\mathbf{ss}_p} \mathbf{u}_{y_p} = 1$, this shows that

$$\phi = \psi = \mathbf{u}_{x_p}^T \tilde{\boldsymbol{\Sigma}}_{\mathbf{rs}_p} \mathbf{u}_{y_p}. \qquad (8.61)$$

Next, we have the following using the definitions in Equations (8.50) and (8.51) through multiplication of Equation (8.57) on the left by $\mathbf{W}_{p-1}^T \tilde{\mathbf{R}}_p^T \tilde{\boldsymbol{\Sigma}}_{\mathbf{rr}_p}^{-1}$ and Equation (8.58) on the left by $\mathbf{Z}_{p-1}^T \tilde{\mathbf{S}}_p^T \tilde{\boldsymbol{\Sigma}}_{\mathbf{ss}_p}^{-1}$:

$$\mathbf{W}_{p-1}^T \tilde{\mathbf{R}}_p^T \tilde{\boldsymbol{\Sigma}}_{\mathbf{rr}_p}^{-1} \tilde{\boldsymbol{\Sigma}}_{\mathbf{rs}_p} \mathbf{u}_{y_p} - \boldsymbol{\Theta}_p \boldsymbol{\theta}_{p-1} = \mathbf{0}, \qquad (8.62)$$

$$\mathbf{Z}_{p-1}^T \tilde{\mathbf{S}}_p^T \tilde{\boldsymbol{\Sigma}}_{\mathbf{ss}_p}^{-1} \tilde{\boldsymbol{\Sigma}}_{\mathbf{sr}_p} \mathbf{u}_{x_p} - \boldsymbol{\Omega}_p \boldsymbol{\omega}_{p-1} = \mathbf{0}. \qquad (8.63)$$

Solving for $\boldsymbol{\theta}_{p-1}$ from Equation (8.62) and $\boldsymbol{\omega}_{p-1}$ from Equation (8.63), and substituting them into Equations (8.57) and (8.58), we have the following based on earlier result $\phi = \psi$ in Equation (8.61) and the definitions in Equations (8.48) and (8.49):

$$\boldsymbol{\Phi}_p \tilde{\boldsymbol{\Sigma}}_{\mathbf{rs}_p} \mathbf{u}_{y_p} = \phi \tilde{\boldsymbol{\Sigma}}_{\mathbf{rr}_p} \mathbf{u}_{x_p}, \quad \boldsymbol{\Psi}_p \tilde{\boldsymbol{\Sigma}}_{\mathbf{sr}_p} \mathbf{u}_{x_p} = \phi \tilde{\boldsymbol{\Sigma}}_{\mathbf{ss}_p} \mathbf{u}_{y_p}. \qquad (8.64)$$

Let $\lambda = \phi^2$ (then $\phi = \sqrt{\lambda}$), we have two eigenvalue problems in Equations (8.46) and (8.47) from Equation (8.64). Because $\phi(= \sqrt{\lambda})$ is the criterion to be maximized, the maximization is achieved by setting $\tilde{\mathbf{u}}_{x_p}$ and $\tilde{\mathbf{u}}_{y_p}$ to be the eigenvectors corresponding to the largest eigenvalue of Equations (8.46) and (8.47), respectively. □

By setting $\boldsymbol{\Phi}_1 = \mathbf{I}$, $\boldsymbol{\Psi}_1 = \mathbf{I}$, we can write Equations (8.38) and (8.39) in the form of Equations (8.46) and (8.47) as well. In this way, Equations (8.46) and (8.47) give unified solutions for $[(\tilde{\mathbf{u}}_{x_p}, \tilde{\mathbf{v}}_{x_p}), (\tilde{\mathbf{u}}_{y_p}, \tilde{\mathbf{v}}_{y_p})]$, $p = 1, ..., P$. They are unique only up to sign [Anderson, 2003].

Algorithm 8.3 Multilinear CCA for matrix sets (Architecture I)

Input: Two paired *zero-mean* matrix sets: $\left\{ \mathbf{X}_m \in \mathbb{R}^{I_1 \times I_2} \right\}$ and $\left\{ \mathbf{Y}_m \in \mathbb{R}^{J_1 \times J_2} \right\}$, $m = 1, ..., M$, the subspace dimension P, and the maximum number of iterations K.

1: **for** $p = 1$ **to** P **do**
2:     Initialize $\mathbf{v}_{x_{p_{(0)}}} = 1/\parallel \mathbf{1} \parallel$, $\mathbf{v}_{y_{p_{(0)}}} = 1/\parallel \mathbf{1} \parallel$.
3:     [**Local optimization:**]
4:     **for** $k = 1$ **to** K **do**
5:         **Mode 1:**
6:         Calculate $\{\tilde{\mathbf{r}}_{m_p}\}$ and $\{\tilde{\mathbf{s}}_{m_p}\}$ according to Equation (8.35) with $\mathbf{v}_{x_{p_{(k-1)}}}$ and $\mathbf{v}_{y_{p_{(k-1)}}}$ for $m = 1, ..., M$.
7:         Calculate $\tilde{\boldsymbol{\Sigma}}_{\mathbf{rr}_p}$, $\tilde{\boldsymbol{\Sigma}}_{\mathbf{ss}_p}$, $\tilde{\boldsymbol{\Sigma}}_{\mathbf{rs}_p}$ and $\tilde{\boldsymbol{\Sigma}}_{\mathbf{sr}_p}$ according to Equations (8.36) and (8.37). Form $\tilde{\mathbf{R}}_p$ and $\tilde{\mathbf{S}}_p$ according to Equation (8.40).
8:         Calculate $\boldsymbol{\Theta}_p$, $\boldsymbol{\Omega}_p$, $\boldsymbol{\Phi}_p$, and $\boldsymbol{\Psi}_p$ according to Equations (8.50), (8.51), (8.48), and (8.49), respectively.
9:         Set $\mathbf{u}_{x_{p_{(k)}}}$ and $\mathbf{u}_{y_{p_{(k)}}}$ to be the eigenvectors of Equations (8.46) and (8.47), respectively, associated with the largest eigenvalue.
10:       **Mode 2:**
11:       Calculate $\{\tilde{\mathbf{r}}_{m_p}\}$ and $\{\tilde{\mathbf{s}}_{m_p}\}$ according to Equation (8.65) with $\mathbf{u}_{x_{p_{(k)}}}$ and $\mathbf{u}_{y_{p_{(k)}}}$ for $m = 1, ..., M$.
12:       Repeat lines 7–9, replacing $\mathbf{u}$ with $\mathbf{v}$ in line 9.
13:     **end for**
14:     Set $\tilde{\mathbf{u}}_{x_p} = \mathbf{u}_{x_{p_{(K)}}}$, $\tilde{\mathbf{u}}_{y_p} = \mathbf{u}_{y_{p_{(K)}}}$, $\tilde{\mathbf{v}}_{x_p} = \mathbf{v}_{x_{p_{(K)}}}$, $\tilde{\mathbf{v}}_{y_p} = \mathbf{v}_{y_{p_{(K)}}}$.
15:     Calculate the projections to get coordinate vectors $\mathbf{w}_p$ and $\mathbf{z}_p$ using $[(\tilde{\mathbf{u}}_{x_p}, \tilde{\mathbf{v}}_{x_p}), (\tilde{\mathbf{u}}_{y_p}, \tilde{\mathbf{v}}_{y_p})]$.
16:     Calculate ρ_p according to Equation (8.34) and form $\mathbf{W}_p$ and $\mathbf{Z}_p$ according to Equations (8.42) and (8.43), respectively.
17: **end for**
Output: $[(\tilde{\mathbf{u}}_{x_p}, \tilde{\mathbf{v}}_{x_p}), (\tilde{\mathbf{u}}_{y_p}, \tilde{\mathbf{v}}_{y_p}), \rho_p]$, $p = 1, ..., P$.

Similarly, to estimate $\mathbf{v}_{x_p}$ and $\mathbf{v}_{y_p}$ conditioned on $\mathbf{u}_{x_p}$ and $\mathbf{u}_{y_p}$, we project input samples to obtain the mode-2 partial multilinear projections

$$\tilde{\mathbf{r}}_{m_p} = \mathbf{X}_m^T \mathbf{u}_{x_p} \in \mathbb{R}^{I_2}, \quad \tilde{\mathbf{s}}_{m_p} = \mathbf{Y}_m^T \mathbf{u}_{y_p} \in \mathbb{R}^{J_2}. \tag{8.65}$$

This conditional subproblem can be solved similarly by following the derivation from Equations (8.36) to (8.51). Algorithm 8.3 summarizes the MCCA algorithm for Architecture I [Lu, 2013a]. It uses the *uniform initialization* in Section 5.2.1 and terminates the iteration by setting a maximum iteration number K.

8.5.3   Discussions on MCCA

Convergence: In step p, each iteration maximizes the correlation between the pth pair of canonical variates $\mathbf{w}_p$ and $\mathbf{z}_p$ subject to the zero-correlation constraint. Thus, the captured correlation ρ_p is expected to increase monotonically. However, due to its iterative nature, MCCA can only converge to a local maximum, giving a suboptimal solution [Lu, 2013a].

Bound for P: Based on a similar argument as for UMPCA and uncorrelated multilinear discriminant analysis (UMLDA), the number of canonical variate pairs that can be extracted by MCCA is upper-bounded as $P \leq \min\{I_1, I_2, J_1, J_2, M\}$.

Feature correlations: MCCA with Architecture I (MCCA1) produces same-set uncorrelated features while MCCA with Architecture II (MCCA2) produces cross-set uncorrelated features. On the other hand, the same-set correlations for MCCA2 and the cross-set correlations for MCCA1 are both low, especially for the first few features [Lu, 2013a]. From UMPCA, UMLDA, and MCCA, we see that TVP could be more suitable for constraint enforcement than TTP because it projects a tensor to a vector so vector-based properties can be enforced.

8.6   Multilinear PLS Algorithms

PLS considers two related datasets and assumes that observed data are generated by a system or process that is driven by a small number of latent variables [Rosipal and Krämer, 2006]. It aims to find projections that maximize the covariance between the two sets. As discussed in Section 2.5, *regression* is one of the most important applications of PLS. Thus, this section discusses two multilinear extensions of PLS that were proposed to solve regression problems for tensor data. One is N-way PLS (N-PLS) [Bro, 1996] based on TVP and the other is higher-order PLS (HOPLS) [Zhao et al., 2011] based on TTP. Because PLS can be considered a special case of CCA, multilinear extensions of CCA studied in previous sections can also be easily modified to get multilinear extensions of PLS.

8.6.1 N-Way PLS

Bro [1996] introduced N-PLS, which can be seen as a multilinear extension of the PLS1 algorithm (Algorithm 2.5). It is a TVP-based method and its solution shares many similarities with the solutions of UMPCA, UMLDA and MCCA.

N-PLS aims to learn the relationships between two datasets. The first dataset is considered the independent variable (the *predictor*) and the second dataset is the dependent variables (the *response* to be predicted). There are many variations depending on the orders (number of dimensions) of the two datasets [Bro, 1996]. Here, we study the *tri-linear PLS1*, where the independent variables form a third-order tensor $\mathcal{X} \in \mathbb{R}^{I_1 \times I_2 \times I_3}$ and the dependent variables form a vector $\mathbf{y} \in \mathbb{R}^{I_1 \times 1}$ (e.g., $\mathbf{y}$ can be the class labels of samples contained in $\mathcal{X}$, and I_1 equals to the number of samples M in this case). They are assumed to have zero-mean with respect to mode 1. Tri-linear PLS1 models dataset $\mathcal{X}$ by extracting P components (*latent factors*) $\{\mathbf{w}_p \in \mathbb{R}^{I_1 \times 1}, p = 1, ..., P\}$ subsequently with deflation from mode 1 to maximize the covariance between $\mathbf{y}$ and $\mathbf{w}_p$.

As in other TVP-based algorithms, the solution for tri-linear PLS1 involves P steps. In each step, we solve for three vectors $\mathbf{w}_p$, $\mathbf{u}_p \in \mathbb{R}^{I_2 \times 1}$, and $\mathbf{v}_p \in \mathbb{R}^{I_3 \times 1}$, and get a *regression coefficients* d_p, as described in the pseudocode of Algorithm 8.4. $\tilde{\mathbf{u}}_p$ and $\tilde{\mathbf{v}}_p$ are obtained from singular value decomposition[2] (SVD) of $\mathbf{Z}$, the projection of $\mathcal{X}$ on $\mathbf{y}$ with the singleton mode (the mode with dimension of 1) removed after projection. The *score vector* $\tilde{\mathbf{w}}_p$ is set to the projection of $\mathcal{X}$ on $\tilde{\mathbf{u}}_p$ and $\tilde{\mathbf{v}}_p$ (with singleton modes removed), and d_p is calculated from $\tilde{\mathbf{w}}_p$ and $\mathbf{y}$. The P components $\{\tilde{\mathbf{w}}_p\}$ form the *score matrix* $\tilde{\mathbf{W}} \in \mathbb{R}^{I_1 \times P}$, with $\tilde{\mathbf{w}}_p$ as its columns. Two component matrices $\tilde{\mathbf{U}} \in \mathbb{R}^{I_2 \times P}$ and $\tilde{\mathbf{V}} \in \mathbb{R}^{I_3 \times P}$ are formed for mode 2 and mode 3 with $\tilde{\mathbf{u}}_p$ and $\tilde{\mathbf{v}}_p$ as their columns, respectively.

8.6.2 Higher-Order PLS

HOPLS [Zhao et al., 2011] is another multilinear regression model based on PLS. It predicts a tensor $\mathcal{Y}$ (the *dependent data*, or *response*) from a tensor $\mathcal{X}$ (the *independent data*, or *predictor*) by projecting the data into low-dimensional *latent spaces* to perform regression of corresponding *latent variables*. Deflation is again employed to solve for the latent spaces. HOPLS can be seen as a multilinear extension of the NIPALS algorithm (Algorithm 2.4). It is a TTP-based method.

In HOPLS, the independent data is an Nth-order tensor $\mathcal{X} \in \mathbb{R}^{I_1 \times ... \times I_N}$, and the dependent data is a Dth-order tensor $\mathcal{Y} \in \mathbb{R}^{J_1 \times ... \times J_D}$. They have the same mode-1 dimension, that is, $I_1 = J_1$. Thus, we consider the covariance of projections in mode 1. The *mode-n cross-covariance* between $\mathcal{X}$ and $\mathcal{Y}$ with $I_n = J_n$ is defined as a tensor $\mathcal{C} \in \mathbb{R}^{I_1 \times ... \times I_{n-1} \times I_{n+1} \times ... \times I_N \times J_1 \times ... \times J_{n-1} \times J_{n+1} \times ... \times J_D}$ of order $N + D - 2$ [Zhao et al., 2011]

$$\mathcal{C} = < \mathcal{X}, \mathcal{Y} >_{\{n;n\}}, \tag{8.66}$$

---

[2]Singular values are all sorted in descending order.

Algorithm 8.4 Tri-linear PLS1 [N-way PLS (N-PLS)]

Input: Two paired zero-mean (with respective to mode 1) datasets $\mathcal{X} \in \mathbb{R}^{I_1 \times I_2 \times I_3}$ and $\mathbf{y} \in \mathbb{R}^{I_1 \times 1}$, the number of latent factors P.

Process:

1: **for** $p = 1$ **to** P **do**
2:     Project $\mathcal{X}$ on $\mathbf{y}$ to get $\mathbf{Z} = \mathcal{X} \times_1 \mathbf{y} \in \mathbb{R}^{I_2 \times I_3}$ (with the singleton mode removed).
3:     Calculate the SVD of $\mathbf{Z}$ as $\mathbf{Z} = \mathbf{U}\mathbf{S}\mathbf{V}^T$.
4:     Set $\tilde{\mathbf{u}}_p = \mathbf{U}(:,1)$ and $\tilde{\mathbf{v}}_p = \mathbf{V}(:,1)$.
5:     Calculate the score vector $\tilde{\mathbf{w}}_p = \mathcal{X} \times_2 \tilde{\mathbf{u}}_p \times_3 \tilde{\mathbf{v}}_p$ (with singleton modes removed).
6:     Calculate the regression coefficient $d_p = \tilde{\mathbf{w}}_p^T \mathbf{y} / (\tilde{\mathbf{w}}_p^T \tilde{\mathbf{w}}_p)$.
7:     Deflation:
8:     **for** $i_1 = 1$ **to** I_1 **do**
9:         Deflation: $\mathcal{X}(i_1,:,:) \leftarrow \mathcal{X}(i_1,:,:) - \tilde{\mathbf{w}}_p(i_1)\tilde{\mathbf{v}}_p\tilde{\mathbf{u}}_p^T$.
10:    **end for**
11:    Deflation: $\mathbf{y} \leftarrow \mathbf{y} - d_p\tilde{\mathbf{w}}_p$.
12: **end for**

Output: P sets of vectors $\{\tilde{\mathbf{w}}_p, \tilde{\mathbf{u}}_p, \tilde{\mathbf{v}}_p\}$, $p = 1, ..., P$ and the regression coefficients $\{d_p\}$.

---

where each element of $\mathcal{C}$ is defined as [Zhao et al., 2011]

$$\mathcal{C}(i_1, ..., i_{n-1}, i_{n+1}, ..., i_N, j_1, ..., j_{n-1}, j_{n+1}, ..., j_D)$$
$$= \sum_{i_n=1}^{I_N} \mathcal{X}(i_1, ..., i_{n-1}, i_n, i_{n+1}, ..., i_N) \cdot \mathcal{Y}(j_1, ..., j_{n-1}, j_n, j_{n+1}, ..., j_D). \quad (8.67)$$

HOPLS aims to find the best rank-$(1, L_2, ..., L_N)$ approximation [De Lathauwer et al., 2000a] of $\mathcal{X}$ and the best rank-$(1, K_2, ..., K_D)$ approximation of $\mathcal{Y}$ such that a set of common *latent vectors* can best approximate both $\mathcal{X}$ and $\mathcal{Y}$. Following Equation (2.77) and (2.78), the HOPLS model is constructed as

$$\mathcal{X} = \sum_{p=1}^{P} \mathcal{W}_p \times_1 \mathbf{u}_{x_p} \times_2 \mathbf{V}_{x_p}^{(2)} \times_3 ... \times_N \mathbf{V}_{x_p}^{(N)} + \mathcal{E}_x, \quad (8.68)$$

$$\mathcal{Y} = \sum_{p=1}^{P} \mathcal{Z}_p \times_1 \mathbf{u}_{x_p} \times_2 \mathbf{V}_{y_p}^{(2)} \times_3 ... \times_D \mathbf{V}_{y_p}^{(D)} + \mathcal{E}_y, \quad (8.69)$$

where P denotes the number of latent vectors, $\mathbf{u}_{x_p} \in \mathbb{R}^{I_1}$ denotes the pth latent vector, $\{\mathbf{V}_{x_p}^{(n)}\}_{n=2}^{N} \in \mathbb{R}^{I_n \times L_n}$ and $\{\mathbf{V}_{y_p}^{(d)}\}_{d=2}^{D} \in \mathbb{R}^{J_d \times K_d}$ are the *loading matrices* for $\mathbf{u}_{x_p}$ on mode n and mode d, respectively, and $\mathcal{W}_p \in \mathbb{R}^{1 \times L_2 \times ... \times L_N}$ and $\mathcal{Z}_p \in \mathbb{R}^{1 \times K_2 \times ... \times K_D}$ are the *core tensors* for $\mathcal{X}$ and $\mathcal{Y}$, respectively. L_n and K_d are the mode-n rank and mode-d rank, respectively. They equal to

---

Algorithm 8.5 Higher-order PLS (HOPLS)

Input: Two tensor datasets $\mathcal{X} \in \mathbb{R}^{I_1 \times \cdots \times I_N}$ and $\mathcal{Y} \in \mathbb{R}^{J_1 \times \cdots \times J_D}$, where $I_1 = J_1$, the number of latent vectors P, the number of loading vectors in each mode $\{L_n\}_{n=2}^N$ and $\{K_d\}_{d=2}^D$, a small number η as the stopping criterion.

Process:

1: **for** $p = 1$ **to** P **do**

2:     Calculate the current mode-1 cross-covariance $\mathcal{C}$ between $\mathcal{X}$ and $\mathcal{Y}$ as $\mathcal{C} = <\mathcal{X}, \mathcal{Y}>_{\{1;1\}}$ from Equation (8.67).

3:     Calculate the best rank-$(L_2, ..., L_N, K_2, ..., K_D)$ approximation of $\mathcal{C}$ using the higher-order orthogonal iteration (HOOI) algorithm [De Lathauwer et al., 2000a] to obtain projection matrices in each mode as $\tilde{\mathbf{V}}_{x_p}^{(2)}, ..., \tilde{\mathbf{V}}_{x_p}^{(N)}, \tilde{\mathbf{V}}_{y_p}^{(2)}, ..., \tilde{\mathbf{V}}_{y_p}^{(D)}$.

4:     Calculate the mode-1 partial multilinear projection of $\mathcal{X}$: $\acute{\mathcal{W}}^{(1)} = \mathcal{X} \times_2 \tilde{\mathbf{V}}_{x_p}^{(2)T} \times_3 ... \times_N \tilde{\mathbf{V}}_{x_p}^{(N)T}$.

5:     Set the latent vector $\tilde{\mathbf{u}}_{x_p}$ as the first leading left singular vector of $\acute{\mathbf{W}}_{(1)}$ (by SVD), the mode-1 unfolding of $\acute{\mathcal{W}}^{(1)}$.

6:     Calculate the core tensor $\tilde{\mathcal{W}}_p = \mathcal{X} \times_1 \tilde{\mathbf{u}}_{x_p}^T \times_2 \tilde{\mathbf{V}}_{x_p}^{(2)T} \times_3 ... \times_N \tilde{\mathbf{V}}_{x_p}^{(N)T}$.

7:     Calculate the core tensor $\tilde{\mathcal{Z}}_p = \mathcal{Y} \times_1 \tilde{\mathbf{u}}_{x_p}^T \times_2 \tilde{\mathbf{V}}_{y_p}^{(2)T} \times_3 ... \times_D \tilde{\mathbf{V}}_{y_p}^{(D)T}$.

8:     Deflation:
$$\mathcal{X} \leftarrow \mathcal{X} - \tilde{\mathcal{W}}_p \times_1 \tilde{\mathbf{u}}_{x_p} \times_2 \tilde{\mathbf{V}}_{x_p}^{(2)} \times_3 ... \times_N \tilde{\mathbf{V}}_{x_p}^{(N)} \text{ and}$$
$$\mathcal{Y} \leftarrow \mathcal{Y} - \tilde{\mathcal{Z}}_p \times_1 \tilde{\mathbf{u}}_{x_p} \times_2 \tilde{\mathbf{V}}_{y_p}^{(2)} \times_3 ... \times_D \tilde{\mathbf{V}}_{y_p}^{(D)}.$$

9:     **if** $\| \mathcal{X} \|_F < \eta$ and $\| \mathcal{Y} \|_F < \eta$ **then**

10:       Break.

11:     **end if**

12: **end for**

Output: P sets of projection matrices, core tensors and latent vectors: $\{\tilde{\mathbf{V}}_{x_p}^{(2)}, ..., \tilde{\mathbf{V}}_{x_p}^{(N)}, \tilde{\mathbf{V}}_{y_p}^{(2)}, ..., \tilde{\mathbf{V}}_{y_p}^{(D)}\}$, $\{\tilde{\mathcal{W}}_p, \tilde{\mathcal{Z}}_p\}$, $\{\tilde{\mathbf{u}}_{x_p}\}$, $p = 1, ..., P$.

---

the number of *loading vectors* for each respective mode, to be specified by the user. Thus, there are many parameters to set, especially for higher-order tensors.

The pseudocode for HOPLS is in Algorithm 8.5. The solution for HOPLS involves P steps, which is similar to TVP-based algorithms but different from other TTP-based algorithms. At each step p, it solves for the pth pair of projection matrix set $(\mathbf{V}_{x_p}^{(2)}, ..., \mathbf{V}_{x_p}^{(N)}, \mathbf{V}_{y_p}^{(2)}, ..., \mathbf{V}_{y_p}^{(D)})$, the pth pair of core tensors $(\mathcal{W}_p, \mathcal{Z}_p)$, and the pth latent vector $\mathbf{u}_{x_p}$.

Chapter 9

Applications of Multilinear Subspace Learning

Multilinear subspace learning (MSL) is very useful for a wide range of application domains that involve tensor-structured datasets. This final chapter starts with an introduction to a typical pattern recognition system as shown in Figure 9.1, and then examines real-world applications of MSL. In particular, we study two biometric applications in greater detail: face recognition and gait recognition, due to their popularity in the development of MSL. Some experimental results are presented for demonstration. For completeness sake, several other applications are also described briefly, such as visual content analysis in computer vision, brain signal/image processing in neuroscience, DNA sequence discovery in bioinformatics, music genre classification in audio signal processing, and data stream monitoring in data mining.

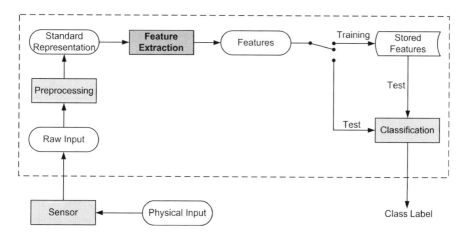

FIGURE 9.1: A typical pattern recognition system.

9.1   Pattern Recognition System

Figure 9.1 shows the block diagram of a typical pattern recognition system. Raw data are acquired from the real world through a certain sensor. Then they are preprocessed to produce standard representations for feature extraction. Common preprocessing steps include noise removal (filtering), enhancement, segmentation/cropping, and normalization (Appendix B gives some examples). Feature extraction is the focus of this book as well as most machine learning algorithms. However, it should be noted that problem formulation, data collection, and data preprocessing are very important factors for a successful machine learning and pattern recognition system, as pointed out in [Wagstaff, 2012].

In the training phase, features are extracted from the *gallery* dataset and stored in a database. In the test phase, features are extracted from the *probe* dataset and compared against the stored gallery features based on a certain similarity measure to make a decision. In the following, we describe several similarity measures for comparison of two feature vectors or two sequences of feature vectors.

Similarity between two feature vectors: The similarity score $S(\mathcal{P}_j, \mathcal{G}_m)$ between a test sample $\mathcal{P}_j \in \mathbb{P}$ from the probe set and a training sample $\mathcal{G}_m \in \mathbb{G}$ from the gallery set can be calculated through measuring the distance between the respective feature vectors $\mathbf{p}_j$ and $\mathbf{g}_m$. Table 9.1 lists six distance measures adapted from [Moon and Phillips, 2001]: the ℓ_1 distance (D_{ℓ_1}), the ℓ_2 distance (D_{ℓ_2}), the angle between feature vectors (D_θ), the modified ℓ_1 distance (M_{ℓ_1}), the modified ℓ_2 distance (M_{ℓ_2}), and the modified angle distance (M_θ), where $\mathbf{w}$ is a weight vector and H is the vector length. The last three measures are the weighted versions of the first three measures. The similarity score between $\mathbf{p}_j$ and $\mathbf{g}_m$ can be obtained as[1]

$$S(\mathcal{P}_j, \mathcal{G}_m) = S(\mathbf{p}_j, \mathbf{g}_m) = -d(\mathbf{p}_j, \mathbf{g}_m), \tag{9.1}$$

using one of the distance measures.

Similarity between two sequences of feature vectors: In applications involving temporal sequences such as gait recognition, a test sequence is often matched against training sequences. Consider a test sequence $\mathcal{P}_j \in \mathbb{P}$ that has K_j samples with corresponding feature vectors $\{\mathbf{p}_{k_j}, k_j = 1, ..., K_j\}$, and a training sequence $\mathcal{G}_m \in \mathbb{G}$ that has K_m samples with corresponding feature vectors $\{\mathbf{g}_{k_m}, k_m = 1, ..., K_m\}$. To obtain the similarity score $S(\mathcal{P}_j, \mathcal{G}_m)$ between sequences $\mathcal{P}_j$ and $\mathcal{G}_m$, the distance calculation process should be symmetric with respect to test and training sequences [Boulgouris et al., 2006]. If the test and training sequences were interchanged, the computed distance

---

[1]Note that a smaller distance indicates higher similarity and a larger distance indicates lower similarity so there is a minus sign in Equation (9.1).

TABLE 9.1: Six distance (dissimilarity) measures $d(\mathbf{a}, \mathbf{b})$ between feature vectors $\mathbf{a} \in \mathbb{R}^H$ and $\mathbf{b} \in \mathbb{R}^H$, with an optional weight vector $\mathbf{w} \in \mathbb{R}^H$.

D_{ℓ_1}	D_{ℓ_2}	D_θ
$\sum_{h=1}^{H} |\mathbf{a}(h) - \mathbf{b}(h)|$	$\sqrt{\sum_{h=1}^{H} [\mathbf{a}(h) - \mathbf{b}(h)]^2}$	$\dfrac{-\sum_{h=1}^{H} \mathbf{a}(h) \cdot \mathbf{b}(h)}{\sqrt{\sum_{h=1}^{H} \mathbf{a}(h)^2 \sum_{h=1}^{H} \mathbf{b}(h)^2}}$
M_{ℓ_1}	M_{ℓ_2}	M_θ
$\sum_{h=1}^{H} \dfrac{|\mathbf{a}(h) - \mathbf{b}(h)|}{\mathbf{w}(h)}$	$\sqrt{\sum_{h=1}^{H} \dfrac{[\mathbf{a}(h) - \mathbf{b}(h)]^2}{\mathbf{w}(h)}}$	$\dfrac{-\sum_{h=1}^{H} \mathbf{a}(h) \cdot \mathbf{b}(h)/\mathbf{w}(h)}{\sqrt{\sum_{h=1}^{H} \mathbf{a}(h)^2 \sum_{h=1}^{H} \mathbf{b}(h)^2}}$

should be identical. With such consideration, each feature $\mathbf{p}_{k_j}$ of a test sequence $\mathcal{P}_j$ can be matched against a training sequence $\mathcal{G}_m$ to find the shortest distance from its features $\{\mathbf{g}_{k_m}\}$ as

$$S(\mathbf{p}_{k_j}, \mathcal{G}_m) = -\min_{k_m} d(\mathbf{p}_{k_j}, \mathbf{g}_{k_m}). \tag{9.2}$$

Similarly, each feature $\mathbf{g}_{k_m}$ of a training sequence $\mathcal{G}_m$ can be matched against a test sequence $\mathcal{P}_j$ as

$$S(\mathbf{g}_{k_m}, \mathcal{P}_j) = -\min_{k_j} d(\mathbf{g}_{k_m}, \mathbf{p}_{k_j}). \tag{9.3}$$

The similarity score between test sequence $\mathcal{P}_j$ and training sequence $\mathcal{G}_m$ can be obtained by summing up the mean matching score of all features from $\mathcal{P}_j$ against $\mathcal{G}_m$ and that of all features from $\mathcal{G}_m$ against $\mathcal{P}_j$:

$$S(\mathcal{P}_j, \mathcal{G}_m) = \frac{1}{K_j} \sum_{k_j=1}^{K_j} S(\mathbf{p}_{k_j}, \mathcal{G}_m) + \frac{1}{K_m} \sum_{k_m=1}^{K_m} S(\mathbf{g}_{k_m}, \mathcal{P}_j). \tag{9.4}$$

9.2 Face Recognition

Face recognition is a popular biometric technology that has received significant attention during the past two decades. It has a large number of commercial security and forensic applications, including video surveillance, access control, mugshot identification, entertainment industry, video communications, and medical diagnosis [Jain et al., 2004a; Chellappa et al., 2005]. Face recognition algorithms are broadly categorized into appearance based [Turk and Pentland, 1991; Belhumeur et al., 1997; Schölkopf et al., 1998] and feature based [Samaria and Young, 1994; Wiskott et al., 1997; Cootes et al., 2001] algorithms and they take digitally captured facial images, mostly in gray-level, as input.

Biometrics refers to the automatic recognition of individuals based on their physiological and/or behavioral characteristics [Jain et al., 2004a]. *Physiological* characteristics are related to the shape of the body, such as fingerprints, faces, hand geometry, and iris. *Behavioral* characteristics are related to the behavior of a person, such as signature, keystroke, voice, and gait. Although biometrics emerged from its extensive use in law enforcement to identify criminals, it is being increasingly used for human recognition in a large number of civilian applications. Biometric systems offer greater security and convenience than traditional methods of personal recognition, such as ID cards and passwords. They give users greater convenience (e.g., no need to remember passwords) while maintaining sufficiently high accuracy and ensuring that the user is present at the point in time of recognition [Jain et al., 2007].

This section presents some face recognition results of linear and multilinear subspace learning algorithms for demonstration. In the experiments, face images are input directly as second-order tensors as shown in Figure 9.2 to multilinear algorithms, while for linear algorithms, they are converted to vectors as input. For each subject in a face recognition experiment, L samples are randomly selected for training and the rest are used for testing. The recognition results report the average recognition rates over twenty random splits. The following presents two sets of face recognition experiments (one set by unsupervised learning algorithms and the other set by supervised learning algorithms) after describing the algorithms compared and their settings briefly. The detailed description and results can be found in [Lu, 2008].

9.2.1 Algorithms and Their Settings

The unsupervised subspace learning algorithms to be studied are principal component analysis (PCA) [Jolliffe, 2002; Turk and Pentland, 1991], two-dimensional PCA (2DPCA) [Yang et al., 2004], concurrent subspaces analysis (CSA) [Xu et al., 2008], tensor rank-one decomposition (TROD) [Shashua and

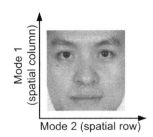

FIGURE 9.2: A face image represented as a second-order tensor (matrix) of *column × row*.

Levin, 2001], multilinear PCA (MPCA) [Lu et al., 2008b], and uncorrelated MPCA (UMPCA) [Lu et al., 2009d].

The supervised subspace learning algorithms to be studied are PCA plus linear discriminant analysis (PCA+LDA) [Belhumeur et al., 1997], uncorrelated linear discriminant analysis (ULDA) [Ye, 2005b][2], regularized direct LDA (R-JD-LDA) [Lu et al., 2003, 2005], discriminant analysis with tensor representation (DATER) [Yan et al., 2007a], general tensor discriminant analysis (GTDA) [Tao et al., 2007b], tensor rank-one discriminant analysis (TR1DA) [Wang and Gong, 2006; Tao et al., 2006], MPCA with discriminative feature selection (MPCA-S) described in Section 5.5.1, MPCA-S plus LDA (MPCA+LDA) [Lu et al., 2008b], regularized uncorrelated multilinear discriminant analysis (R-UMLDA) [Lu et al., 2009c], and R-UMLDA with aggregation (R-UMLDA-A) [Lu et al., 2009c].

The *correct recognition rate* (CRR), or the *rank-1 identification rate*, is calculated through similarity measurement between feature vectors for up to 600 features. The algorithms 2DPCA, CSA, MPCA, DATER, and GTDA produce tensor-valued features, which need to be converted to vector-valued features for classification and comparison. For the conversion, the unsupervised feature selection scheme in Section 5.5.2 is used for unsupervised methods 2DPCA, CSA, and MPCA, while the supervised feature selection scheme in Section 5.5.1 is used for supervised methods DATER and GTDA. In addition, we need to determine the best dimensions of the feature vectors for input to LDA for PCA+LDA and MPCA+LDA.

9.2.2 Recognition Results for Supervised Learning Algorithms

We show face recognition results of supervised learning algorithms on a subset from the Pose, Illumination, and Expression (PIE) database [Sim et al., 2003]. Seven poses (C05, C07, C09, C27, C29, C37, C11) are chosen, with at most 45 degrees of pose variation and under 21 illumination conditions (02 to 22). Thus, there are about 147 (7×21) samples per subject and there are a total number of 9,987 face images (with nine faces missing). All face images are normalized to 32×32 pixels, with 256 gray levels per pixel (preprocessing details are described in Appendix B.1.3).

Figure 9.3 depicts the detailed results for $L = 2, 6, 10$, and 40 by the supervised algorithms, where the horizontal axis is shown in log scale and the best results from using D_{ℓ_1}, D_{ℓ_2}, and D_θ distance measures in Table 9.1 are plotted. For $L = 2, 10, 40$, MPCA+LDA gives the best results. For $L = 6$, R-UMLDA-A gives the best result. For $L = 10$, the first four features extracted by R-UMLDA are the most effective features in recognition. For $L = 40$, the

---

[2]This ULDA algorithm is different from the ULDA in [Jin et al., 2001a] so it is different from the classical LDA.

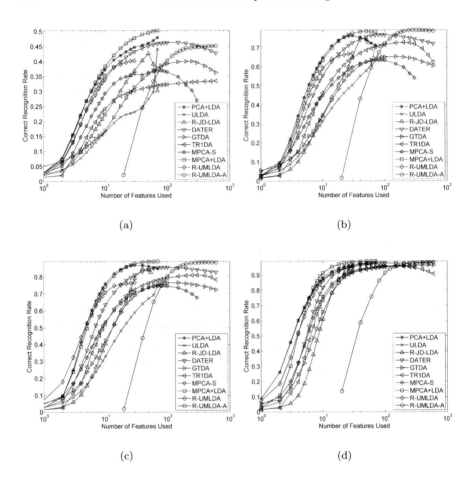

(a)　　　　　　　　　　　　　　　　(b)

(c)　　　　　　　　　　　　　　　　(d)

FIGURE 9.3: Face recognition results by supervised subspace learning algorithms on the PIE database: correct recognition rate against the number of features used for L training samples per subject with (a) $L = 2$, (b) $L = 6$, (c) $L = 10$, and (d) $L = 40$.

first five features extracted by PCA+LDA and ULDA are the most effective features in recognition.

9.2.3　Recognition Results for Unsupervised Learning Algorithms

In classification tasks, unsupervised learning algorithms are usually inferior to supervised ones. However, because no class specific information is required in the learning process, unsupervised subspace learning algorithms have wider

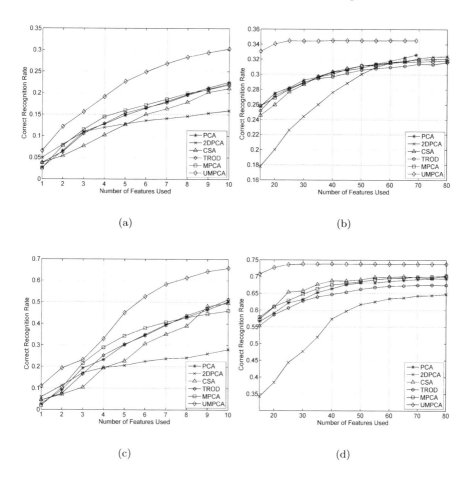

FIGURE 9.4: Face recognition results by unsupervised subspace learning algorithms on the FERET database: correct recognition rate against the number of features used P for L training samples per subject with (a) $P = 1, ..., 10$ for $L = 1$; (b) $P = 15, ..., 80$ for $L = 1$; (c) $P = 1, ..., 10$ for $L = 7$; and (d) $P = 15, ..., 80$ for $L = 7$.

applications. In particular, in the so-called one training sample case [Tan et al., 2006; Wang et al., 2006], an extreme small sample size (SSS) scenario where only one sample per class is available for training (i.e., $L = 1$), the supervised subspace learning algorithms studied here cannot be applied because it is impossible to measure the within-class scatter with only one sample per class available. In contrast, unsupervised subspace learning algorithms are still applicable even in this difficult scenario.

This set of experiments studies the recognition performance of unsupervised subspace learning algorithms on higher resolution face images and particularly in low-dimensional subspace. A subset of the Facial Recognition Technology (FERET) database [Phillips et al., 2000] is selected to consist of those subjects having at least eight images with at most 15 degrees of pose variation, resulting in 721 face images from 70 subjects. All face images are preprocessed and normalized to 80×80 pixels, with 256 gray levels per pixel. For the unsupervised subspace learning algorithms studied, extracted features are all arranged in descending variation captured (measured by respective total scatter), and the D_{ℓ_2} (Euclidean) distance measure is tested.

Figure 9.4 shows the results for $L = 1$ and $L = 7$. $L = 1$ is the one training sample (per class) case, and $L = 7$ is the maximum number of training samples (per subject) that can be used in this set of experiments. For PCA and UMPCA, there are at most 69 (uncorrelated) features when $L = 1$ because there are only 70 faces for training. Figures 9.4(a) and 9.4(c) plot CRRs against the number of features used P (the dimension of the subspace), for $P = 1, ..., 10$. Figures 9.4(b) and 9.4(d) plot CRRs for P ranging from 15 to 80. UMPCA performs best in these figures, indicating that uncorrelated features extracted directly from tensor (matrix) face data are effective in these settings. In particular, for smaller values of P, UMPCA outperforms the other algorithms significantly. The figures also show that for UMPCA, the recognition rate saturates around $P = 30$, due to decreasing variation (scatter) captured by individual features [Lu et al., 2009d].

9.3   Gait Recognition

Gait, a person's walking style, is a complex spatio-temporal biometric [Nixon and Carter, 2006, 2004; Jain et al., 2004a]. It is considered the only true remote biometric [Jain et al., 2007]. The interest in gait recognition is strongly motivated by the need for an automated human identification system at a distance in visual surveillance and monitoring applications in security-sensitive environments, for example, banks, parking lots, museums, malls, and transportation hubs such as airports and train stations [Wang et al., 2003]. Other biometrics such as fingerprint, face, or iris information are usually not available at high enough resolution for recognition in these circumstances [Chellappa et al., 2005]. Furthermore, night vision capability (an important component in surveillance) is usually not possible with other biometrics due to limited biometric details in an IR image at large distances [Kale, 2003]. Therefore, gait recognition became a popular research area due to its advantages of unobtrusiveness, hard-to-hide, and recognition at a distance [Kale et al., 2004; Sarkar et al., 2005; Boulgouris et al., 2005].

Similar to face recognition, gait recognition algorithms are either appearance based [Liu et al., 2004; Boulgouris et al., 2004; Vega and Sarkar, 2003; Little and Boyd, 1998; Tolliver and Collins, 2003; Foster et al., 2003; Boyd,

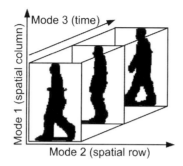

FIGURE 9.5: A gait silhouette sequence as a third-order tensor of *column* × *row* × *time*.

2001; Grant et al., 2004] or model based [Wang et al., 2004; Yam et al., 2004; Cunado et al., 2003; Wagg and Nixon, 2004; Lu et al., 2006a, 2008a]. Their inputs are usually binary gait silhouette sequences because color or texture is not reliable for recognition.

This section studies recognition performance on third-order tensor-valued gait objects using the similarity measure between two sequences of feature vectors in Equation (9.4). Gait samples are input directly as third-order tensors to multilinear algorithms with three modes of spatial column, spatial row, and time, as shown in Figure 9.5. While for linear algorithms, they are converted to vectors for input. The experiments follow the standard procedures described in Section B.2. Here we only study supervised linear and multilinear subspace learning algorithms. Their detailed settings can be found in [Lu, 2008].

We show the results on the University of South Florida (USF) gait database V.1.7 [Sarkar et al., 2005] (with details described in Section B.2) with probes of varying difficulty. The original resolution $128 \times 88 \times 20$ results in vectors of $225,280 \times 1$, which leads to high memory and computational demand for linear subspace learning algorithms. Therefore, in this set of experiments, each normalized gait sample of $128 \times 88 \times 20$ is down-sampled to a tensor of $32 \times 22 \times 10$ so that all linear subspace learning algorithms can be applied more efficiently.

Figure 9.6 plots the CRRs for supervised subspace learning algorithms (with the best results from the D_{ℓ_1}, D_{ℓ_2}, and D_θ distance measures). Figures 9.6(a) and 9.6(b) depict the average CRRs for probes A, B, and C (the easier probes) as well as the average CRRs for all seven probes, respectively. The horizontal axis is shown in log scale. From the figures, MPCA+LDA achieves the best overall performance. The overall results of MPCA-S is competitive as well.

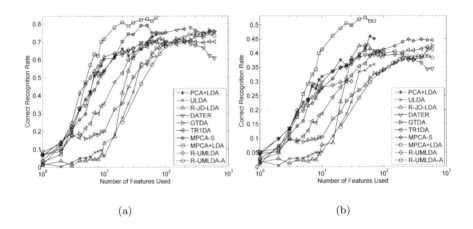

(a)   (b)

FIGURE 9.6: The correct recognition rates of supervised subspace learning algorithms on the $32 \times 22 \times 10$ USF gait database V.1.7: (a) the average over probes A, B, and C; and (b) the average over all seven probes.

9.4   Visual Content Analysis in Computer Vision

As a dimensionality reduction technique, MSL can be used for various visual content analysis tasks. For example, Gao et al. [2009] built an optical flow tensor and used GTDA to reduce the dimensionality for further semantic analysis of video sequences. For 3D facial data modeling, Tao et al. [2008b] employed the Bayesian tensor analysis (BTA) algorithm for 3D facial expression retargeting task. Wen et al. [2009] applied a modified version of the dynamic tensor analysis (DTA) algorithm (with mean and variance updating) on a weighted tensor representation for visual tracking of human faces, and Ruiz-Hernandez et al. [2010a] developed an automatic human age estimation method using MPCA for dimensionality reduction of tensor-valued Gaussian receptive maps. In the following, we discuss several applications of MSL in visual content analysis in greater detail.

9.4.1   Crowd Event Visualization and Clustering

Crowd video content analysis is difficult because detection and tracking of individual subjects become complex tasks due to large numbers of subjects in the scene [Wang et al., 2009]. In [Lu et al., 2010b], MPCA was used to analyze crowd activities and events without detection and tracking. A video segment is viewed as a basic video element to be mapped to a point in the MPCA subspace. Thus, the entire video sequence is visualized as a trajectory in the

MPCA subspace. This visualization enables visual clustering of events and provides a valuable tool in analyzing video contents for video summarization, anomaly detection, and behavior understanding.

The method was tested on four sequences from the dataset S3 "Event Recognition" of the Performance Evaluation of Tracking and Surveillance (PETS) 2009 database with dense crowd [Ferryman and Shahrokni, 2009]. The sequences contain walking, running, evacuation (rapid dispersion), local dispersions, crowd formation/merging and splitting. In experiments, the MPCA-based method gives much better visualization of the four PETS2009 crowd video sequences and produces more visible clusters of events than Isomap [Tenenbaum et al., 2000], Laplacian Eigenmaps [Belkin and Niyogi, 2003], and diffusion map [Coifman and Lafon, 2006].

9.4.2 Target Tracking in Video

The online MPCA approach in [Wang et al., 2011a] detects local spatial structures in the target tensor space through fusing information from different feature spaces using tensor representation. It can learn a more discriminative appearance model in case of significant target appearance variations and cluttered background. The tracking is done using a likelihood model to measure the similarity between a test sample and the learned appearance model, and a particle filter [Arulampalam et al., 2002] to recursively estimate the target state over time.

The incremental tensor biased discriminant analysis (ITBDA) in [Wen et al., 2010] extracts features from color objects represented as third-order tensors (*column* × *row* × *color*) for tracking. The method is able to adapt to appearance change in both the object and the background. It can track objects precisely undergoing partial occlusion, and large changes in pose, scale, and lighting changes.

9.4.3 Action, Scene, and Object Recognition

Jia et al. [2011] developed the multilinear discriminant analysis of canonical correlations (MDCC) and incremental MDCC (IMDCC) for action recognition, where each action sample, a short video sequence, is represented as a high-order tensor. The methods have advantages in accuracy, time complexity, and robustness against partial occlusion. In addition, MPCA was also applied to action recognition in [Sun et al., 2011].

Han et al. [2012] proposed the multilinear supervised neighborhood embedding (MSNE) algorithm for scene and object recognition. They used the Scale-Invariant Feature Transform (SIFT) [Lowe, 2004] features or the histogram of orientation weighted with a normalized gradient (NHOG) for local region representation. These local descriptors form a tensor, from which features are further extracted using MSNE. This way of combining features is shown to be more efficient than the popular bag-of-feature (BOF) model [Sivic and Zisserman, 2003].

9.5   Brain Signal/Image Processing in Neuroscience

This section discusses two applications of MSL in processing brain signals and images in neuroscience.

9.5.1   EEG Signal Analysis

Electroencephalography (EEG), as shown in Figure 9.7, records brain activities as multichannel time series from multiple electrodes placed on the scalp of a subject to provide a direct communication channel between brain and computer [Blankertz et al., 2008]. It is widely used in noninvasive brain-computer interface (BCI) applications [Lu et al., 2009b, 2010a; Lotte et al., 2007].

Acar et al. [2007] employs the N-way partial least squares (N-PLS) for automated visual analysis of EEG data and patient-specific epileptic seizure recognition [Khan and Gotman, 2003; Lu et al., 2013]. Several features in both the time and frequency domains are extracted from multichannel EEG signals for each epoch to detect distinct trends during seizure and non-seizure periods. They are arranged into an epilepsy feature tensor with modes: time epochs, features, and electrodes. N-PLS is then used to model this feature tensor to facilitate EEG data analysis and identify important features for seizure recognition.

Li et al. [2009a] proposed a tensor-based EEG classification scheme using GTDA. The wavelet transform [Daubechies, 1990] is applied to EEG signals

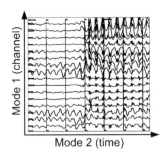

FIGURE 9.7: Multichannel electroencephalography (EEG) signals with each channel as a time series recorded by an electrode placed on the scalp ("Electroencephalography," Wikipedia, the free encyclopedia, http://en.wikipedia.org/wiki/Electroencephalography).

to result in third-order tensor representations in the spatial-spectral-temporal domain. GTDA is then applied to obtain low-dimensional tensors, from which discriminative features are selected for support vector machine (SVM) classification [Burges, 1998]. This approach has been shown to outperform many other existing EEG signal classification schemes in motor imagery experiments on three datasets, especially when there is no prior neurophysiologic knowledge available for EEG signal preprocessing.

Washizawa et al. [2010] introduced a weighted MPCA method for EEG signal classification in motor imagery BCI. This method has been shown to outperform the standard common spatial pattern (CSP) method [Ramoser et al., 2000] and a combination of PCA and CSP methods.

Zhao et al. [2011] proposed the higher-order PLS (HOPLS) to decode human electrocorticogram (ECoG)[3] from simultaneously recorded scalp EEG, assuming that both EEG and ECoG are related to the same brain activities. Common latent components between EEG and ECoG are extracted to examine whether ECoG can be decoded from the corresponding EEG using HOPLS.

9.5.2 fMRI Image Analysis

Functional magnetic resonance image (fMRI) scan sequences are fourth-order tensors as shown in Figure 9.8. Barnathan et al. [2010] applied tensor analysis to computer-aided diagnosis (CAD) of fMRIs of the brain and proposed a TWave method based on multi-step wavelet and tensor representation. A sixth-order tensor is formed with three spatial modes, one temporal mode,

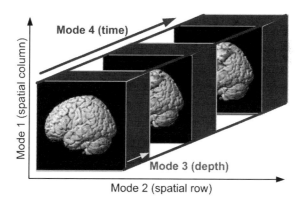

FIGURE 9.8: A functional magnetic resonance imaging (fMRI) scan sequence with three spatial modes and one temporal mode [Pantano et al., 2005].

---

[3]ECoG is also called intracranial EEG (iEEG); it records electrical activity from the cerebral cortex through electrodes placed directly on the exposed surface of the brain.

and two categorical modes (subject and motor task). Compared with TWave, the combination of TWave and MPCA+LDA can improve subject-based classification accuracy from 96% to 100% and task-based classification accuracy from 74% to 93% [Barnathan, 2010]. Compared with PARAFAC, the combination of TWave and MPCA+LDA can improve subject-based classification accuracy from 88% to 100% and task-based classification accuracy from 52% to 93% [Barnathan, 2010].

9.6 DNA Sequence Discovery in Bioinformatics

Bioinformatics studies methods for biological data processing, including nucleic acid (DNA/RNA) and protein sequence [Saeys et al., 2007; Rozas et al., 2003]. To identify regulatory circuits controlling gene expression, biologists often need to map some functional regions in genomic sequence, such as promoters, which are DNA regions that initiate transcription of a particular gene. To analyze and predict promoter regions in DNA sequences, it is assumed that sub-sequences from different functional regions differ in local term composition [Li et al., 2008].

To capture both the term composition and position, DNA sequences can be represented as binary matrices $\{\mathbf{A}\}$ with the terms as mode 1 and the positions within the sequence as mode 2. An element of the binary tensor (matrix) $\mathbf{A}(i_1, i_2)$ is an indicator of whether the sequence represented by $\mathbf{A}$ has a term i_1 at the position i_2. The MPCA extension for binary tensors in [Mažgut et al., 2010, 2013] has been shown to reveal natural groupings of sub-sequences from different functional regions in a large-scale dataset of DNA sequences through subspace learning and visualization.

9.7 Music Genre Classification in Audio Signal Processing

Music genre is a popular description of music content for music database organization. The nonnegative MPCA (NMPCA) algorithm proposed in [Panagakis et al., 2010] was designed for music genre classification, combined with the nearest neighbor or SVM classifiers. A 2D joint acoustic and modulation frequency representation of music signals is utilized in this work to capture slow temporal modulations [Sukittanon et al., 2004], inspired by psychophysiological investigations of the human auditory system. An acoustic signal is first converted to a time-frequency auditory spectrogram [Pitton et al., 1996], as il-

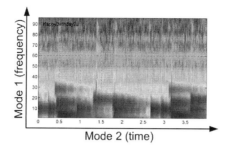

FIGURE 9.9: A 2D auditory spectrogram representation of music signals.

lustrated in Figure 9.9. Then, a wavelet transform is applied to each row of this spectrogram via a bank of Gabor filters [Jain and Farrokhnia, 1991] to estimate its temporal modulation content through modulation scale analysis [Qian, 2002]. This analysis results in a 3D representation of $rate \times time \times frequency$. The power of this 3D representation is next obtained by integrating across the wavelet translation axis to get a joint rate-frequency representation. In three different sets of experiments on public music genre datasets, NMPCA has achieved the state-of-the-art classification results.

9.8 Data Stream Monitoring in Data Mining

In streaming data applications such as network forensics, large volumes of high-order data continuously arrive incrementally [Han and Kamber, 2006]. It is challenging to perform incremental pattern discovery on such data streams.

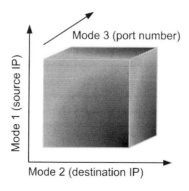

FIGURE 9.10: Network traffic data organized as a third-order tensor of source IP×destination IP×port number.

The incremental tensor analysis (ITA) framework in [Sun et al., 2008a] was devised to tackle this problem using tensor representation of network traffic data, as illustrated in Figure 9.10. It aims to find suspicious Internet activity in a real trace. The abnormality is modeled by the reconstruction error of incoming tensor data streams. A large reconstruction error often indicates an anomaly. This method has been illustrated on network flow data of 100GB over 1 month, where three types of anomalies (abnormal source hosts, abnormal destination hosts, and abnormal ports) can be detected with very high precision [Sun et al., 2006].

9.9   Other MSL Applications

Other MSL applications include palmprint recognition [Zhang et al., 2008; Guo et al., 2011], gait-based ethnicity classification [Zhang et al., 2010], color face recognition [Wang et al., 2011b], handwritten digit recognition [Ye, 2005a; Xu et al., 2008; Barnathan et al., 2010], image compression [Ye et al., 2004a; Xu et al., 2008; Shashua and Levin, 2001], image/video retrieval [Ye et al., 2004a; Gao et al., 2010b], head-pose estimation [Guo et al., 2012], object categorization and recognition [Wang and Gong, 2006], hyperspectral image classification [Zhang et al., 2013], reactor of thermal anneal (RTA) batch process monitoring [Guo et al., 2010], and target classification in synthetic aperture radar (SAR) images [Porges and Favier, 2011].

For more general tensor data applications, some good references are [Kolda and Bader, 2009], [Aja-Fernández et al., 2009], [Cichocki et al., 2009], and [MMD, 2006, 2008, 2010].

Appendix A

Mathematical Background

This appendix presents the basic results and definitions from linear algebra, probability theory, optimization, and matrix calculus that serve as the mathematical foundations for multilinear subspace learning. Systematic studies of these results can be found in many textbooks. In particular, [Moon and Stirling, 2000] has comprehensive coverage of these mathematical foundations. Note that in this book, we are only concerned about real numbers.

A.1 Linear Algebra Preliminaries

In this book, we denote vectors by lowercase boldface letters such as **a** and matrices by uppercase boldface such as **A**.

A.1.1 Transpose

A column vector **a** of size $d \times 1$ consists of d *entries*. Its *transpose* $\mathbf{a}^T$ is a row vector of size $1 \times d$. They can be written as

$$\mathbf{a} = \begin{bmatrix} a_1 \\ a_2 \\ \vdots \\ a_d \end{bmatrix} \text{ and } \mathbf{a}^T = \begin{bmatrix} a_1 \, a_2 \, \ldots \, a_d \end{bmatrix}, \tag{A.1}$$

respectively. An $r \times c$ matrix **A** consists of r rows and c columns. Its transpose $\mathbf{A}^T$ consists of c rows and r columns. They can be written as

$$\mathbf{A} = \begin{bmatrix} a_{11} & a_{12} & \cdots & a_{1c} \\ a_{21} & a_{22} & \cdots & a_{2c} \\ \vdots & \vdots & \ddots & \vdots \\ a_{r1} & a_{r2} & \cdots & a_{rc} \end{bmatrix} \text{ and } \mathbf{A}^T = \begin{bmatrix} a_{11} & a_{21} & \cdots & a_{r1} \\ a_{12} & a_{22} & \cdots & a_{r2} \\ \vdots & \vdots & \ddots & \vdots \\ a_{1c} & a_{2c} & \cdots & a_{rc} \end{bmatrix}, \tag{A.2}$$

respectively.

A *symmetric* matrix is a square matrix **A** of size $d \times d$ with its entries $a_{ij} = a_{ji}$ for $i, j = 1, ..., d$, that is, $\mathbf{A} = \mathbf{A}^T$. A *nonnegative* matrix is a matrix with $a_{ij} \geq 0$ for all i and j. A *diagonal* matrix is a $d \times d$ square matrix $\mathbf{A}$ with all nondiagonal entries equal to 0. A *rectangular diagonal* matrix is an $r \times c$ matrix $\mathbf{A}$ with only the entries of the form a_{ii} possibly nonzero.

A.1.2   Identity and Inverse Matrices

An *identity* matrix $\mathbf{I}$ is a square (diagonal) matrix of size $d \times d$ with its diagonal entries equal to 1 while all other entries equal to 0:

$$
\mathbf{I} = \begin{bmatrix} 1 & 0 & \cdots & 0 \\ 0 & 1 & \cdots & 0 \\ \vdots & \vdots & \ddots & \vdots \\ 0 & 0 & \cdots & 1 \end{bmatrix}.
\tag{A.3}
$$

A $d \times d$ square matrix $\mathbf{A}$ is *invertible* (or *nonsingular*) if there exists an *inverse matrix* $\mathbf{A}^{-1}$ such that their product is an identity matrix:

$$
\mathbf{A}^{-1}\mathbf{A} = \mathbf{A}\mathbf{A}^{-1} = \mathbf{I}.
\tag{A.4}
$$

For a non-square matrix $\mathbf{A}$ (or if $\mathbf{A}^{-1}$ does not exist), we can define the *left inverse* (or *pseudoinverse*) $\mathbf{A}^{+}$ if $\mathbf{A}^{T}\mathbf{A}$ is nonsingular:

$$
\mathbf{A}^{+} = (\mathbf{A}^{T}\mathbf{A})^{-1}\mathbf{A}^{T},
\tag{A.5}
$$

such that $\mathbf{A}^{+}\mathbf{A} = \mathbf{I}$.

Inverse of product: The inverse of the product of two matrices is equivalent to the product of their inverses in reverse order:

$$
(\mathbf{A}\mathbf{B})^{-1} = \mathbf{B}^{-1}\mathbf{A}^{-1}.
\tag{A.6}
$$

Transpose of inverse: The transpose of the inverse matrix is equivalent to the inverse of the transpose:

$$
(\mathbf{A}^{-1})^{T} = (\mathbf{A}^{T})^{-1}.
\tag{A.7}
$$

Inverse of inverse: The inverse of the inverse matrix is equivalent to the matrix itself:

$$
(\mathbf{A}^{-1})^{-1} = \mathbf{A}.
\tag{A.8}
$$

A.1.3   Linear Independence and Vector Space Basis

A set of vectors $\mathbf{x}_1, \mathbf{x}_2, \ldots, \mathbf{x}_p$ are *linearly dependent* if there exist constants $a_1, a_2, \ldots, a_p$ not all zero such that

$$
a_1\mathbf{x}_1 + a_2\mathbf{x}_2 + \cdots + a_p\mathbf{x}_p = 0.
\tag{A.9}
$$

Otherwise, no vector in this set can be written as a linear combination of other vectors so these vectors are *linearly independent*. A set of p linearly independent vectors spans a p-dimensional *vector space* [Moon and Stirling, 2000], which means that any vector in that space can be written as a linear combination of such spanning vectors. These spanning vectors form the *basis* of that vector space.

A.1.4 Products of Vectors and Matrices

The *inner product* of two vectors of the same dimension d results in a scalar

$$\mathbf{a}^T\mathbf{b} = \mathbf{b}^T\mathbf{a} = \sum_{i=1}^{d} x_i y_i. \tag{A.10}$$

It is also called the *scalar product* or the *dot product*. The Euclidean norm of a vector is

$$\| \mathbf{a} \| = \sqrt{\mathbf{a}^T\mathbf{a}}, \tag{A.11}$$

which is the length of the vector. If $\| \mathbf{a} \| = 1$, we call it a *normalized vector* or a *unit vector*. The angle θ between two vectors of the same dimension d is determined by

$$\cos\theta = \frac{\mathbf{a}^T\mathbf{b}}{\| \mathbf{a} \| \| \mathbf{b} \|}. \tag{A.12}$$

If $\mathbf{a}^T\mathbf{b} = 0$, $\cos\theta = 0$ so $\mathbf{a}$ and $\mathbf{b}$ are *orthogonal*. Two vectors are *orthonormal vectors* if they are orthogonal and normalized vectors.

A $d \times d$ square matrix $\mathbf{U}$ is an *orthogonal matrix* if its columns and rows are orthonormal vectors. For an orthogonal matrix $\mathbf{U}$, its transpose is equal to its inverse:

$$\mathbf{U}^T = \mathbf{U}^{-1} \tag{A.13}$$

so we have

$$\mathbf{U}\mathbf{U}^T = \mathbf{U}^T\mathbf{U} = \mathbf{I}, \tag{A.14}$$

where $\mathbf{I}$ is an identity matrix.

The *outer product* of two vectors $\mathbf{a}$ $(r \times 1)$ and $\mathbf{b}$ $(c \times 1)$ results in a matrix of size $r \times c$ so it is also called the *matrix product* or *tensor product*[1]:

$$\mathbf{a} \circ \mathbf{b} = \mathbf{a}\mathbf{b}^T = \begin{bmatrix} a_1 \\ a_2 \\ \vdots \\ a_r \end{bmatrix} \begin{bmatrix} b_1 & b_2 & \ldots & b_c \end{bmatrix} = \begin{bmatrix} a_1b_1 & a_1b_2 & \ldots & a_1b_c \\ a_2b_1 & a_2b_2 & \ldots & a_2b_c \\ \vdots & \vdots & \ddots & \vdots \\ a_rb_1 & a_rb_2 & \ldots & a_rb_c \end{bmatrix}. \tag{A.15}$$

The *Kronecker product* (also called the *tensor product*) of two matrices $\mathbf{A}$

---

[1]Here, we use a notation '$\circ$' different from the conventional notation '$\otimes$' to better differentiate outer product of vectors from Kronecker product of matrices.

$(r \times s)$ and $\mathbf{B}$ $(c \times d)$ results in a block matrix of size $rc \times sd$, which can also be viewed as a fourth-order tensor of size $r \times s \times c \times d$:

$$
\mathbf{A} \otimes \mathbf{B} = \begin{bmatrix} a_{11}\mathbf{B} & a_{12}\mathbf{B} & \cdots & a_{1s}\mathbf{B} \\ a_{21}\mathbf{B} & a_{22}\mathbf{B} & \cdots & a_{2s}\mathbf{B} \\ \vdots & \vdots & \ddots & \vdots \\ a_{r1}\mathbf{B} & a_{r2}\mathbf{B} & \cdots & a_{rs}\mathbf{B} \end{bmatrix}. \tag{A.16}
$$

It is a generalization of outer product from vectors to matrices. On the other hand, the outer product of vectors can be considered a special case of the Kronecker product with $s = d = 1$. They have the following properties:

$$
\mathbf{A} \otimes (\mathbf{B} + \mathbf{C}) = \mathbf{A} \otimes \mathbf{B} + \mathbf{A} \otimes \mathbf{C}, \tag{A.17}
$$

$$
(\mathbf{A} \otimes \mathbf{B}) \otimes \mathbf{C} = \mathbf{A} \otimes (\mathbf{B} \otimes \mathbf{C}), \tag{A.18}
$$

$$
(\mathbf{A} \otimes \mathbf{B})^T = \mathbf{A}^T \otimes \mathbf{B}^T, \tag{A.19}
$$

$$
(\mathbf{A} \otimes \mathbf{B})^{-1} = \mathbf{A}^{-1} \otimes \mathbf{B}^{-1}, \tag{A.20}
$$

$$
(\mathbf{A} \otimes \mathbf{B})(\mathbf{C} \otimes \mathbf{D}) = (\mathbf{AC} \otimes \mathbf{BD}). \tag{A.21}
$$

We can multiply a matrix $\mathbf{A}$ of size $r \times c$ with a vector $\mathbf{x}$ of size $c \times 1$ to obtain a vector $\mathbf{y}$ of size $r \times 1$:

$$
\begin{bmatrix} a_{11} & a_{12} & \cdots & a_{1c} \\ a_{21} & a_{22} & \cdots & a_{2c} \\ \vdots & \vdots & \ddots & \vdots \\ a_{r1} & a_{r2} & \cdots & a_{rc} \end{bmatrix} \begin{bmatrix} x_1 \\ x_2 \\ \vdots \\ x_c \end{bmatrix} = \begin{bmatrix} y_1 \\ y_2 \\ \vdots \\ y_r \end{bmatrix}, \tag{A.22}
$$

which can be written as

$$
\mathbf{Ax} = \mathbf{y}. \tag{A.23}
$$

Each entry of $\mathbf{y}$ is

$$
y_i = \sum_{j=1}^{c} a_{ij} x_j, \quad i = 1, ..., r. \tag{A.24}
$$

We can multiply a matrix $\mathbf{A}$ of size $r \times c$ with another matrix $\mathbf{B}$ of size $c \times d$ to obtain a matrix $\mathbf{C}$ of size $r \times d$:

$$
\mathbf{C} = \mathbf{AB}. \tag{A.25}
$$

If $\mathbf{B}$ (or $\mathbf{A}$) is an identity matrix, then $\mathbf{C} = \mathbf{A}$ (or $\mathbf{C} = \mathbf{B}$).

A symmetric $d \times d$ matrix $\mathbf{A}$ is *positive definite* if $\mathbf{x}^T \mathbf{Ax}$ is positive for any nonzero column vector $\mathbf{x}$.

Product of sum: The product of a matrix $\mathbf{A}$ with the sum of two matrices $\mathbf{B}$ and $\mathbf{C}$ is equivalent to the sum of their products $\mathbf{AB}$ and $\mathbf{AC}$:

$$
\mathbf{A}(\mathbf{B} + \mathbf{C}) = \mathbf{AB} + \mathbf{AC}. \tag{A.26}
$$

Transpose of sum: The transpose of the sum of two matrices is equivalent to the sum of their transposes:

$$(\mathbf{A} + \mathbf{B})^T = \mathbf{A}^T + \mathbf{B}^T. \tag{A.27}$$

Transpose of product: The transpose of the product of two matrices is equivalent to the product of their transposes in reverse order:

$$(\mathbf{AB})^T = \mathbf{B}^T \mathbf{A}^T. \tag{A.28}$$

A.1.5 Vector and Matrix Norms

The ℓ_1 norm of a $d \times 1$ vector is

$$\| \mathbf{a} \|_1 = \sum_{i=1}^{d} |a_i|. \tag{A.29}$$

The ℓ_2 norm of a $d \times 1$ vector is

$$\| \mathbf{a} \|_2 = \sqrt{\sum_{i=1}^{d} a_i^2}, \tag{A.30}$$

which is the Euclidean norm in Equation (A.11).

The ℓ_p norm of a $d \times 1$ vector is

$$\| \mathbf{a} \|_p = \left(\sum_{i=1}^{d} |a_i|^p \right)^{1/p}. \tag{A.31}$$

The *Frobenius norm* of an $r \times c$ matrix is defined as the square root of the sum of the squares of its elements:

$$\| \mathbf{A} \|_F = \sqrt{\sum_{i=1}^{r} \sum_{j=1}^{c} a_{ij}^2}. \tag{A.32}$$

A.1.6 Trace

There is lots of information contained in matrices and we often want to extract useful information from them. *Trace* is a scalar quantity that can provide some useful information about a matrix. The trace of a $d \times d$ square matrix $\mathbf{A}$ is the sum of its diagonal entries:

$$\mathrm{tr}(\mathbf{A}) = \sum_{i=1}^{d} a_{ii}. \tag{A.33}$$

The trace has several interesting properties:

Trace of sum:

$$\text{tr}(\mathbf{A} + \mathbf{B}) = \text{tr}(\mathbf{A}) + \text{tr}(\mathbf{B}). \tag{A.34}$$

Trace of scalar multiplication:

$$\text{tr}(c\mathbf{A}) = c \cdot \text{tr}(\mathbf{A}). \tag{A.35}$$

Trace of transpose:

$$\text{tr}(\mathbf{A}) = \text{tr}(\mathbf{A}^T). \tag{A.36}$$

Trace of product:

$$\text{tr}(\mathbf{AB}) = \text{tr}(\mathbf{BA}). \tag{A.37}$$

Trace of cyclic products:

$$\text{tr}(\mathbf{ABC}...) = \text{tr}(\mathbf{BC}...\mathbf{A}) = \text{tr}(\mathbf{C}...\mathbf{AB}). \tag{A.38}$$

Trace and the Frobenius norm:

$$\| \mathbf{A} \|_F = \sqrt{\text{tr}(\mathbf{A}^T \mathbf{A})}. \tag{A.39}$$

A.1.7   Determinant

The *determinant* of a $d \times d$ square matrix $\mathbf{A}$ is another scalar quantity that can provide some useful information about a matrix. We first consider a simple 2×2 matrix:

$$\mathbf{A} = \begin{bmatrix} a & b \\ c & d \end{bmatrix}. \tag{A.40}$$

We can verify that its inverse is

$$\mathbf{A}^{-1} = \frac{1}{ad - bc} \begin{bmatrix} d & -b \\ -c & a \end{bmatrix}. \tag{A.41}$$

The denominator $ad - bc$ is the determinant of this matrix, denoted as $\det(\mathbf{A})$ or $|\mathbf{A}|$:

$$\det(\mathbf{A}) = |\mathbf{A}| = \begin{vmatrix} a & b \\ c & d \end{vmatrix}. \tag{A.42}$$

The determinant for a 3×3 matrix can be calculated as

$$\begin{vmatrix} a & b & c \\ d & e & f \\ g & h & i \end{vmatrix} = a \begin{vmatrix} e & f \\ h & i \end{vmatrix} - b \begin{vmatrix} d & f \\ g & i \end{vmatrix} + c \begin{vmatrix} d & e \\ g & h \end{vmatrix} = aei + bfg + cdh - ceg - bdi - afh. \tag{A.43}$$

The calculation of the determinant for matrices of higher dimensions is more involved; for example, it can be done by expansion of cofactors/minors [Moon and Stirling, 2000].

If the column vectors of $\mathbf{A}$ are not linearly independent, then $|\mathbf{A}| = 0$. In this case, $\mathbf{A}$ is a *singular* matrix and its inverse does not exist. If $|\mathbf{A}| \neq 0$, $\mathbf{A}$ is *nonsingular* and its inverse exists.

Determinant of inverse:

$$|\mathbf{A}^{-1}| = \frac{1}{|\mathbf{A}|}. \tag{A.44}$$

Determinant of transpose:

$$|\mathbf{A}| = |\mathbf{A}^T|. \tag{A.45}$$

Determinant of product:

$$|\mathbf{AB}| = |\mathbf{B}||\mathbf{A}|. \tag{A.46}$$

Determinant of an identity matrix:

$$|\mathbf{I}| = 1. \tag{A.47}$$

A.1.8   Eigenvalues and Eigenvectors

For a $d \times d$ square matrix $\mathbf{A}$, the following form represents a very important class of linear equations

$$\mathbf{Ax} = \lambda\mathbf{x}, \tag{A.48}$$

where λ is a scalar and $\mathbf{x}$ is usually the unknown vector to be solved. It can also be written as

$$(\mathbf{A} - \lambda\mathbf{I})\mathbf{x} = \mathbf{0}, \tag{A.49}$$

where $\mathbf{I}$ is an identity matrix and $\mathbf{0}$ is a zero vector (all entries are zero). This is called the *eigenvalue problem* or *eigendecomposition*. The ith (nonzero) solution $\mathbf{x}_i$ and the corresponding ith scalar λ_i are the ith *eigenvector* and *eigenvalue*, respectively. For real and symmetric $\mathbf{A}$, there are d eigenvectors $\mathbf{x}_1, \mathbf{x}_2, \ldots, \mathbf{x}_d$ associated with d eigenvalues $\lambda_1, \lambda_2, \ldots, \lambda_d$, respectively.

Multiplication of $\mathbf{A}$ with its eigenvector $\mathbf{x}_i$ results in a scaled version of $\mathbf{x}_i$ (by λ_i):

$$\mathbf{Ax}_i = \lambda_i\mathbf{x}_i. \tag{A.50}$$

In other words, the linear transformation $\mathbf{A}$ only elongates or shrinks the eigenvectors by the amount indicated by the eigenvalues and it does not change the direction of the eigenvectors.

The system in Equation (A.49) has nonzero solutions if and only if the matrix $\mathbf{A} - \lambda\mathbf{I}$ is singular so its determinant should be zero:

$$|\mathbf{A} - \lambda\mathbf{I}| = 0. \tag{A.51}$$

This gives the *characteristic equation*, a dth order polynomial equation in the unknown λ. The roots of this characteristic equation are the eigenvalues.

Trace and eigenvalues: The trace of a matrix is the sum of its eigenvalues:

$$\text{tr}(\mathbf{A}) = \sum_{i=1}^{d} \lambda_i. \tag{A.52}$$

Determinant and eigenvalues: The determinant of a matrix is the product of its eigenvalues:

$$|\mathbf{A}| = \prod_{i=1}^{d} \lambda_i. \tag{A.53}$$

Trace maximization: For a $d \times d$ symmetric matrix $\mathbf{A}$ and some arbitrary $d \times p$ matrix $\mathbf{U}$ with orthonormal columns, the trace of $\mathbf{U}^T \mathbf{A} \mathbf{U}$ is maximized when $\mathbf{U}$ consists of eigenvectors associated with the largest p eigenvalues [Kokiopoulou et al., 2011].

A.1.9   Generalized Eigenvalues and Eigenvectors

For a pair of matrices $\mathbf{A}$ and $\mathbf{B}$, we have the following *generalized eigenvalue problem* similar to Equation (A.48):

$$\mathbf{A}\mathbf{x} = \lambda \mathbf{B}\mathbf{x}. \tag{A.54}$$

For this problem, the solutions λ and $\mathbf{x}$ are the *generalized eigenvalue* and *generalized eigenvector* of $\mathbf{A}$ and $\mathbf{B}$, respectively. The corresponding characteristic equation is

$$|\mathbf{A} - \lambda \mathbf{B}\mathbf{I}| = 0. \tag{A.55}$$

When $\mathbf{B}$ is invertible (nonsingular), we can write the generalized eigenvalue problem into a standard eigenvalue problem as

$$\mathbf{B}^{-1}\mathbf{A}\mathbf{x} = \lambda \mathbf{x}. \tag{A.56}$$

However, when $\mathbf{A}$ and $\mathbf{B}$ are symmetric matrices, the conversion often results in a nonsymmetric matrix $\mathbf{B}^{-1}\mathbf{A}$.

When $\mathbf{A}$ and $\mathbf{B}$ are both symmetric matrices and $\mathbf{B}$ is a positive-definite matrix, the generalized eigenvalues are real and any two generalized eigenvectors $\mathbf{x}_i$ and $\mathbf{x}_j$ with distinct generalized eigenvalues are $\mathbf{B}$-orthogonal, that is, $\mathbf{x}_i^T \mathbf{B} \mathbf{x}_j = 0$.

A.1.10   Singular Value Decomposition

The *singular value decomposition* (SVD) is a matrix factorization closely related with the eigenvalue problem. The SVD of an $r \times c$ matrix $\mathbf{A}$ is a factorization of the following form:

$$\mathbf{A} = \mathbf{U}\boldsymbol{\Sigma}\mathbf{V}^T, \tag{A.57}$$

where $\mathbf{U}$ is an $r \times r$ orthogonal matrix, $\boldsymbol{\Sigma}$ is an $r \times c$ rectangular diagonal matrix with nonnegative numbers on the diagonal, and $\mathbf{V}$ is a $c \times c$ orthogonal matrix. The diagonal entries of $\boldsymbol{\Sigma}$, denoted as $\{\sigma_i\}$, are the *singular values* of $\mathbf{A}$. The r columns of $\mathbf{U}$ and the c columns of $\mathbf{V}$ are the *left-singular vectors* and *right-singular vectors* of $\mathbf{A}$, respectively.

The *condition number* of a matrix $\mathbf{A}$ is the ratio of its largest singular value to its smallest singular value:

$$\kappa(\mathbf{A}) = \frac{\sigma_{max}}{\sigma_{min}}. \tag{A.58}$$

A matrix is said to be *ill-conditioned* if its condition number is too large.

Relationships between singular vectors/values and eigenvectors/eigenvalues:

- The left-singular vectors of $\mathbf{A}$ (columns of $\mathbf{U}$) are the eigenvectors of $\mathbf{A}\mathbf{A}^T$.

- The right-singular vectors of $\mathbf{A}$ (columns of $\mathbf{V}$) are the eigenvectors of $\mathbf{A}^T\mathbf{A}$.

- The nonzero singular values of $\mathbf{A}$ (the diagonal entries of $\boldsymbol{\Sigma}$) are the square roots of the nonzero eigenvalues of both $\mathbf{A}\mathbf{A}^T$ and $\mathbf{A}^T\mathbf{A}$.

A.1.11 Power Method for Eigenvalue Computation

The *power method* is a simple iterative algorithm for finding the largest eigenvalue and its associated eigenvector of a $d \times d$ matrix $\mathbf{A}$. It starts with an initial vector $\mathbf{x}_0$ and does the following iteration

$$\mathbf{x}_{k+1} = \frac{\mathbf{A}\mathbf{x}_k}{\| \mathbf{A}\mathbf{x}_k \|}. \tag{A.59}$$

This method may only converge slowly.

A.2 Basic Probability Theory

This section only reviews those basic theories needed in this book.

A.2.1 One Random Variable

First, we consider only one discrete random variable x that can assume any of d different values in the set $\mathbb{X} = \{c_1, c_2, \ldots, c_d\}$. The probability that x assumes c_i is denoted by p_i as

$$p_i = \Pr[x = c_i], \ for \ i = 1, \ldots d, \tag{A.60}$$

where

$$p_i \geq 0, \ and \ \sum_{i=1}^{d} p_i = 1. \tag{A.61}$$

The *mean* or *expected value* of the random variable x is defined as

$$E[x] = \mu = \sum_{i=1}^{d} c_i p_i. \tag{A.62}$$

The *variance* of the random variable x is defined as

$$\text{Var}[x] = \sigma^2 = E[(x - \mu)^2] = \sum_{i=1}^{d} (c_i - \mu)^2 p_i. \tag{A.63}$$

The square root of the variance, σ, is the *standard deviation* of x.

A.2.2   Two Random Variables

We consider two discrete random variables x and y that can assume different values in the set $\mathbb{X} = \{c_1, c_2, \ldots, c_d\}$ and $\mathbb{Y} = \{e_1, e_2, \ldots, e_f\}$, respectively. The probability that x or y assumes a particular value c_i in $\mathbb{X}$ or e_j in $\mathbb{Y}$ is denoted by the *probability mass function* $P(x)$ and $P(y)$, respectively. The probability that x assumes a particular value c_i in $\mathbb{X}$ and y assumes a particular value e_j in $\mathbb{Y}$ is denoted by the *joint probability mass function* $P(x, y)$, where

$$P(x, y) \geq 0, \ and \ \sum_{x \in \mathbb{X}} \sum_{y \in \mathbb{Y}} P(x, y) = 1. \tag{A.64}$$

The means and variances for x and y are

$$\mu_x = E[x] = \sum_{x \in \mathbb{X}} \sum_{y \in \mathbb{Y}} x P(x, y), \tag{A.65}$$

$$\mu_y = E[y] = \sum_{x \in \mathbb{X}} \sum_{y \in \mathbb{Y}} y P(x, y), \tag{A.66}$$

$$\sigma_x^2 = \text{Var}[x] = E[(x - \mu_x)^2] = \sum_{x \in \mathbb{X}} \sum_{y \in \mathbb{Y}} (x - \mu_x)^2 P(x, y), \tag{A.67}$$

$$\sigma_y^2 = \text{Var}[y] = E[(y - \mu_y)^2] = \sum_{x \in \mathbb{X}} \sum_{y \in \mathbb{Y}} (y - \mu_y)^2 P(x, y). \tag{A.68}$$

The *covariance* of x and y is defined as

$$\sigma_{xy} = E[(x - \mu_x)(y - \mu_y)] = \sum_{x \in \mathbb{X}} \sum_{y \in \mathbb{Y}} (x - \mu_x)(y - \mu_y) P(x, y). \tag{A.69}$$

The *correlation* coefficient is defined as

$$\rho = \frac{\sigma_{xy}}{\sigma_x \sigma_y}. \tag{A.70}$$

If $\sigma_{xy} = 0$, $\rho = 0$, so x and y are *uncorrelated*.

A.3   Basic Constrained Optimization

Constrained optimization is employed in this book at several points to get some desired property. Here, we present the basic concepts.

Let $f(\mathbf{x})$ be a function to be minimized or maximized, subject to a constraint $g(\mathbf{x}) = 0$. The constraint is introduced by forming the *Lagrangian function*

$$L(\mathbf{x}, \lambda) = f(\mathbf{x}) + \lambda g(\mathbf{x}), \qquad (A.71)$$

where λ is a *Lagrange multiplier*. The conditions for an optimal solution are obtained by setting the derivatives to zero:

$$\frac{\partial L(\mathbf{x}, \lambda)}{\partial \mathbf{x}} = 0, \quad \frac{\partial L(\mathbf{x}, \lambda)}{\partial \lambda} = 0. \qquad (A.72)$$

When there are several variables, a partial derivative is introduced for each. When there are several constraints, each constraint is introduced using its own Lagrange multiplier. For example, if there are P constraints: $g_1(\mathbf{x}) = 0$, $g_2(\mathbf{x}) = 0$, ..., $g_P(\mathbf{x}) = 0$, then we form

$$L(\mathbf{x}, \{\lambda_p, p = 1, ..., P\}) = f(\mathbf{x}) + \lambda_1 g_1(\mathbf{x}) + \lambda_2 g_2(\mathbf{x}) + ... + \lambda_p g_P(\mathbf{x}). \quad (A.73)$$

The conditions for an optimal solution are obtained by setting:

$$\frac{\partial L(\mathbf{x}, \{\lambda_i\})}{\partial \mathbf{x}} = 0, \quad \frac{\partial L(\mathbf{x}, \{\lambda_i\})}{\partial \lambda_1} = 0, \quad \frac{\partial L(\mathbf{x}, \{\lambda_i\})}{\partial \lambda_2} = 0, \quad ..., \quad \frac{\partial L(\mathbf{x}, \{\lambda_i\})}{\partial \lambda_p} = 0.$$
$$(A.74)$$

A.4   Basic Matrix Calculus

As reviewed in the previous section, the solution for an optimization problem often involves the calculation of derivatives. A handy reference is available in [Roweis, 1999]. Here, we review some basic matrix calculus.

A.4.1   Basic Derivative Rules

Constant rule:

$$\frac{\mathrm{d}f(x)}{\mathrm{d}x} = 0, \text{ if } f(x) \text{ is constant.} \qquad (A.75)$$

Sum rule:

$$\frac{\mathrm{d}(\alpha f(x) + \beta g(x))}{\mathrm{d}x} = \alpha \frac{\mathrm{d}f(x)}{\mathrm{d}x} + \beta \frac{\mathrm{d}g(x)}{\mathrm{d}x}. \qquad (A.76)$$

Product rule:

$$\frac{\mathrm{d}(f(x)g(x))}{\mathrm{d}x} = g(x)\frac{\mathrm{d}f(x)}{\mathrm{d}x} + f(x)\frac{\mathrm{d}g(x)}{\mathrm{d}x}. \tag{A.77}$$

Quotient rule:

$$\frac{\mathrm{d}(f(x)/g(x))}{\mathrm{d}x} = \frac{g(x)\frac{\mathrm{d}f(x)}{\mathrm{d}x} + f(x)\frac{\mathrm{d}g(x)}{\mathrm{d}x}}{(g(x))^2}, \quad \text{if } g(x) \neq 0. \tag{A.78}$$

Chain rule:

$$\frac{\mathrm{d}(f(x))}{\mathrm{d}x} = \frac{\mathrm{d}g(x)}{\mathrm{d}x} \cdot \frac{\mathrm{d}f(g(x))}{\mathrm{d}g(x)}, \quad \text{for } f(x) = f(g(x)). \tag{A.79}$$

A.4.2 Derivative of Scalar/Vector with Respect to Vector

In the following, $\mathbf{a}$ or $\mathbf{A}$ is not a function of $\mathbf{x}$:

$$\frac{\partial \mathbf{x}}{\partial \mathbf{x}} = \mathbf{I}. \tag{A.80}$$

$$\frac{\partial \mathbf{a}}{\partial \mathbf{x}} = \mathbf{0}. \tag{A.81}$$

$$\frac{\partial \mathbf{A}\mathbf{x}}{\partial \mathbf{x}} = \mathbf{A}^T. \tag{A.82}$$

$$\frac{\partial \mathbf{x}^T \mathbf{A}}{\partial \mathbf{x}} = \mathbf{A}. \tag{A.83}$$

$$\frac{\partial \mathbf{x}^T \mathbf{x}}{\partial \mathbf{x}} = 2\mathbf{x}. \tag{A.84}$$

$$\frac{\partial \mathbf{x}^T \mathbf{A}\mathbf{x}}{\partial \mathbf{x}} = (\mathbf{A} + \mathbf{A}^T)\mathbf{x}. \tag{A.85}$$

A.4.3 Derivative of Trace with Respect to Matrix

In the following, $\mathbf{A}$ or $\mathbf{B}$ is not a function of $\mathbf{x}$:

$$\frac{\partial \mathrm{tr}(\mathbf{X})}{\partial \mathbf{X}} = \mathbf{I}. \tag{A.86}$$

$$\frac{\partial \mathrm{tr}(\mathbf{A}\mathbf{X})}{\partial \mathbf{X}} = \frac{\partial \mathrm{tr}(\mathbf{X}\mathbf{A})}{\partial \mathbf{X}} = \mathbf{A}^T. \tag{A.87}$$

$$\frac{\partial \mathrm{tr}(\mathbf{A}\mathbf{X}^T)}{\partial \mathbf{X}} = \frac{\partial \mathrm{tr}(\mathbf{X}^T \mathbf{A})}{\partial \mathbf{X}} = \mathbf{A}. \tag{A.88}$$

$$\frac{\partial \mathrm{tr}(\mathbf{X}^T \mathbf{A}\mathbf{X})}{\partial \mathbf{X}} = (\mathbf{A} + \mathbf{A}^T)\mathbf{X}. \tag{A.89}$$

$$\frac{\partial \text{tr}(\mathbf{AXB})}{\partial \mathbf{X}} = \mathbf{A}^T \mathbf{B}^T. \tag{A.90}$$

$$\frac{\partial \text{tr}(\mathbf{X}^{-1}\mathbf{A})}{\partial \mathbf{X}} = -\mathbf{X}^{-1}\mathbf{A}^T\mathbf{X}^{-1}. \tag{A.91}$$

A.4.4 Derivative of Determinant with Respect to Matrix

In the following, $\mathbf{A}$ or $\mathbf{B}$ is not a function of $\mathbf{x}$:

$$\frac{\partial |\mathbf{X}|}{\partial \mathbf{X}} = |\mathbf{X}|(\mathbf{X}^{-1})^T. \tag{A.92}$$

$$\frac{\partial \ln |\mathbf{X}|}{\partial \mathbf{X}} = (\mathbf{X}^{-1})^T. \tag{A.93}$$

$$\frac{\partial |\mathbf{AXB}|}{\partial \mathbf{X}} = |\mathbf{AXB}|(\mathbf{X}^{-1})^T. \tag{A.94}$$

Appendix B

Data and Preprocessing

This appendix covers three popular face and gait databases for evaluating and developing multilinear subspace learning (MSL) algorithms. As raw data are usually noisy and/or not in a standard representation for learning, *preprocessing* steps are often needed before extracting or learning features, which can play an important role in the successfulness of a learning-based system [Wagstaff, 2012]. Therefore, we also discuss related preprocessing steps to obtain standard tensor data for input to learning algorithms.

B.1 Face Databases and Preprocessing

We consider two widely used public face databases here: the Pose, Illumination, and Expression (PIE) database [Sim et al., 2003]; and the Facial Recognition Technology (FERET) database [Phillips et al., 2000].

B.1.1 PIE Database

Visually perceived human faces are significantly affected by three factors: the pose, which is the angle they are viewed from, the illumination (or lighting) condition, and the facial expression such as happy, sad, and anger. The collection of the PIE database was motivated by a need for a database with a fairly large number of subjects imaged a large number of times to cover these three significant factors, that is, from a variety of different poses, under a wide range of illumination variations, and with several expressions [Sim et al., 2003].

This database was collected at the Carnegie Mellon University (CMU) between October 2000 and December 2000 using the CMU 3D room. It contains 41,368 face images from 68 individuals, with a total size of about 40GB of data. The captured images have a size of 640 × 486. Face images with 13 different poses were captured using 13 synchronized cameras. For the illumination variation, the 3D room was augmented with a flash system having 21 flashes. Images were captured with and without background lighting, resulting in $21 \times 2 + 1 = 43$ different illumination conditions. In addition, the subjects were asked to pose with four different expressions.

The PIE database can be used for a variety of purposes, including evaluating the robustness of face recognition systems against the three variations and three-dimensional modeling. In particular, this database has a large number (around 600 on average) of facial images available for each subject, allowing us to study the effects of the number of training samples (per subject) on the recognition performance. In practice, a subset is usually selected with a specific range of pose, illumination, and expression for experiments so that datasets with various degrees of difficulty can be obtained, such as in Section 9.2.2. A wider range of the three variations leads to a more difficult recognition task.

B.1.2 FERET Database

The FERET database is a widely used database for face recognition performance evaluation. It was constructed through the FERET program, which aimed to develop automatic face recognition systems to assist security, intelligence, and law enforcement personnel in the performance of their duties [Phillips et al., 2000]. The face images in this database cover a wide range of variations in pose (viewpoint), illumination, facial expression, acquisition time, ethnicity, and age.

The FERET database was collected in 15 sessions between August 1993 and July 1996, and it contains a total of 14,126 images from 1,199 individuals with views ranging from frontal to left and right profiles. The face images were collected under relatively unconstrained conditions. The same physical setup and location were used in each session to maintain a degree of consistency throughout the database. However, because the equipment was reassembled for each session, images collected on different dates have some minor variation. Sometimes, a second set of images of an individual was captured on a later date, resulting in variations in scale, pose, expression, and illumination of the face. Furthermore, for some people, over 2 years elapsed between their first and last capturing in order to study changes in a subject's facial appearance over a year.

The experimental studies in Section 9.2.3 were based on the color FERET database. The images have size of 786×512 and they are encoded with 24 bits. The total data size is around 8GB. This database has a large number of subjects and it becomes the de facto standard for evaluating face recognition technologies after release [Liu, 2006], especially in the small sample size (SSS) scenario, where a smaller number of training samples per subject and a large number of total subjects lead to a more difficult recognition task.

B.1.3 Preprocessing of Face Images for Recognition

Here, we discuss preprocessing steps involved in the face recognition experiments in Section 9.2. We consider only gray-level facial images, without taking color information into account. Moreover, because the focus is on the recog-

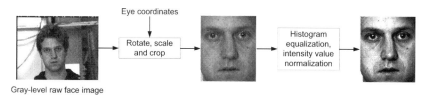

Gray-level raw face image

FIGURE B.1: Illustration of face image preprocessing.

FIGURE B.2: Sample face images of one subject from the CMU PIE database.

nition of faces rather than their detection, all face images from the PIE and FERET databases were manually aligned with manually annotated coordinate information of eyes, cropped, and normalized. A practical system will need face detection [Viola and Jones, 2004; Hsu et al., 2002] and eye detection [Zhou and Geng, 2004; Feng and Yuen, 1998, 2001] modules to perform such preprocessing automatically. The detailed preprocessing procedures are described below and illustrated in Figure B.1.

First, all color images are transformed to gray-level images by taking the luminance component in the YC_bC_r color space [Plataniotis and Venetsanopoulos, 2000]. Then, all face images are rotated and scaled so that the centers of the eyes are placed on specific pixels. Next, the images are cropped and normalized to a standard size, followed by histogram equalization, and image intensity values are normalized to have zero mean and unit standard deviation. Finally, each image is represented with 256 gray levels (eight bits) per pixel. Figure B.2 shows 160 near-frontal face images for one subject in the PIE database, and Figure B.3 shows some sample face images from two subjects in the FERET database.

FIGURE B.3: Examples of face images from two subjects in the FERET database.

B.2 Gait Database and Preprocessing

The HumanID Gait Challenge datasets from the University of South Florida (USF) [Sarkar et al., 2005] is a standard testbed for gait recognition algorithms [Liu and Sarkar, 2004, 2005; Lee et al., 2003]. This database captures the variations of a number of covariates for a large group of people. This section describes this USF database and then introduces how to obtain gait samples from a gait silhouette sequence for MSL.

B.2.1 USF Gait Challenge Database

Gait recognition has many open questions to answer [Nixon and Carter, 2004]. It is important to investigate the conditions under which this problem is "solvable," and to find out what factors affect gait recognition and to what extent. The HumanID Gait Challenge Problem was thus introduced by USF in order to assess the potential of gait recognition by providing a means for measuring progress and characterizing the properties [Sarkar et al., 2005].

This challenge problem consists of a baseline algorithm, a large dataset, and a set of twelve experiments. The baseline algorithm extracts silhouettes through background subtraction and performs recognition via temporal correlation of silhouettes. The data were collected outdoors because gait, as a biometric, is most appropriate in outdoor at-a-distance settings where other biometrics are difficult to capture. Figure B.4 shows two sample frames from this database.

The twelve experiments, in increasing difficulty, examine the effects of five covariates on recognition performance: change in viewing angle, change in shoe type, change in walking surface, carrying or not carrying a briefcase, and temporal (time) differences, where the time covariate implicitly includes other changes naturally occurring between video acquisition sessions such as change of shoes and cloths, change in the outdoor lighting conditions, and

FIGURE B.4: Sample frames from the Gait Challenge datasets.

inherent variation in gait over time. These covariates either affect gait or affect the extraction of gait features from images. They were selected, based on logistical issues and collection feasibility, from a list of factors compiled through discussions with researchers at CMU, Maryland, MIT, Southampton, and Georgia Tech about potentially important covariates for gait analysis. It was shown in [Sarkar et al., 2005] that the shoe type has the least impact on the performance, next is the viewpoint, the third is briefcase, then surface type (flat concrete surface and typical grass lawn surface), and time (6 months) difference has the greatest impact. The latter two are the most "difficult" covariates to deal with. In particular, it was found that the surface covariate impacts the gait period more than the other covariates.

The experimental studies in Section 9.3 were conducted on the USF gait database V.1.7 collected in May 2001. This database consists of 452 sequences from 74 subjects walking in elliptical paths in front of the camera. The raw video frames are of size 720×480 in 24-bit RGB and a subject's size in the back portion of the ellipse is, on average, 100 pixels in height. The total data is around 300GB. For each subject, there are three covariates: viewpoint (left or right), shoe type (two different types, A or B), and surface type (grass or concrete). The gallery set contains seventy-one, sequences (subjects) and seven experiments (probe sets) were designed for human identification as shown in Table B.1. The capturing condition for each probe set is summarized in the parentheses after the probe name in the table, where C, G, A, B, L, and R stand for concrete surface, grass surface, shoe type A, shoe type B, left view, and right view, respectively. For instance, the capturing condition of the gallery set is GAR (Grass surface, shoe type A and Right view). Each set has only one sequence for a subject. Subjects are unique in the gallery and each probe set. There are no common sequences between the gallery set and any of the probe sets. In addition, all the probe sets are distinct.

TABLE B.1: Characteristics of the gait data from the USF Gait Challenge datasets version 1.7.

Gait Dataset	Number of Sequences	Difference from the Gallery
Gallery (GAR)	71 (731)	-
Probe A(GAL)	71 (727)	View
Probe B(GBR)	41 (423)	Shoe
Probe C(GBL)	41 (420)	Shoe, view
Probe D(CAR)	70 (682)	Surface
Probe E(CBR)	44 (435)	Surface, shoe
Probe F(CAL)	70 (685)	Surface, view
Probe G(CBL)	44 (424)	Surface, shoe, view

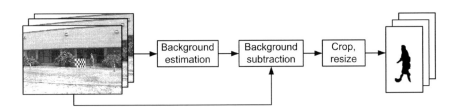

FIGURE B.5: Illustration of the silhouette extraction process.

B.2.2 Gait Silhouette Extraction

Figure B.5 illustrates the process of silhouette extraction through background subtraction, where a background model is estimated from the input raw gait sequences and then it is subtracted to get the silhouettes [Elgammal et al., 1999; Lu et al., 2006c]. The extracted silhouettes are then cropped and resized to a standard size. The silhouettes data extracted through the baseline algorithm are provided by the USF and are widely used in the gait recognition literature [Boulgouris et al., 2005]. Thus, this silhouettes data were used in Section 9.3.

B.2.3 Normalization of Gait Samples

While in many recognition problems, an input sample is unambiguously defined, such as iris, face, or fingerprint images, there is no obvious definition of a gait sample. In [Lu et al., 2008b], each half gait cycle is treated as a data sample.

To obtain half cycles, a gait silhouette sequence is partitioned following the approach in [Sarkar et al., 2005]. The number of foreground pixels is counted

FIGURE B.6: Three gait samples from the USF Gait database V.1.7, shown by concatenating frames in rows.

in the bottom half of each silhouette as legs are the major visible moving body parts from a distance. This number will reach a maximum when the two legs are farthest apart and drop to a minimum when the legs overlap. The sequence of these numbers is smoothed with a moving average filter [Oppenheim et al., 1983], and the minimums in this number sequence partition the sequence into several half gait cycles. In this way, 731 gait samples were obtained in the gallery set, and each subject has roughly ten samples available on average. The number of samples for each set is indicated in parentheses following the number of sequences in the second column of Table B.1.

Each frame of the gait silhouette sequences from the USF datasets is of standard size 128×88, but the number of frames in each gait sample obtained through half cycle partition has some variation. Before feeding the gait samples to a subspace learning algorithm, they need to be normalized to the same dimensions. Because the row and column dimensions (mode-1 and mode-2 dimensions) are normalized already, only the time mode (mode-3), that is, the number of frames in each gait sample, is subject to normalization. The normalized time mode dimension was chosen to be 20, roughly the average number of frames in each gait sample. Thus, each gait sample has a standard representation of $I_1 \times I_2 \times I_3 = 128 \times 88 \times 20$. The following describes a simple procedure for this time-mode normalization.

Conventional interpolation algorithms, such as linear interpolation [Oppenheim et al., 1983], can be applied to time-mode normalization. Consider one gait sample of size $I_1 \times I_2 \times D_3$. Each mode-3 (time-mode) vector is interpolated linearly from the original size $D_3 \times 1$ to the normal size $I_3 \times 1$, followed by binarization to get binary silhouettes. Figure B.6 shows three gait samples obtained in this way from the USF gait database V.1.7 by concatenating the frames of one sample in a row.

Appendix C

Software

This final appendix provides information on multilinear subspace learning (MSL) software, and discusses the importance of open-source software and some good software practices.

C.1 Software for Multilinear Subspace Learning

The *MATLAB Tensor Toolbox* developed by Bader and Kolda [2006, 2007] is widely used for tensor manipulations and fast algorithm prototyping in MSL and tensor-based computation in general. This toolbox includes a set of classes for mathematical operations on tensors that extends the functionality of MATLAB's multidimensional arrays to support operations such as tensor matrix/vector multiplications, unfolding, and many other operations. The current version also supports efficient manipulation of *sparse tensors*, where only a small fraction of the elements are nonzero. This toolbox is freely available for research and evaluation purposes at the following website:

http://www.sandia.gov/~tgkolda/TensorToolbox/

Since the development of our first MSL algorithm, multilinear principal component analysis (MPCA), we have put in efforts to make sure that our algorithms are available for those interested. We have made the MATLAB source code for our algorithms available at the following website:

http://www.mathworks.com/matlabcentral/fileexchange/authors/80417

Furthermore, we also set up a page to include useful information on this topic, including software, data, related sites, and important publications at

http://www.comp.hkbu.edu.hk/~haiping/MSL.html

or

http://www.dsp.toronto.edu/~haiping/MSL.html

or

https://sites.google.com/site/tensormsl/

In the future, we will continue to make more learning algorithms, education materials, and other relevant sources available in the public domain.

C.2 Benefits of Open-Source Software

Open-source software is important for advancing the state-of-the-art [Sonnenburg et al., 2007]. To utilize the true potential of MSL algorithms that is not utilized, it is important to openly share their implementations. Their true impact will be judged eventually by their success in solving real scientific and technological problems. Therefore, it is crucial for the progress of this field to make algorithm replication and application to new tasks easy. Four key benefits are summarized below based on [Sonnenburg et al., 2007]:

- Increase reproducibility and reusability of scientific research and reduce reimplementation efforts.

- Facilitate fair comparisons and evaluations of algorithms.

- Uncover problems with the algorithms (including bugs) and develop enhancements and extensions.

- Build on each other's open-source software tools to develop more complex machine learning algorithms.

C.3 Software Development Tips

We found that the following machine learning software development practices suggested in [Sonnenburg et al., 2007] and from online resources such as blogs (e.g. http://hunch.net/ by John Langford) to be very useful, especially for new comers:

- The software should be structured well and logically such that its usability is high.

- It is good to keep code, input data, and results in separate directories for easy management, archiving, debugging, and distribution.

- Parameters and/or program options used should be saved together with the results (and even documentation on observations) for easy tracing and regeneration.

- The development cycle can be reduced significantly for big datasets by working on smaller datasets at an early stage to find bugs earlier and faster. For example, we can first work on a small portion of the full database, a smaller problem of the big problem, or a lower-resolution/dimension of the full resolution/dimension.

- It should have good documentations such that others can learn to use the software quickly, for example, including tutorials, examples, references (e.g., the corresponding paper with key calculation steps indicated in the code).

- It should provide testing routines to verify automatically whether the code is correct and tolerate incorrect inputs by providing relevant error messages rather than breaking down silently.

Bibliography

Workshop on Algorithms for Modern Massive Data Sets, 2006, 2008, 2010. URL http://www.stanford.edu/group/mmds/.

NSF Workshop on Future Directions in Tensor-Based Computation and Modeling, 2009. URL http://www.cs.cornell.edu/cv/TenWork/FinalReport.pdf.

Acar, E. and Yener, B. Unsupervised multiway data analysis: A literature survey. *IEEE Transactions on Knowledge and Data Engineering*, 21(1): 6–20, October 2009.

Acar, E., Bingol, C.A., Bingol, H., Bro, R., and Yener, B. Seizure recognition on epilepsy feature tensor. In *Proc. 29th International Conference of the IEEE Engineering in Medicine and Biology Society*, pp. 4273–4276, 2007.

Aja-Fernández, S., d. L. García, R., Tao, D., and Li (Eds.), X. *Tensors in Image Processing and Computer Vision*. Springer, 2009.

Akaike, H. Information theory and an extension of the maximum likelihood principle. In *Breakthroughs in statistics*, pp. 610–624. Springer, 1992.

Allwein, E. L., Schapire, R. E., and Singer, Y. Reducing multiclass to binary: A unifying approach for margin classifiers. *Journal of Machine Learning Research*, 1:113–141, September 2000.

Amari, S.I. Natural gradient works efficiently in learning. *Neural Computation*, 10(2):251–276, 1998.

Anderson, T. W. *An Introduction to Multivariate Statistical Analysis*. Wiley, New York, third edition, 2003.

Antonini, M., Barlaud, M., Mathieu, P., and Daubechies, I. Image coding using wavelet transform. *IEEE Transactions on Image Processing*, 1(2): 205–220, 1992.

Armbrust, M., Fox, A., Griffith, R., Joseph, A., Katz, R., Konwinski, A., Lee, G., Patterson, D., Rabkin, A., Stoica, I., et al. A view of cloud computing. *Communications of the ACM*, 53(4):50–58, 2010.

Arulampalam, M. S., Maskell, S., Gordon, N., and Clapp, T. A tutorial on particle filters for online nonlinear/non-Gaussian Bayesian tracking. *IEEE Transactions on Signal Processing*, 50(2):174–188, 2002.

Ayers, J., Davis, J.L., and Rudolph, A. *Neurotechnology for Biomimetic Robots*. MIT Press, 2002.

Bader, B. W. and Kolda, T. G. Algorithm 862: MATLAB tensor classes for fast algorithm prototyping. *ACM Transactions on Mathematical Software*, 32(4):635–653, December 2006.

Bader, B. W. and Kolda, T. G. Efficient MATLAB computations with sparse and factored tensors. *SIAM Journal on Scientific Computing*, 30(1):205–231, December 2007.

Barker, M. and Rayens, W. Partial least squares for discrimination. *Journal of Chemometrics*, 17(3):166–173, 2003.

Barnathan, M. Mining Complex High-Order Datasets. PhD thesis, Temple University, 2010.

Barnathan, M., Megalooikonomou, V., Faloutsos, C., Mohamed, F., and Faro, S. TWave: High-order analysis of spatiotemporal data. In *Proc. Pacific-Asia Conference on Knowledge Discovery and Data Mining*, pp. 246–253, 2010.

Bartlett, M. S., Movellan, J. R., and Sejnowski, T. J. Face recognition by independent component analysis. *IEEE Transactions on Neural Networks*, 13(6):1450–1464, November 2002.

Beckmann, C. F. and Smith, S. M. Tensorial extensions of independent component analysis for multisubject fMRI analysis. *NeuroImage*, 25(1):294–311, 2005.

Beckmann, C. F. and Smith, S. M. Probabilistic independent component analysis for functional magnetic resonance imaging. *IEEE Transactions on Medical Imaging*, 23(2):137–152, 2004.

Belhumeur, P. N., Hespanha, J. P., and Kriegman, D. J. Eigenfaces vs. fisherfaces: Recognition using class specific linear projection. *IEEE Transactions on Pattern Analysis and Machine Intelligence*, 19(7):711–720, July 1997.

Belkin, M. and Niyogi, P. Laplacian eigenmaps and spectral techniques for embedding and clustering. In *Advances in Neural Information Processing Systems (NIPS)*, pp. 585–591, 2001.

Belkin, M. and Niyogi, P. Laplacian eigenmaps for dimensionality reduction and data representation. *Neural Computation*, 15(6):1373–1396, June 2003.

Belkin, M. and Niyogi, P. Semi-supervised learning on Riemannian manifolds. *Machine Learning*, 56(1):209–239, 2004.

Bell, A. J. and Sejnowski, T. J. An information-maximization approach to blind separation and blind deconvolution. *Neural Computation*, 7(6):1129–1159, 2000.

Bie, T., Cristianini, N., and Rosipal, R. Eigenproblems in pattern recognition. *Handbook of Geometric Computing*, pp. 129–167, 2005.

Bishop, C. M. Bayesian PCA. In *Advances in Neural Information Processing Systems (NIPS)*, pp. 382–388, 1999.

Blankertz, B., Tomioka, R., Lemm, S., Kawanabe, M., and Müller, K.-R. Optimizing spatial filters for robust EEG single-trial analysis. *IEEE Signal Processing Magazine*, 25(1):41–56, January 2008.

Boulgouris, N. V., Plataniotis, K. N., and Hatzinakos, D. An angular transform of gait sequences for gait assisted recognition. In *Proc. IEEE International Conference on Image Processing*, volume 2, pp. 857–860, October 2004.

Boulgouris, N. V., Hatzinakos, D., and Plataniotis, K. N. Gait recognition: A challenging signal processing technology for biometrics. *IEEE Signal Processing Magazine*, 22(6):78–90, November 2005.

Boulgouris, N. V., Plataniotis, K. N., and Hatzinakos, D. Gait recognition using linear time normalization. *Pattern Recognition*, 39(5):969–979, 2006.

Bowyer, K. W., Chang, K., and Flynn, P. A survey of approaches and challenges in 3D and multi-modal 3D + 2D face recognition. *Computer Vision and Image Understanding*, 101(1):1–15, January 2006.

Boyd, J. E. Video phase-locked loops in gait recognition. In *Proc. IEEE Conference on Computer Vision*, volume 1, pp. 696–703, July 2001.

Brand, M. Incremental singular value decomposition of uncertain data with missing values. In *Proc. European Conference on Computer Vision*, pp. 707–720, 2002.

Breiman, L. Arcing classifiers. *The Annals of Statistics*, 26(3):801–849, 1998.

Breiman, L. Bagging predictors. *Machine Learning*, 24(2):123–140, 1996.

Bro, R. Multi-way calibration. Multi-linear PLS. *Journal of Chemometrics*, 10(1):47–61, 1996.

Bro, R. and De Jong, S. A fast non-negativity-constrained least squares algorithm. *Journal of Chemometrics*, 11(5):393–401, 1997.

Burges, C. J. C. Dimension reduction: A guided tour. *Foundations and Trends in Machine Learning*, 2(4):275–365, 2010.

Burges, C.J.C. A tutorial on support vector machines for pattern recognition. *Data Mining and Knowledge Discovery*, 2(2):121–167, 1998.

Cai, D., He, X., Han, J., and Zhang, H. J. Orthogonal Laplacianfaces for face recognition. *IEEE Transactions on Image Processing*, 15(11):3608–3614, November 2006.

Cardoso, J.-F. Super-symmetric decomposition of the fourth-order cumulant tensor. Blind identification of more sources than sensors. In *Proc. IEEE International Conference on Acoustics, Speech and Signal Processing*, pp. 3109–3112, 1991.

Cardoso, J.-F. Blind signal separation: statistical principles. *Proceedings of the IEEE*, 86(10):2009–2025, 1998.

Cardoso, J.-F. and Souloumiac, A. Blind beamforming for non-Gaussian signals. *IEE Proceedings-F*, 140(6):362–370, December 1993.

Carroll, J. D. and Chang, J. J. Analysis of individual differences in multidimensional scaling via an N-way generalization of "Eckart-Young" decomposition. *Psychometrika*, 35:283–319, 1970.

Cattell, R.B. "Parallel proportional profiles" and other principles for determining the choice of factors by rotation. *Psychometrika*, 9(4):267–283, 1944.

Cattell, R.B. The three basic factor-analytic research designs — Their interrelations and derivatives. *Psychological Bulletin*, 49(5):499–520, 1952.

Chang, C.C. and Lin, C.J. LIBSVM: A library for support vector machines. *ACM Transactions on Intelligent Systems and Technology (TIST)*, 2(3): 1–27, 2011.

Chapelle, O., Schölkopf, B., Zien, A., et al. *Semi-Supervised Learning*, volume 2. MIT Press, Cambridge, MA, 2006.

Chaudhuri, K., Kakade, S. M., Livescu, K., and Sridharan, K. Multi-view clustering via canonical correlation analysis. In *Proc. International Conference on Machine Learning*, pp. 129–136, 2009.

Chellappa, R., Roy-Chowdhury, A., and Zhou, S. *Recognition of Humans and Their Activities Using Video*. Morgan & Claypool Publishers, San Rafael, CA, 2005.

Chen, C., Zhang, J., and Fleischer, R. Distance approximating dimension reduction of Riemannian manifolds. *IEEE Transactions on Systems, Man, and Cybernetics, Part B: Cybernetics*, 40(1):208–217, 2010.

Cichocki, A., Zdunek, R., Phan, A. H., and Amari, S. *Nonnegative Matrix and Tensor Factorizations: Applications to Exploratory Multi-Way Data Analysis and Blind Source Separation.* John Wiley & Sons Ltd., Chichester West Sussex, United Kingdom, 2009.

Coifman, R. R. and Lafon, S. Diffusion maps. *Applied and Computational Harmonic Analysis*, 21(1):5–30, July 2006.

Colombo, A., Cusano, C., and Schettini, R. 3D face detection using curvature analysis. *Pattern Recognition*, 39(3):444–455, March 2006.

Comon, P. Independent component analysis, a new concept? *Signal Processing*, 36(3):287–314, 1994.

Comon, P. and Mourrain, B. Decomposition of quantics in sums of powers of linear forms. *Signal Processing*, 53:93–108, 1996.

Comon, P., Luciani, X., and De Almeida, A.L.F. Tensor decompositions, alternating least squares and other tales. *Journal of Chemometrics*, 23(7-8):393–405, 2009.

Cootes, T. F., Edwards, G. J., and Taylor, C. J. Active appearance models. *IEEE Transactions on Pattern Analysis and Machine Intelligence*, 23(6): 681–685, June 2001.

Cristianini, N. and Shawe-Taylor, J. *An Introduction to Support Vector Machines and Other Kernel-Based Learning Methods.* Cambridge University Press, United Kingdom, 2000.

Cunado, D., Nixon, M. S., and Carter, J. N. Automatic extraction and description of human gait models for recognition purposes. *Computer Vision and Image Understanding*, 90(1):1–41, January 2003.

Dai, G. and Yeung, D. Y. Tensor embedding methods. In *Proc. Twenty-First National Conference on Artificial Intelligence*, pp. 330–335, July 2006.

Dalal, N. and Triggs, B. Histograms of oriented gradients for human detection. In *Proc. IEEE Conference on Computer Vision and Pattern Recognition*, volume 1, pp. 886–893, 2005.

Daubechies, I. The wavelet transform, time-frequency localization and signal analysis. *IEEE Transactions on Information Theory*, 36(5):961–1005, 1990.

De la Torre, F. A least-squares framework for component analysis. *IEEE Transactions on Pattern Analysis and Machine Intelligence*, (34):1041–1055, 2012.

De Lathauwer, L. *Signal Processing Based on Multilinear Algebra.* PhD thesis, Katholieke Universiteit Leuven, 1997. URL `ftp://ftp.esat.kuleuven.ac.be/sista/delathauwer/reports/PHD.pdf`.

De Lathauwer, L., De Moor, B., and Vandewalle, J. On the best rank-1 and rank-$(R_1, R_2, ..., R_N)$ approximation of higher-order tensors. *SIAM Journal of Matrix Analysis and Applications*, 21(4):1324–1342, 2000a.

De Lathauwer, L., De Moor, B., and Vandewalle, J. A multilinear singular value decomposition. *SIAM Journal of Matrix Analysis and Applications*, 21(4):1253–1278, 2000b.

De Lathauwer, L., De Moor, B., and Vandewalle, J. Dimensionality reduction in higher-order-only ICA. In *Proceedings of the IEEE Signal Processing Workshop on Higher-Order Statistics*, pp. 316–320, 1997.

De Lathauwer, L., De Moor, B., and Vandewalle, J. Independent component analysis and (simultaneous) third-order tensor diagonalization. *IEEE Transactions on Signal Processing*, 49(10):2262–2271, 2001.

De Lathauwer, L. and Vandewalle, J. Dimensionality reduction in higher-order signal processing and rank-$(R_1, R_2, ..., R_N)$ reduction in multilinear algebra. *Linear Algebra and its Applications*, 391:31–55, November 2004.

Dean, J. and Ghemawat, S. MapReduce: Simplified data processing on large clusters. *Communications of the ACM*, 51(1):107–113, 2008.

Delorme, A. and Makeig, S. EEGLAB: An open source toolbox for analysis of single-trial EEG dynamics including independent component analysis. *Journal of Neuroscience Methods*, 134(1):9–21, 2004.

Deng, J., Dong, W., Socher, R., Li, L.J., Li, K., and Fei-Fei, L. Imagenet: A large-scale hierarchical image database. In *Proc. IEEE Conference on Computer Vision and Pattern Recognition*, pp. 248–255, 2009.

Dhillon, P. S., Foster, D., and Ungar, L. Multi-view learning of word embeddings via CCA. In *Advances in Neural Information Processing Systems (NIPS)*, pp. 199–207, 2011.

Domingos, P. A few useful things to know about machine learning. *Communications of the ACM*, 55(10):78–87, 2012.

Domingos, P. The role of occam's razor in knowledge discovery. *Data Mining and Knowledge Discovery*, 3(4):409–425, 1999.

Dorogovtsev, S. N., Goltsev, A. V., and Mendes, J. F. Pseudofractal scale-free web. *Physical Review E*, 65(6):066122, 2002.

Duda, R. O., Hart, P. E., and Stork, D. G. *Pattern Classification*. Wiley Interscience, New York, second edition, 2001.

Edelman, A., Arias, T.A., and Smith, S.T. The geometry of algorithms with orthogonality constraints. *SIAM Journal on Matrix Analysis and Applications*, 20(2):303–353, 1998.

Elgammal, A., Harwood, D., and Davis, L.S. Non-parametric model for background subtraction. In *Proc. IEEE International Conference on Computer Vision FRAME-RATE Workshop*, 1999.

Faber, N. M., Bro, R., and Hopke, P. K. Recent developments in CANDECOMP/PARAFAC algorithms: A critical review. *Chemometrics and Intelligent Laboratory Systems*, 65(1):119–137, January 2003.

Faloutsos, C., Kolda, T. G., and Sun, J. Mining large time-evolving data using matrix and tensor tools. *International Conference on Machine Learning 2007 Tutorial*, 2007. URL http://www.cs.cmu.edu/~christos/TALKS/ICML-07-tutorial/ICMLtutorial.pdf.

Feng, G. C. and Yuen, P. C. Variance projection function and its application to eye detection for human face recognition. *Pattern Recognition Letters*, 19(9):899–906, 1998.

Feng, G. C. and Yuen, P. C. Multi-cues eye detection on gray intensity image. *Pattern Recognition*, 34(5):1033–1046, 2001.

Ferryman, J. and Shahrokni, A. Pets2009: Dataset and challenge. In *Proc. 2009 Twelfth IEEE International Workshop on Performance Evaluation of Tracking and Surveillance (PETS-Winter)*, pp. 1–6, December 2009.

Fisher, R.A. The use of multiple measurements in taxonomic problems. *Annals of Human Genetics*, 7(2):179–188, 1936.

Forman, S.D., Cohen, J.D., Fitzgerald, M., Eddy, W.F., Mintun, M.A., and Noll, D.C. Improved assessment of significant activation in functional magnetic resonance imaging (fMRI): Use of a cluster-size threshold. *Magnetic Resonance in Medicine*, 33(5):636–647, 1995.

Foster, J. P., Nixon, M. S., and Prügel-Bennett, A. Automatic gait recognition using area-based metrics. *Pattern Recognition Letters*, 24(14):2489–2497, October 2003.

Freeman, W.T. Where computer vision needs help from computer science. In *Proc. Twenty-Second Annual ACM-SIAM Symposium on Discrete Algorithms*, pp. 814–819, 2011.

Freund, Y. and Schapire, R. E. Experiments with a new boosting algorithm. In *Proc. of the Thirteenth International Conference on Machine Learning*, pp. 148–156, 1996.

Freund, Y. and Schapire, R. E. A decision-theoretic generalization of on-line learning and an application to boosting. *Journal of Computer and System Sciences*, 55(1):119–139, 1997.

Freund, Y. and Schapire, R. E. A short introducion to boosting. *Journal of Japanese Society for Artificial Intelligence*, 14(5):771–780, 1999.

Friedlander, M.P. and Hatz, K. Computing non-negative tensor factorizations. *Optimisation Methods and Software*, 23(4):631–647, 2008.

Friedman, J., Hastie, T., and Tibshirani, R. *The Elements of Statistical Learning*, volume 1. Springer Series in Statistics, 2001.

Friedman, J. H. Regularized discriminant analysis. *Journal of the American Statistical Association*, 84(405):165–175, March 1989.

Fukunaga, K. *Introduction to Statistical Pattern Recognition*. Academic Press, Boston, MA, 1990.

Gang, L., Yong, Z., Liu, Y.-L., and Jing, D. Three dimensional canonical correlation analysis and its application to facial expression recognition. In *Proc. International Conference on Intelligent Computing and Information Science*, pp. 56–61, 2011.

Gao, Q., Zhang, L., Zhang, D., and Xu, H. Independent components extraction from image matrix. *Pattern Recognition Letters*, 31(3):171–178, February 2010a.

Gao, X., Yang, Y., Tao, D., and Li, X. Discriminative optical flow tensor for video semantic analysis. *Computer Vision and Image Understanding*, 113 (3):372–383, March 2009.

Gao, X., Li, X., Feng, J., and Tao, D. Shot-based video retrieval with optical flow tensor and HMMs. *Pattern Recognition Letters*, 30(2):140–147, January 2010b.

Gold, B., Morgan, N., and Ellis, D., *Speech and Audio Signal Processing*. Wiley Online Library, 2011.

Grant, M. G., Shutler, J. D., Nixon, M. S., and Carter, J. N. Analysis of a human extraction system for deploying gait biometrics. In *Proc. IEEE Southwest Symposium on Image Analysis and Interpretation*, pp. 46–50, March 2004.

Green, R. D. and Guan, L. Quantifying and recognizing human movement patterns from monocular video images Part II: Applications to biometrics. *IEEE Transactions on Circuits and Systems for Video Technology*, 14(2): 191–198, February 2004.

Greub, W. H. *Multilinear Algebra*. Springer-Verlag, Berlin, 1967.

Guan, Z., Wang, C., Chen, Z., Bu, J., and Chen, C. Efficient face recognition using tensor subspace regression. *Neurocomputing*, 73(13):2744–2753, 2010.

Guo, J., Li, Y., Wang, G., and Zeng, J. Batch process monitoring based on multilinear principal component analysis. In *International Conference on Intelligent System Design and Engineering Application (ISDEA)*, volume 1, pp. 413–416, 2010.

Guo, J., Liu, Y., and Yuan, W. Palmprint recognition based on phase congruency and two-dimensional principal component analysis. In *Proc. 4th International Congress on Image and Signal Processing*, volume 3, pp. 1527–1530, 2011.

Guo, W., Kotsia, I., and Patras, I. Tensor learning for regression. *IEEE Transactions on Image Processing*, 21(2):816–827, 2012.

Hackbusch, W. *Tensor Spaces and Numerical Tensor Calculus*, volume 42. Springer, Berlin, 2012.

Hamming, R. You and your research. In *Kaiser JF Transcription of the Bell Communications Research Colloquium Seminar*, 1986.

Hampel, F.R., Ronchetti, E.M., Rousseeuw, P.J., and Stahel, W.A. *Robust Statistics: The Approach Based on Influence Functions*, volume 114. Wiley, New York, 2011.

Han, J. and Kamber, M. *Data Mining: Concepts and Techniques*. Morgan Kaufmann, 2006.

Han, X.H., Chen, Y.W., and Ruan, X. Multilinear supervised neighborhood embedding of a local descriptor tensor for scene/object recognition. *IEEE Transactions on Image Processing*, 21(3):1314–1326, 2012.

Hao, Z., He, L., Chen, B., and Yang, X. A linear support higher-order tensor machine for classification. *IEEE Transactions on Image Processing*, 22(7): 2911–2920, 2013.

Hardoon, D., Szedmak, S., and Shawe-Taylor, J. Canonical correlation analysis: An overview with application to learning methods. *Neural Computation*, 16(12):2639–2664, 2004.

Harshman, R. Generalization of canonical correlation to N-way arrays. In *Poster at the 34th Annual Meeting of the Statistical Society Canada*, May 2006.

Harshman, R. A. Foundations of the PARAFAC procedure: Models and conditions for an "explanatory" multi-modal factor analysis. *UCLA Working Papers in Phonetics*, 16:1–84, 1970.

Hazan, T., Polak, S., and Shashua, A. Sparse image coding using a 3D non-negative tensor factorization. In *Proc. IEEE Conference on Computer Vision*, volume 1, pp. 50–57, 2005.

He, X. Incremental semi-supervised subspace learning for image retrieval. In *ACM Conference on Multimedia 2004*, pp. 2–8, October 2004.

He, X., Cai, D., and Niyogi, P. Tensor subspace analysis. In *Advances in Neural Information Processing Systems 18 (NIPS)*, pp. 499–506, 2005a.

He, X., Yan, S., Hu, Y., Niyogi, P., and Zhang, H. Face recognition using Laplacianfaces. *IEEE Transactions on Pattern Analysis and Machine Intelligence*, 27(3):328–340, March 2005b.

Hillis, D. M. and Bull, J. J. An empirical test of bootstrapping as a method for assessing confidence in phylogenetic analysis. *Systematic Biology*, 42(2): 182–192, 1993.

Hinton, G.E. and Salakhutdinov, R.R. Reducing the dimensionality of data with neural networks. *Science*, 313(5786):504–507, 2006.

Hinton, G.E., Osindero, S., and Teh, Y.W. A fast learning algorithm for deep belief nets. *Neural Computation*, 18(7):1527–1554, 2006.

Hitchcock, F. L. Multiple invariants and generalized rank of a p-way matrix or tensor. *Journal of Mathematical Physics*, 7(1):39–79, 1927a.

Hitchcock, F. L. The expression of a tensor or a polyadic as a sum of products. *Journal of Mathematical Physics*, 6(1):164–189, 1927b.

Ho, T. K. The random subspace method for constructing decision forests. *IEEE Transactions on Pattern Analysis and Machine Intelligence*, 20(8): 832–844, August 1998.

Höskuldsson, A. PLS regression methods. *Journal of Chemometrics*, 2(3): 211–228, 1988.

Hotelling, H. Analysis of a complex of statistical variables into principal components. *Journal of Educational Psychology*, 24(6):417–441, 1933.

Hotelling, H. Relations between two sets of variables. *Biometrika*, 28(3/4): 312–377, 1936.

Howe, D., Costanzo, M., Fey, P., Gojobori, T., Hannick, L., Hide, W., Hill, D. P, Kania, R., Schaeffer, M., St. Pierre, S., et al. Big data: The future of biocuration. *Nature*, 455(7209):47–50, 2008.

Hsu, R.-L., Abdel-Mottaleb, M., and Jain, A. K. Face detection in color images. *IEEE Transactions on Pattern Analysis and Machine Intelligence*, 24(5):696–706, 2002.

Hua, G., Viola, P. A., and Drucker, S. M. Face recognition using discriminatively trained orthogonal rank one tensor projections. In *Proc. IEEE Conference on Computer Vision and Pattern Recognition*, pp. 1–8, June 2007.

Hyvärinen, A. Fast and robust fixed-point algorithms for independent component analysis. *IEEE Transactions on Neural Networks*, 10(3):626–634, May 1999.

Hyvärinen, A. and Oja, E. Independent component analysis: Algorithms and applications. *Neural Networks*, 13(4-5):411–430, 2000.

Hyvärinen, A., Karhunen, J., and Oja, E. *Independent Component Analysis*. John Wiley & Sons, 2001.

Inoue, K., Hara, K., and Urahama, K. Robust multilinear principal component analysis. In *Proc. IEEE Conference on Computer Vision*, pp. 591–597, 2009.

Inoue, K. and Urahama, K. Non-iterative two-dimensional linear discriminant analysis. In *Proc. International Conference on Pattern Recognition*, volume 2, pp. 540–543, 2006.

Jackson, J.E. *A User's Guide to Principal Components*. Wiley, New York, 1991.

Jain, A. K., Ross, A., and Prabhakar, S. An introduction to biometric recognition. *IEEE Transactions on Circuits and Systems for Video Technology*, 14(1):4–20, January 2004a.

Jain, A. K., Chellappa, R., Draper, S. C., Memon, N., Phillips, P. J., and Vetro, A. Signal processing for biometric systems. *IEEE Signal Processing Magazine*, 24(6):146–152, November 2007.

Jain, A.K. and Farrokhnia, F. Unsupervised texture segmentation using Gabor filters. *Pattern Recognition*, 24(12):1167–1186, 1991.

Jia, C.C., Wang, S.J., Peng, X.J., Pang, W., Zhang, C.Y., Zhou, C.G., and Yu, Z.Z. Incremental multi-linear discriminant analysis using canonical correlations for action recognition. *Neurocomputing*, 83:56–63, 2011.

Jin, Z., Yang, J. Y., Hu, Z. S., and Lou, Z. Face recognition based on the uncorrelated discriminant transformation. *Pattern Recognition*, 34:1405–1416, 2001a.

Jin, Z., Yang, J. Y., Tang, Z. M., and Hu, Z. S. A theorem on the uncorrelated optimal discriminant vectors. *Pattern Recognition*, 34(10):2041–2047, October 2001b.

Jolliffe, I. T. *Principal Component Analysis*. Springer Series in Statistics, second edition, 2002.

Jørgensen, B. and Goegebeur, Y. Module 7: Partial least squares regression i, 2007. URL http://statmaster.sdu.dk/courses/ST02/module07/module.pdf.

Kaelbling, L.P., Littman, M.L., and Moore, A.W. Reinforcement learning: A survey. *Journal of Artificial Intelligence Research*, 4:237–285, 1996.

Kale, A. *Algorithms for Gait-Based Human Identification from a Monocular Video Sequences.* PhD thesis, Department of Electrical and Computer Engineering, University of Maryland, College Park, 2003. URL http://www.cs.uky.edu/~amit/thesis.pdf.

Kale, A., Rajagopalan, A. N., Sunderesan, A., Cuntoor, N., Roy-Chowdhury, A., Krueger, V., and Chellappa, R. Identification of humans using gait. *IEEE Transactions on Image Processing*, 13(9):1163–1173, September 2004.

Kang, U., Papalexakis, E., Harpale, A., and Faloutsos, C. Gigatensor: Scaling tensor analysis up by 100 times-algorithms and discoveries. In *Proceedings of the 18th ACM SIGKDD International Conference on Knowledge discovery and Data Mining*, pp. 316–324, 2012.

Kearns, M. and Valiant, L. G. Crytographic limitations on learning Boolean formulae and finite automata. *Journal of the Association for Computing Machinery*, 41(1):67–95, January 1994.

Khan, Y. and Gotman, J. Wavelet based automatic seizure detection in intracerebral electroencephalogram. *Clinical Neurophysiology*, 114(5):898–908, 2003.

Kim, T.-K. and Cipolla, R. Canonical correlation analysis of video volume tensors for action categorization and detection. *IEEE Transactions on Pattern Analysis and Machine Intelligence*, 31(8):1415–1428, 2009.

Kim, T.K., Kittler, J., and Cipolla, R. Discriminative learning and recognition of image set classes using canonical correlations. *IEEE Transactions on Pattern Analysis and Machine Intelligence*, 29(6):1005–1018, 2007.

Kittler, J., Hatef, M., Duin, R. P. W., and Matas, J. On combining classifiers. *IEEE Transactions on Pattern Analysis and Machine Intelligence*, 20(3): 226–239, March 1998.

Kivinen, J., Smola, A.J., and Williamson, R.C. Online learning with kernels. *IEEE Transactions on Signal Processing*, 52(8):2165–2176, 2004.

Kofidis, E. and Regalia, P.A. Tensor approximation and signal processing applications. *Contemporary Mathematics*, 280:103–134, 2001.

Kofidis, E. and Regalia, P.A. On the best rank-1 approximation of higher-order supersymmetric tensors. *SIAM Journal on Matrix Analysis and Applications*, 23(3):863–884, 2002.

Kokiopoulou, E., Chen, J., and Saad, Y. Trace optimization and eigenproblems in dimension reduction methods. *Numerical Linear Algebra with Applications*, 18(3):565–602, 2011.

Kolda, T. G. Orthogonal tensor decompositions. *SIAM Journal of Matrix Analysis and Applications*, 23(1):243–255, 2001.

Kolda, T. G. and Bader, B. W. Tensor decompositions and applications. *SIAM Review*, 51(3):455–500, September 2009.

Kolda, T.G. A counterexample to the possibility of an extension of the Eckart–Young low-rank approximation theorem for the orthogonal rank tensor decomposition. *SIAM Journal on Matrix Analysis and Applications*, 24(3): 762–767, 2003.

Kolda, T.G. and Sun, J. Scalable tensor decompositions for multi-aspect data mining. In *Proc. IEEE International Conference on Data Mining*, pp. 363–372, 2008.

Kong, H., Wang, L., Teoh, E. K., Wang, J.-G., and Venkateswarlu, R. A framework of 2D Fisher discriminant analysis: application to face recognition with small number of training samples. In *Proc. IEEE Conference on Computer Vision and Pattern Recognition*, volume 2, pp. 1083–1088, 2005.

Kotsia, I. and Patras, I. Multiplicative update rules for multilinear support tensor machines. In *Proceedings of the 20th International Conference on Pattern Recognition*, pp. 33–36, 2010a.

Kotsia, I. and Patras, I. Relative margin support tensor machines for gait and action recognition. In *Proceedings of the ACM International Conference on Image and Video Retrieval*, pp. 446–453, 2010b.

Kroonenberg, P. *Three-Mode Principal Component Analysis: Theory and Applications*. Leiden: DSWO Press, 1983.

Kroonenberg, P. and Leeuw, J. Principal component analysis of three-mode data by means of alternating least squares algorithms. *Psychometrika*, 45 (1):69–97, 1980.

Kruskal, J.B. Three-way arrays: Rank and uniqueness of trilinear decompositions, with application to arithmetic complexity and statistics. *Linear Algebra and Its Applications*, 18(2):95–138, 1977.

Kwak, N. Principal component analysis based on l1-norm maximization. *IEEE Transactions on Pattern Analysis and Machine Intelligence*, 30(9):1672–1680, 2008.

Lang, S. *Algebra*. Addison Wesley, Reading, MA, 1984.

Law, M. H. C. and Jain, A. K. Incremental nonlinear dimensionality reduction by manifold learning. *IEEE Transactions on Pattern Analysis and Machine Intelligence*, 28(3):377–391, March 2006.

Lawrence, N.D. Gaussian process latent variable models for visualisation of high dimensional data. *Advances in Neural Information Processing Systems (NIPS)*, 16:329–336, 2004.

Lebedev, L. P. and Cloud, M. J. *Tensor Analysis*. World Scientific, Singapore, 2003.

Lee, C. S. and Elgammal, A. Towards scalable view-invariant gait recognition: Multilinear analysis for gait. In *Proc. International Conference on Audio and Video-Based Biometric Person Authentication*, pp. 395–405, July 2005.

Lee, D.D. and Seung, H.S. Learning the parts of objects by non-negative matrix factorization. *Nature*, 401(6755):788–791, 1999.

Lee, K.C. and Kriegman, D. Online learning of probabilistic appearance manifolds for video-based recognition and tracking. In *Proc. IEEE Conference on Computer Vision and Pattern Recognition*, volume 1, pp. 852–859, 2005.

Lee, L., Dalley, G., and Tieu, K. Learning pedestrian models for silhouette refinement. In *Proc. IEEE Conference on Computer Vision*, pp. 663–670, October 2003.

Lee, S. H. and Choi, S. Two-dimensional canonical correlation analysis. *IEEE Signal Processing Letters*, 14(10):735–738, October 2007.

Lenhart, A., Purcell, K., Smith, A., and Zickuhr, K.. *Social Media & Mobile Internet Use among Teens and Young Adults*. Pew Internet & American Life Project Washington, DC, 2010.

Levey, A. and Lindenbaum, M. Sequential Karhunen-Loeve basis extraction and its application to images. *IEEE Transactions on Image Processing*, 9(8):1371–1374, 2000.

Li, J., Zhang, L., Tao, D., Sun, H., and Zhao, Q. A prior neurophysiologic knowledge free tensor-based scheme for single trial EEG classification. *IEEE Transactions on Neural Systems and Rehabilitation Engineering*, 17(2):107–115, April 2009a.

Li, M. and Yuan, B.. 2D-LDA: A statistical linear discriminant analysis for image matrix. *Pattern Recognition Letters*, 26(5):527–532, 2005.

Li, S. Z., Zhao, C., Ao, M., and Lei, Z. Learning to fuse 3D+2D based face recognition at both feature and decision levels. In *Proc. IEEE International Workshop on Analysis and Modeling of Faces and Gestures*, pp. 43–53, October 2005.

Li, X., Zeng, J., and Yan, H. PCA-HPR: A principle component analysis model for human promoter recognition. *Bioinformation*, 2(9):373–378, 2008.

Li, X., Pang, Y., and Yuan, Y. L1-norm-based 2DPCA. *IEEE Transactions on Systems, Man, and Cybernetics, Part B: Cybernetics*, 40(4):1170–1175, 2009b.

Librado, P. and Rozas, J. DNASP v5: a software for comprehensive analysis of DNA polymorphism data. *Bioinformatics*, 25(11):1451–1452, 2009.

Lindeberg, T. Scale-space theory: A basic tool for analyzing structures at different scales. *Journal of Applied Statistics*, 21(1-2):225–270, 1994.

Little, J. J. and Boyd, J. E. Recognizing people by their gait: The shape of motion. *Videre*, 1(2):1–32, 1998.

Liu, C. Capitalize on dimensionality increasing techniques for improving face recognition grand challenge performance. *IEEE Transactions on Pattern Analysis and Machine Intelligence*, 28(5):725–737, May 2006.

Liu, J., Chen, S., Zhou, Z.H., and Tan, X. Generalized low-rank approximations of matrices revisited. *IEEE Transactions on Neural Networks*, 21(4): 621–632, 2010.

Liu, N., Zhang, B., Yan, J., Chen, Z., Liu, W., Bai, F., and Chien, L. Text representation: From vector to tensor. In *Fifth IEEE International Conference on Data Mining*, pp. 725–728. IEEE, 2005.

Liu, Q., Tang, X., Lu, H., and Ma, S. Face recognition using kernel scatter-difference-based discriminant analysis. *IEEE Transactions on Neural Networks*, 17(4):1081–1085, July 2006.

Liu, Y., Collins, R. T., and Tsin, Y. A computational model for periodic pattern perception based on frieze and wallpaper groups. *IEEE Transactions on Pattern Analysis and Machine Intelligence*, 26(3):354–371, March 2004.

Liu, Z. and Sarkar, S. Simplest representation yet for gait recognition: Averaged silhouette. In *Proc. International Conference on Pattern Recognition*, volume 4, pp. 211–214, August 2004.

Liu, Z. and Sarkar, S. Effect of silhouette quality on hard problems in gait recognition. *IEEE Transactions on Systems, Man, and Cybernetics—Part B: Cybernetics*, 35(2):170–178, 2005.

Loog, M., Duin, R. P. W., and Haeb-Umbach, R. Multiclass linear dimension reduction by weighted pairwise Fisher criteria. *IEEE Transactions on Pattern Analysis and Machine Intelligence*, 23(7):762–766, July 2001.

Lotte, F., Congedo, M., Lécuyer, A., Lamarche, F., Arnaldi, B., et al. A review of classification algorithms for EEG-based brain–computer interfaces. *Journal of Neural Engineering*, 4(2):R1–R13, 2007.

Lowe, D.G. Distinctive image features from scale-invariant keypoints. *International Journal of Computer Vision*, 60(2):91–110, 2004.

Lu, H. *Multilinear Subspace Learning for Face and Gait Recognition*. PhD thesis, University of Toronto, 2008. URL https://tspace.library.utoronto.ca/handle/1807/16750.

Lu, H. Learning canonical correlations of paired tensor sets via tensor-to-vector projection. In *Proc. 23rd International Joint Conference on Artificial Intelligence (IJCAI 2013)*, pp. 1516–1522, 2013a.

Lu, H. Learning modewise independent components from tensor data using multilinear mixing model. In *Proc. European Conference on Machine Learning and Principles and Practice of Knowledge Discovery in Databases (ECML PKDD'13)*, pp. 288–303, 2013b.

Lu, H., Kot, A. C., and Shi, Y. Q. Distance-reciprocal distortion measure for binary document images. *IEEE Signal Processing Letters*, 11(2):228–231, February 2004.

Lu, H., Plataniotis, K. N., and Venetsanopoulos, A. N. A layered deformable model for gait analysis. In *Proc. IEEE International Conference on Automatic Face and Gesture Recognition*, pp. 249–254, April 2006a.

Lu, H., Plataniotis, K. N., and Venetsanopoulos, A. N. Gait recognition through MPCA plus LDA. In *Proc. Biometrics Symposium 2006*, pp. 1–6, September 2006b. doi:10.1109/BCC.2006.4341613.

Lu, H., Plataniotis, K. N., and Venetsanopoulos, A. N. Coarse-to-fine pedestrian localization and silhouette extraction for the gait challenge data sets. In *Proc. IEEE Conference on Multimedia and Expo*, pp. 1009–1012, July 2006c.

Lu, H., Plataniotis, K. N., and Venetsanopoulos, A. N. Multilinear principal component analysis of tensor objects for recognition. In *Proc. International Conference on Pattern Recognition*, volume 2, pp. 776 – 779, August 2006d.

Lu, H., Plataniotis, K. N., and Venetsanopoulos, A. N. Boosting LDA with regularization on MPCA features for gait recognition. In *Proc. Biometrics Symposium 2007*, September 2007a. doi:10.1109/BCC.2007.4430542.

Lu, H., Plataniotis, K. N., and Venetsanopoulos, A. N. Uncorrelated multilinear discriminant analysis with regularization for gait recognition. In *Proc. Biometrics Symposium 2007*, September 2007b. doi:10.1109/BCC.2007.4430540.

Lu, H., Plataniotis, K. N., and Venetsanopoulos, A. N. A full-body layered deformable model for automatic model-based gait recognition. *EURASIP Journal on Advances in Signal Processing: Special Issue on Advanced Signal Processing and Pattern Recognition Methods for Biometrics*, 2008, 2008a. Article ID 261317, 13 pages, doi:10.1155/2008/261317.

Lu, H., Plataniotis, K. N., and Venetsanopoulos, A. N. MPCA: Multilinear principal component analysis of tensor objects. *IEEE Transactions on Neural Networks*, 19(1):18–39, January 2008b.

Lu, H., Plataniotis, K. N., and Venetsanopoulos, A. N. Uncorrelated multilinear principal component analysis through successive variance maximization. In *Proc. International Conference on Machine Learning*, pp. 616–623, July 2008c.

Lu, H., Plataniotis, K. N., and Venetsanopoulos, A. N. Boosting discriminant learners for gait recognition using MPCA features. *EURASIP Journal on Image and Video Processing*, 2009, 2009a. Article ID 713183, 11 pages, doi:10.1155/2009/713183.

Lu, H., Plataniotis, K. N., and Venetsanopoulos, A. N. Regularized common spatial patterns with generic learning for EEG signal classification. In *Proc. 31st International Conference of the IEEE Engineering in Medicine and Biology Society*, pp. 6599–6602, September 2009b.

Lu, H., Plataniotis, K. N., and Venetsanopoulos, A. N. Uncorrelated multilinear discriminant analysis with regularization and aggregation for tensor object recognition. *IEEE Transactions on Neural Networks*, 20(1):103–123, January 2009c.

Lu, H., Plataniotis, K. N., and Venetsanopoulos, A. N. Uncorrelated multilinear principal component analysis for unsupervised multilinear subspace learning. *IEEE Transactions on Neural Networks*, 20(11):1820–1836, November 2009d.

Lu, H., Eng, H.-L., Guan, C., Plataniotis, K. N., and Venetsanopoulos, A. N. Regularized common spatial pattern with aggregation for EEG classification in small-sample setting. *IEEE Transactions on Biomedical Engineering*, 57 (12):2936–2946, December 2010a.

Lu, H., Eng, H.-L., Thida, M., and Plataniotis, K. N. Visualization and clustering of crowd video content in MPCA subspace. In *Proc. 19st ACM Conference on Information and Knowledge Management*, pp. 1777–1780, October 2010b.

Lu, H., Plataniotis, K. N., and Venetsanopoulos, A. N. A survey of multilinear subspace learning for tensor data. *Pattern Recognition*, 44(7):1540–1551, July 2011.

Lu, H., Pan, Y., Mandal, B., Eng, H, Guan, C., and Chan, D. Quantifying limb movements in epileptic seizures through color-based video analysis. *IEEE Transactions on Biomedical Engineering*, 60(2):461–469, February 2013.

Lu, J., Plataniotis, K. N., and Venetsanopoulos, A. N. Face recognition using LDA based algorithms. *IEEE Transactions on Neural Networks*, 14(1):195–200, January 2003.

Lu, J., Plataniotis, K. N., and Venetsanopoulos, A. N. Regularization studies of linear discriminant analysis in small sample size scenarios with application to face recognition. *Pattern Recognition Letters*, 26(2):181–191, 2005.

Lu, J., Plataniotis, K. N., Venetsanopoulos, A. N., and Li, S. Z. Ensemble-based discriminant learning with boosting for face recognition. *IEEE Transactions on Neural Networks*, 17(1):166–178, January 2006e.

Luo, D., Ding, C., and Huang, H. Symmetric two dimensional linear discriminant analysis (2DLDA). In *Proc. IEEE Conference on Computer Vision and Pattern Recognition*, pp. 2820–2827, 2009.

Ma, Y., Niyogi, P., Sapiro, G., and Vidal, R. Dimensionality reduction via subspace and submanifold learning [from the guest editors]. *IEEE Signal Processing Magazine*, 28(2):14–126, January 2011.

MacKay, D.J.C. Introduction to Gaussian processes. *NATO ASI Series F Computer and Systems Sciences*, 168:133–166, 1998.

Mahoney, M.W., Maggioni, M., and Drineas, P. Tensor-CUR decompositions for tensor-based data. *SIAM Journal on Matrix Analysis and Applications*, 30(3):957–987, 2008.

Mairal, J., Bach, F., Ponce, J., and Sapiro, G. Online learning for matrix factorization and sparse coding. *The Journal of Machine Learning Research*, 11:19–60, 2010.

Manning, C.D. and Schütze, H. *Foundations of Statistical Natural Language Processing*. MIT Press, Cambridge, MA, 1999.

Marden, J.I. Multivariate statistical analysis old school, 2011. URL http://istics.net/pdfs/multivariate.pdf.

Mažgut, J., Tiňo, P., Bodén, M., and Yan, H. Multilinear decomposition and topographic mapping of binary tensors. *Proceedings of International Conference on Artificial Neural Networks*, pp. 317–326, 2010.

Mažgut, J., Tiňo, P., Bodén, M., and Yan, H. Dimensionality reduction and topographic mapping of binary tensors. *Pattern Analysis and Applications*, pp. 1–19, February 2013.

McLachlan, G.J. *Discriminant Analysis and Statistical Pattern Recognition*. Wiley, New York, 1992.

McLachlan, G.J. and Krishnan, T. *The EM Algorithm and Extensions*, volume 382. Wiley-Interscience, New York, 2007.

Meng, J. and Zhang, W. Volume measure in 2DPCA-based face recognition. *Pattern Recognition Letters*, 28(10):1203–1208, 2007.

Meng, X.L. and Van Dyk, D. The EM algorithman old folk-song sung to a fast new tune. *Journal of the Royal Statistical Society: Series B (Statistical Methodology)*, 59(3):511–567, 2002.

Moon, H. and Phillips, P. J. Computational and performance aspects of PCA-based face recognition algorithms. *Perception*, 30:303–321, 2001.

Moon, T. K. and Stirling, W. C. *Mathematical methods and Algorithms for Signal Processing*. Prentice Hall, Englewood Cliffs, NJ, 2000.

Mørup, M., Hansen, L.K., and Arnfred, S.M. Algorithms for sparse nonnegative Tucker decompositions. *Neural Computation*, 20(8):2112–2131, 2008.

Muja, M. and Lowe, D.G. Fast approximate nearest neighbors with automatic algorithm configuration. In *International Conference on Computer Vision Theory and Applications (VISSAPP09)*, pp. 331–340, 2009.

Müller, K.-R., Mika, S., Rätsch, G., Tsuda, K., and Schölkopf, B. An introduction to kernel-based learning algorithms. *IEEE Transactions on Neural Networks*, 12(2):181–201, March 2001.

Muti, D. and Bourennane, S. Survey on tensor signal algebraic filtering. *Signal Processing*, 87(2):237–249, February 2007.

Nie, F., Xiang, S., Song, Y., and Zhang, C. Extracting the optimal dimensionality for local tensor discriminant analysis. *Pattern Recognition*, 42(1): 105–114, 2009.

Nixon, M. S. and Carter, J. N. Advances in automatic gait recognition. In *Proc. IEEE International Conference on Automatic Face and Gesture Recognition*, pp. 139–144, May 2004.

Nixon, M. S. and Carter, J. N. Automatic recognition by gait. *Proceedings of the IEEE*, 94(11):2013–2024, November 2006.

Nolker, C. and Ritter, H. Visual recognition of continuous hand postures. *IEEE Transactions on Neural Networks*, 13(4):983–994, July 2002.

Oja, E. and Yuan, Z. The FastICA algorithm revisited: Convergence analysis. *IEEE Transactions on Neural Networks*, 17(6):1370–1381, November 2006.

Ojala, T., Pietikainen, M., and Maenpaa, T. Multiresolution gray-scale and rotation invariant texture classification with local binary patterns. *IEEE Transactions on Pattern Analysis and Machine Intelligence*, 24(7):971–987, 2002.

Oppenheim, A.V, Willsky, A.S, and Nawab, S.H. *Signals and Systems*, volume 2. Prentice-Hall, Englewood Cliffs, NJ, 1983.

Oweiss, K.G. *Statistical Signal Processing for Neuroscience and Neurotechnology*. Academic Press, Burlington, MA, 2010.

Paatero, P. A weighted non-negative least squares algorithm for three-way PARAFAC factor analysis. *Chemometrics and Intelligent Laboratory Systems*, 38(2):223–242, 1997.

Paatero, P. and Tapper, U. Positive matrix factorization: A non-negative factor model with optimal utilization of error estimates of data values. *Environmetrics*, 5(2):111–126, 1994.

Pan, S.J. and Yang, Q. A survey on transfer learning. *IEEE Transactions on Knowledge and Data Engineering*, 22(10):1345–1359, 2010.

Panagakis, Y., Kotropoulos, C., and Arce, G. R. Non-negative multilinear principal component analysis of auditory temporal modulations for music genre classification. *IEEE Transactions on Audio, Speech, and Language Processing*, 18(3):576–588, March 2010.

Pang, Y., Li, X., and Yuan, Y. Robust tensor analysis with L1-norm. *IEEE Transactions on Circuits and Systems for Video Technology*, 20(2):172–178, 2010.

Pantano, P., Mainero, C., Lenzi, D., Caramia, F., Iannetti, G.D., Piattella, M.C., Pestalozza, I., Di Legge, S., Bozzao, L., and Pozzilli, C. A longitudinal fMRI study on motor activity in patients with multiple sclerosis. *Brain*, 128 (9):2146–2153, 2005.

Papadimitriou, S., Sun, J., and Faloutsos, C. Streaming pattern discovery in multiple time-series. In *Proc. 31st International Conference on Very Large Data Bases*, pp. 697–708, 2005.

Pearson, K. On lines and planes of closest fit to systems of points in space. *Philosophical Magazine*, 2(6):559–572, 1901.

Phillips, P. J., Moon, H., Rizvi, S. A., and Rauss, P. The FERET evaluation method for face recognition algorithms. *IEEE Transactions on Pattern Analysis and Machine Intelligence*, 22(10):1090–1104, October 2000.

Phillips, P. J., Flynn, P.J., Scruggs, T., Bowyer, K.W., Chang, J., Hoffman, K., Marques, J., Min, J., , and Worek, W. Overview of the face recognition grand challenge. In *Proc. IEEE Conference on Computer Vision and Pattern Recognition*, volume 1, pp. 947–954, June 2005.

Pitton, J.W., Wang, K., and Juang, B.H. Time-frequency analysis and auditory modeling for automatic recognition of speech. *Proceedings of the IEEE*, 84(9):1199–1215, 1996.

Plataniotis, K. N. and Venetsanopoulos, A. N. *Color Image Processing and Applications*. Springer Verlag, Berlin, 2000.

Porges, T. and Favier, G. Automatic target classification in SAR images using MPCA. In *Proc. IEEE International Conference on Acoustics, Speech and Signal Processing*, pp. 1225–1228, 2011.

Porro-Muñoz, D., Duin, R.P.W., Talavera, I., and Orozco-Alzate, M. Classification of three-way data by the dissimilarity representation. *Signal Processing*, 91(11):2520–2529, 2011.

Qi, L., Sun, W., and Wang, Y. Numerical multilinear algebra and its applications. *Frontiers of Mathematics in China*, 2(4):501–526, 2007.

Qian, S. *Introduction to Time-Frequency and Wavelet Transforms*, volume 68. Prentice Hall PTR, Englewood Cliffs, NJ, 2002.

Rai, P. and Daumé III, H. Multi-label prediction via sparse infinite CCA. In *Advances in Neural Information Processing Systems (NIPS)*, pp. 1518–1526, 2009.

Raina, R., Battle, A., Lee, H., Packer, B., and Ng, A.Y. Self-taught learning: Transfer learning from unlabeled data. In *Proc. 24th International Conference on Machine Learning*, pp. 759–766, 2007.

Raj, R. G. and Bovik, A. C. MICA: A multilinear ICA decomposition for natural scene modeling. *IEEE Transactions on Image Processing*, 17(3): 259–271, March 2009.

Rajih, M., Comon, P., and Harshman, R.A. Enhanced line search: A novel method to accelerate PARAFAC. *SIAM Journal on Matrix Analysis and Applications*, 30(3):1128–1147, 2008.

Ramoser, H., Muller-Gerking, J., and Pfurtscheller, G. Optimal spatial filtering of single trial EEG during imagined hand movement. *IEEE Transactions on Rehabilitation Engineering*, 8(4):441–446, 2000.

Rasmussen, C.E. and Williams, C.K.I. *Gaussian Processes for Machine Learning*, volume 1. MIT Press, Cambridge, MA, 2006.

Renard, N. and Bourennane, S. Dimensionality reduction based on tensor modeling for classification methods. *IEEE Transactions on Geoscience and Remote Sensing*, 47(4):1123–1131, April 2009.

Rissanen, J. A universal prior for integers and estimation by minimum description length. *The Annals of Statistics*, pp. 416–431, 1983.

Rosipal, R. and Krämer, N. Overview and recent advances in partial least squares. *Subspace, Latent Structure and Feature Selection*, Springer-Verlag, Berlin, pp. 34–51, 2006.

Ross, A. and Govindarajan, R. Feature level fusion of hand and face biometrics. In *Proc. SPIE Conference on Biometric Technology for Human Identification II*, pp. 196–204, March 2005.

Ross, A., Jain, A. K., and Qian, J. Z. Information fusion in biometrics. *Pattern Recognition Letters*, 24:2115–2125, 2003.

Ross, D.A., Lim, J., Lin, R.S., and Yang, M.H. Incremental learning for robust visual tracking. *International Journal of Computer Vision*, 77(1):125–141, 2008.

Rosten, E., Porter, R., and Drummond, T. Faster and better: A machine learning approach to corner detection. *IEEE Transactions on Pattern Analysis and Machine Intelligence*, 32(1):105–119, 2010.

Roweis, S. and Saul, L. Nonlinear dimensionality reduction by locally linear embedding. *Science*, 290(22):2323–2326, December 2000.

Roweis, S. Matrix identities. Note available on the Internet at http://www.cs.toronto.edu/ roweis/notes/matrixid.pdf, 1999.

Rozas, J., Sánchez-DelBarrio, J. C., Messeguer, X., and Rozas, R. DNASP, DNA polymorphism analyses by the coalescent and other methods. *Bioinformatics*, 19(18):2496–2497, 2003.

Ruiz-Hernandez, J. A., Crowley, J. L., and Lux, A. "How old are you?": Age estimation with tensors of binary Gaussian receptive maps. In *Proceedings of the British Machine Vision Conference*, pp. 1–11, 2010a.

Ruiz-Hernandez, J.A., Crowley, J.L., and Lux, A. Tensor-jet: A tensorial representation of local binary Gaussian jet maps. In *Proc. IEEE Conference on Computer Vision and Pattern Recognition Workshops*, pp. 41–47, 2010b.

Saeys, Y., Inza, I., and Larrañaga, P. A review of feature selection techniques in bioinformatics. *Bioinformatics*, 23(19):2507–2517, 2007.

Sahambi, H. S. and Khorasani, K. A neural-network appearance-based 3-D object recognition using independent component analysis. *IEEE Transactions on Neural Networks*, 14(1):138–149, January 2003.

Samaria, F. and Young, S. HMM based architecture for face identification. *Image and Vision Computing*, 12:537–583, 1994.

Sarkar, S., Phillips, P. J., Liu, Z., Robledo, I., Grother, P., and Bowyer, K. W. The human ID gait challenge problem: Data sets, performance, and analysis. *IEEE Transactions on Pattern Analysis and Machine Intelligence*, 27(2): 162–177, February 2005.

Schapire, R. E. Using output codes to boost multiclass learning problems. In *Proc. Fourteenth International Conference on Machine Learning*, pp. 313–321, 1997.

Schapire, R. E. The boosting approach to machine learning: An overview. In Denison, D. D., Hansen, M. H., Holmes, C., Mallick, B., and Yu, B. (Eds.), *MSRI Workshop on Nonlinear Estimation and Classification*. Springer, Berlin, 2003. URL http://stat.haifa.ac.il/~goldensh/DM/msri.pdf.

Schapire, R. E. and Singer, Y. Improved boosting algorithms using confidence-rated predictions. *Machine Learning*, 37(3):297–336, December 1999.

Schapire, R. E., Freund, Y., Bartlett, P., and Lee, W. S. Boosting the margin: A new explanation for the effectiveness of voting methods. In *Proc. Fourteenth International Conference on Machine Learning*, pp. 322–330, 1997.

Schein, A.I., Saul, L.K., and Ungar, L.H. A generalized linear model for principal component analysis of binary data. In *Proc. Ninth International Workshop on Artificial Intelligence and Statistics*, pp. 14–21, 2003.

Schölkopf, B., Smola, A., and Müller, K. R. Nonlinear component analysis as a kernel eigenvalue problem. *Neural Computation*, 10(5):1299–1319, 1998.

Schwarz, G. Estimating the dimension of a model. *The Annals of Statistics*, 6(2):461–464, 1978.

Shakhnarovich, G. and Moghaddam, B. Face recognition in subspaces. In Li, Stan Z. and Jain, Anil K. (Eds.), *Handbook of Face Recognition*, pp. 141–168. Springer-Verlag, Berlin, 2004.

Shashua, A. and Hazan, T. Non-negative tensor factorization with applications to statistics and computer vision. In *Proc. International Conference on Machine Learning*, pp. 792–799, 2005.

Shashua, A. and Levin, A. Linear image coding for regression and classification using the tensor-rank principle. In *Proc. IEEE Conference on Computer Vision and Pattern Recognition*, volume I, pp. 42–49, 2001.

Shechtman, E., Caspi, Y., and Irani, M. Space-time super-resolution. *IEEE Transactions on Pattern Analysis and Machine Intelligence*, 27(4):531–545, April 2005.

Sim, T., Baker, S., and Bsat, M. The CMU pose, illumination, and expression database. *IEEE Transactions on Pattern Analysis and Machine Intelligence*, 25(12):1615–1618, December 2003.

Sivic, J. and Zisserman, A. Video Google: A text retrieval approach to object matching in videos. In *Proc. IEEE Conference on Computer Vision*, pp. 1470–1477, 2003.

Skurichina, M. and Duin, R. P. W. Bagging, boosting and the random subspace method for linear classifiers. *Pattern Analysis & Applications*, 5(2): 121–135, 2002.

Sonnenburg, S., Braun, M. L., Ong, C. S., Bengio, S., Bottou, L., Holmes, G., Y. LeCun, K.-R. Müller, F. Pereira, Rasmussen, C. E., Rätsch, G., Schölkopf, B., Smola, A., Vincent, P., Weston, J., and Williamson, R. C. The need for open-source software in machine learning. *Journal of Machine Learning Research*, 8:2443–2466, October 2007.

Steinwart, I. and Christmann, A. *Support Vector Machines*. Springer, Berlin, 2008.

Sukittanon, S., Atlas, L.E., and Pitton, J.W. Modulation-scale analysis for content identification. *IEEE Transactions on Signal Processing*, 52(10): 3023–3035, 2004.

Sun, J., Tao, D., and Faloutsos, C. Beyond streams and graphs: Dynamic tensor analysis. In *Proc. 12th ACM SIGKDD International Conference on Knowledge Discovery and Data Mining*, pp. 374–383, August 2006.

Sun, J., Tao, D., Papadimitriou, S., Yu, P. S., and Faloutsos, C. Incremental tensor analysis: Theory and applications. *ACM Transactions on Knowledge Discovery from Data*, 2(3):11:1–11:37, October 2008a.

Sun, J., Xie, Y., Zhang, H., and Faloutsos, C. Less is more: Sparse graph mining with compact matrix decomposition. *Statistical Analysis and Data Mining*, 1(1):6–22, February 2008b.

Sun, J. *Incremental Pattern Discovery on Streams, Graphs and Tensors*. PhD thesis, Carnegie Mellon University, 2007.

Sun, L., Ji, S., and Ye, J. A least squares formulation for canonical correlation analysis. In *Proc. International Conference on Machine Learning*, pp. 1024–1031, July 2008c.

Sun, M., Wang, S., Liu, X., Jia, C., and Zhou, C. Human action recognition using tensor principal component analysis. In *Proc. 4th IEEE International Conference on Computer Science and Information Technology*, pp. 487–491, 2011.

Sutton, R.S. and Barto, A.G. *Reinforcement Learning: An Introduction*, volume 1. Cambridge Univ Press, 1998.

Szeliski, R. *Computer vision: Algorithms and applications*. Springer, 2010.

Tan, X., Chen, S., Zhou, Z.-H., and Zhang, F. Face recognition from a single image per person: A survey. *Pattern Recognition*, 39(9):1725–1745, 2006.

Tao, D., Li, X., Wu, X., and Maybank, S. J. Elapsed time in human gait recognition: A new approach. In *Proc. IEEE International Conference on Acoustics, Speech and Signal Processing*, volume 2, pp. 177–180, April 2006.

Tao, D., Li, X., Wu, X., Hu, W., and Maybank, S. J. Supervised tensor learning. *Knowledge and Information Systems*, 13(1):1–42, January 2007a.

Tao, D., Li, X., Wu, X., and Maybank, S. J. General tensor discriminant analysis and gabor features for gait recognition. *IEEE Transactions on Pattern Analysis and Machine Intelligence*, 29(10):1700–1715, October 2007b.

Tao, D., Li, X., Wu, X., and Maybank, S. J. Tensor rank one discriminant analysis: A convergent method for discriminative multilinear subspace selection. *Neurocomputing*, 71(10-12):1866–1882, June 2008a.

Tao, D., Song, M., Li, X., Shen, J., Sun, J., Wu, X., Faloutsos, C., and Maybank, S. J. Bayesian tensor approach for 3-D face modeling. *IEEE Transactions on Circuits and Systems for Video Technology*, 18(10):1397–1410, October 2008b.

Tenenbaum, J. B., de Silva, V., and Langford, J.C. A global geometric framework for nonlinear dimensionality reduction. *Science*, 290(22):2319–2323, December 2000.

Tibshirani, R. Regression shrinkage and selection via the lasso. *Journal of the Royal Statistical Society. Series B (Methodological)*, pp. 267–288, 1996.

Tipping, M.E. and Bishop, C.M. Mixtures of probabilistic principal component analyzers. *Neural Computation*, 11(2):443–482, 1999a.

Tipping, M.E. and Bishop, C.M. Probabilistic principal component analysis. *Journal of the Royal Statistical Society: Series B (Statistical Methodology)*, 61(3):611–622, 1999b.

Tolliver, D. and Collins, R. T. Gait shape estimation for identification. In *Proc. International Conference on Audio and Video-Based Biometric Person Authentication*, pp. 734–742, June 2003.

Tong, S. and Chang, E. Support vector machine active learning for image retrieval. In *Proceedings of the Ninth ACM International Conference on Multimedia*, pp. 107–118, 2001.

Tong, S. and Koller, D. Support vector machine active learning with applications to text classification. *The Journal of Machine Learning Research*, 2: 45–66, 2002.

Torralba, A., Fergus, R., and Freeman, W.T. 80 million tiny images: A large data set for nonparametric object and scene recognition. *IEEE Transactions on Pattern Analysis and Machine Intelligence*, 30(11):1958–1970, 2008.

Tsourakakis, C.E. MACH: Fast randomized tensor decompositions. arXiv preprint arXiv:0909.4969, 2009.

Tsybakov, A. B. Optimal aggregation of classifiers in statistical learning.(English summary). *Annals of Statistics*, 32(1):135–166, 2004.

Tucker, L. R. Some mathematical notes on three-mode factor analysis. *Psychometrika*, 31:279–311, 1966.

Turk, M. and Pentland, A. Eigenfaces for recognition. *Journal of Cognitive Neurosicence*, 3(1):71–86, 1991.

van de Ven, V.G., Formisano, E., Prvulovic, D., Roeder, C.H., and Linden, D.E.J. Functional connectivity as revealed by spatial independent component analysis of fMRI measurements during rest. *Human Brain Mapping*, 22(3):165–178, 2004.

Vapnik, V and Sterin, A. On structural risk minimization or overall risk in a problem of pattern recognition. *Automation and Remote Control*, 10(3): 1495–1503, 1977.

Vapnik, V. N. *The Nature of Statistical Learning Theory*. Springer-Verlag, New York, 1995.

Vasilescu, M. A. O. Human motion signatures: analysis, synthesis, recognition. In *Proc. International Conference on Pattern Recognition*, volume 3, pp. 456–460, August 2002.

Vasilescu, M. A. O. and Terzopoulos, D. Multilinear image analysis for facial recognition. In *Proc. International Conference on Pattern Recognition*, volume 2, pp. 511–514, August 2002a.

Vasilescu, M. A. O. and Terzopoulos, D. Multilinear analysis of image ensembles: Tensorfaces. In *Proc. Seventh European Conference on Computer Vision*, pp. 447–460, May 2002b.

Vasilescu, M. A. O. and Terzopoulos, D. Multilinear subspace analysis of image ensembles. In *Proc. IEEE Conference on Computer Vision and Pattern Recognition*, volume II, pp. 93–99, June 2003.

Vasilescu, M. A. O. and Terzopoulos, D. Multilinear independent components analysis. In *Proc. IEEE Conference on Computer Vision and Pattern Recognition*, volume I, pp. 547–553, June 2005.

Vega, I. R. and Sarkar, S. Statistical motion model based on the change of feature relationships: Human gait-based recognition. *IEEE Transactions on Pattern Analysis and Machine Intelligence*, 25(10):1323–1328, October 2003.

Vidal, R., Ma, Y., and Sastry, S. Generalized principal component analysis (gpca). *IEEE Transactions on Pattern Analysis and Machine Intelligence*, 27(12):1945–1959, 2005.

Viola, P. and Jones, M. J. Robust real-time face detection. *International Journal of Computer Vision*, 57(2):137–154, May 2004.

Wagg, D. K. and Nixon, M. S. On automated model-based extraction and analysis of gait. In *Proc. IEEE International Conference on Automatic Face and Gesture Recognition*, pp. 11–16, May 2004.

Wagstaff, K. Machine learning that matters. In *Proc. International Conference on Machine Learning*, pp. 529–536, 2012.

Wang, H. Local two-dimensional canonical correlation analysis. *IEEE Signal Processing Letters*, 17(11):921–924, November 2010.

Wang, H. and Ahuja, N. A tensor approximation approach to dimensionality reduction. *International Journal of Computer Vision*, 76(3):217–229, March 2008.

Wang, J., Plataniotis, K. N., and Venetsanopoulos, A. N. Selecting discriminant eigenfaces for face recognition. *Pattern Recognition Letters*, 26(10): 1470–1482, 2005.

Wang, J., Plataniotis, K. N., Lu, J., and Venetsanopoulos, A. N. On solving the face recognition problem with one training sample per subject. *Pattern Recognition*, 39(9):1746–1762, 2006.

Wang, L., Tan, T., Ning, H., and Hu, W. Silhouette analysis-based gait recognition for human identification. *IEEE Transactions on Pattern Analysis and Machine Intelligence*, 25(12):1505–1518, December 2003.

Wang, L., Ning, H., Tan, T., and Hu, W. Fusion of static and dynamic body biometrics for gait recognition. *IEEE Transactions on Circuits and Systems for Video Technology*, 14(2):149–158, February 2004.

Wang, Q., Chen, F., and Xu, W. Tracking by third-order tensor representation. *IEEE Transactions on Systems, Man, and Cybernetics, Part B: Cybernetics*, 41(2):385–396, 2011a.

Wang, S.J., Yang, J., Zhang, N., and Zhou, C.G. Tensor discriminant color space for face recognition. *IEEE Transactions on Image Processing*, 20(9): 2490–2501, 2011b.

Wang, S.J., Zhou, C.G., Zhang, N., Peng, X.J., Chen, Y.H., and Liu, X. Face recognition using second-order discriminant tensor subspace analysis. *Neurocomputing*, 74(12):2142–2156, 2011c.

Wang, S.J., Yang, J., Sun, M.F., Peng, X.J., Sun, M.M., and Zhou, C.G. Sparse tensor discriminant color space for face verification. *IEEE Transactions on Neural Networks and Learning Systems*, 23(6):876–888, 2012.

Wang, X., Ma, X., and Grimson, W.E.L. Unsupervised activity perception in crowded and complicated scenes using hierarchical bayesian models. *IEEE Transactions on Pattern Analysis and Machine Intelligence*, 31(3):539–555, 2009.

Wang, Y. and Gong, S. Tensor discriminant analysis for view-based object recognition. In *Proc. International Conference on Pattern Recognition*, volume 3, pp. 33–36, August 2006.

Wang, Z., Chen, S., Liu, J., and Zhang, D. Pattern representation in feature extraction and classifier design: matrix versus vector. *IEEE Transactions on Neural Networks*, 19(5):758–769, 2008.

Washizawa, Y., Higashi, H., Rutkowski, T., Tanaka, T., and Cichocki, A. Tensor based simultaneous feature extraction and sample weighting for EEG classification. *Neural Information Processing. Models and Applications*, pp. 26–33, 2010.

Weenink, D. J. M. Canonical correlation analysis. *Proceedings of the Institute of Phonetic Sciences of the University of Amsterdam*, 25:81–99, 2003.

Welling, M. and Weber, M. Positive tensor factorization. *Pattern Recognition Letters*, 22(12):1255–1261, 2001.

Wen, J., Li, X., Gao, X., and Tao, D. Incremental learning of weighted tensor subspace for visual tracking. In *Proc. 2009 IEEE International Conference on Systems, Man and Cybernetics*, pp. 3688–3693, October 2009.

Wen, J., Gao, X., Yuan, Y., Tao, D., and Li, J. Incremental tensor biased discriminant analysis: A new color-based visual tracking method. *Neurocomputing*, 73(4-6):827–839, January 2010.

Williams, C.K.I. Prediction with Gaussian processes: From linear regression to linear prediction and beyond. *NATO ASI Series D Behavioural And Social Sciences*, 89:599–621, 1998.

Wiskott, L., Fellous, J. M., Kruger, N., and von der Malsburg, C. Face recognition by elastic bunch graph matching. *IEEE Transactions on Pattern Analysis and Machine Intelligence*, 19(7):775–779, July 1997.

Witten, I.H. and Frank, E. *Data Mining: Practical Machine Learning Tools and Techniques*. Morgan Kaufmann, 2005.

Wold, H. Path models with latent variables: The NIPALS approach. In et al., H.M. Blalock (Ed.), *Quantitative Sociology: International Perspectives on Mathematical and Statistical Model Building*, pp. 307–357. Academic Press, 1975.

Wold, S., Sjöström, M., L., and Eriksson. PLS-regression: A basic tool of chemometrics. *Chemometrics and Intelligent Laboratory Systems*, 58(2): 109–130, 2001.

Wright, J., Yang, A.Y., Ganesh, A., Sastry, S.S., and Ma, Y. Robust face recognition via sparse representation. *IEEE Transactions on Pattern Analysis and Machine Intelligence*, 31(2):210–227, 2009.

Wu, S., Li, W., Wei, Z., and Yang, J. Local discriminative orthogonal rank-one tensor projection for image feature extraction. In *First Asian Conference on Pattern Recognition (ACPR)*, pp. 367–371, 2011.

Xu, D., Yan, S., Tao, D., Zhang, L., Li, X., and Zhang, H.-J. Human gait recognition with matrix representation. *IEEE Transactions on Circuits and Systems for Video Technology*, 16(7):896–903, July 2006.

Xu, D., Lin, S., Yan, S., and Tang, X. Rank-one projections with adaptive margins for face recognition. *IEEE Transactions on Systems, Man, and Cybernetics—Part B: Cybernetics*, 37(5):1226–1236, October 2007.

Xu, D., Yan, S., Zhang, L., Lin, S., Zhang, H.-J., and Huang, T. S. Reconstruction and recognition of tensor-based objects with concurrent subspaces analysis. *IEEE Transactions on Circuits and Systems for Video Technology*, 18(1):36–47, January 2008.

Xu, R. and Wunsch II, D. C. Survey of clustering algorithms. *IEEE Transactions on Neural Networks*, 16(3):645–678, May 2005.

Yam, C. Y., Nixon, M. S., and Carter, J. N. Automated person recognition by walking and running via model-based approaches. *Pattern Recognition*, 37(5):1057–1072, May 2004.

Yan, J., Zheng, W., Zhou, X., and Zhao, Z. Sparse 2-D canonical correlation analysis via low rank matrix approximation for feature extraction. *IEEE Signal Processing Letters*, 19(1):51–54, January 2012.

Yan, S., Xu, D., Yang, Q., Zhang, L., Tang, X., and Zhang, H.-J. Discriminant analysis with tensor representation. In *Proc. IEEE Conference on Computer Vision and Pattern Recognition*, volume I, pp. 526–532, June 2005.

Yan, S., Xu, D., Yang, Q., Zhang, L., Tang, X., and Zhang, H. Multilinear discriminant analysis for face recognition. *IEEE Transactions on Image Processing*, 16(1):212–220, January 2007a.

Yan, S., Xu, D., Zhang, B., Zhang, H. J., Yang, Q., and Lin, S. Graph embedding and extensions: A general framework for dimensionality reduction. *IEEE Transactions on Pattern Analysis and Machine Intelligence*, 29(1): 40–51, January 2007b.

Yang, J., Zhang, D., Frangi, A. F., and Yang, J. Two-dimensional PCA: A new approach to appearance-based face representation and recognition. *IEEE Transactions on Pattern Analysis and Machine Intelligence*, 26(1):131–137, January 2004.

Yang, J., Zhang, D., Yong, X., and Yang, J.-Y. Two-dimensional discriminant transform for face recognition. *Pattern Recognition*, 38(7):1125–1129, 2005.

Ye, J. Generalized low rank approximations of matrices. *Machine Learning*, 61(1-3):167–191, 2005a.

Ye, J. Characterization of a family of algorithms for generalized discriminant analysis on undersampled problems. *Journal of Machine Learning Research*, 6:483–502, April 2005b.

Ye, J., Janardan, R., and Li, Q. GPCA: An efficient dimension reduction scheme for image compression and retrieval. In *Tenth ACM SIGKDD International Conference on Knowledge Discovery and Data Mining*, pp. 354–363, 2004a.

Ye, J., Janardan, R., and Li, Q. Two-dimensional linear discriminant analysis. In *Advances in Neural Information Processing Systems (NIPS)*, pp. 1569–1576, 2004b.

Ye, J., Janardan, R., Li, Q., and Park, H. Feature reduction via generalized uncorrelated linear discriminant analysis. *IEEE Transactions on Knowledge and Data Engineering*, 18(10):1312–1322, October 2006.

Zafeiriou, S. Algorithms for nonnegative tensor factorization. In Aja-Fernández, S., d. L. García, R., Tao, D., and Li, X. (Eds.), *Tensors in Image Processing and Computer Vision*, pp. 105–124. Springer, Berlin, 2009a.

Zafeiriou, S. Discriminant nonnegative tensor factorization algorithms. *IEEE Transactions on Neural Networks*, 20(2):217–235, 2009b.

Zafeiriou, S. and Petrou, M. Nonnegative tensor factorization as an alternative Csiszar–Tusnady procedure: Algorithms, convergence, probabilistic interpretations and novel probabilistic tensor latent variable analysis algorithms. *Data Mining and Knowledge Discovery*, 22(3):419–466, 2011.

Zangwill, W.I. *Nonlinear Programming: A Unified Approach*. Prentice-Hall Englewood Cliffs, NJ, 1969.

Zhang, D., Wang, Y., and Bhanu, B. Ethnicity classification based on gait using multi-view fusion. In *Proc. IEEE Conference on Computer Vision and Pattern Recognition Workshops*, pp. 108–115, 2010.

Zhang, J., Li, S. Z., and Wang, J. Manifold learning and applications in recognition. In Tan, Y. P., Yap, K. H., and Wang, L. (Eds.), *Intelligent Multimedia Processing with Soft Computing*, pp. 281–300. Springer-Verlag, Berlin, 2004.

Zhang, L., Gao, Q., and Zhang, D. Directional independent component analysis with tensor representation. In *Proc. IEEE Conference on Computer Vision and Pattern Recognition*, pp. 1–7, June 2008.

Zhang, L., Tao, D, and Huang, X. Tensor discriminative locality alignment for hyperspectral image spectral–spatial feature extraction. *IEEE Transactions on Neural Networks*, 51(1):242–256, 2013.

Zhang, T. and Golub, G.H. Rank-one approximation to high order tensors. *SIAM Journal on Matrix Analysis and Applications*, 23(2):534–550, 2001.

Zhang, Y., Zhou, G., Zhao, Q., Onishi, A., Jin, J., Wang, X., and Cichocki, A. Multiway canonical correlation analysis for frequency components recognition in SSVEP-based BCIs. In *Proc. International Conference on Neural Information Processing (ICONIP)*, pp. 287–295, 2011.

Zhao, J., Yu, P.L.H., and Kwok, J.T. Bilinear probabilistic principal component analysis. *IEEE Transactions on Neural Networks and Learning Systems*, 23(3):492–503, 2012.

Zhao, Q., Caiafa, C. F., Mandic, D. P., Zhang, L., Ball, T., Schulze-Bonhage, A., and Cichocki, A.. Multilinear subspace regression: An orthogonal tensor decomposition approach. In *Advances in Neural Information Processing Systems (NIPS)*, pp. 1269–1277, 2011.

Zhou, X.S. and Huang, T.S. Comparing discriminating transformations and SVM for learning during multimedia retrieval. In *Proceedings of the Ninth ACM International Conference on Multimedia*, pp. 137–146, 2001a.

Zhou, X.S. and Huang, T.S. Small sample learning during multimedia retrieval using biasmap. In *Proc. IEEE Conference on Computer Vision and Pattern Recognition*, volume 1, pp. I–11, 2001b.

Zhou, Z.-H. *Ensemble Methods: Foundations and Algorithms*. Chapman & Hall, Boca Raton, FL, 2012.

Zhou, Z.-H. and Geng, X. Projection functions for eye detection. *Pattern Recognition*, 37(5):1049–1056, 2004.

Zou, H. and Hastie, T. Regression shrinkage and selection via the elastic net, with applications to microarrays. *Journal of the Royal Statistical Society: Series B. v67*, pp. 301–320, 2003.

Zuo, W., Zhang, D., and Wang, K. An assembled matrix distance metric for 2DPCA-based image recognition. *Pattern Recognition Letters*, 27(3): 210–216, 2006.

Index